Information Sources in Science and Technology

A practical guide to traditional and online use

C. C. Parker C.Chem., M.R.S.C., M.Phil., M.I.Inf.Sci.
Information Officer (Physical Sciences and Engineering),
Southampton University Library

R. V. Turley B.Sc., Ph.D., C.Phys., M.Inst.P., A.F.I.M.A.
Divisional Librarian (Physical Sciences and Engineering),
Southampton University Library

Butterworths
London Boston Durban Singapore Sydney Toronto Wellington

First published 1975
Second edition 1986

© Butterworth & Co (Publishers) Ltd, 1986

British Library Cataloguing in Publication Data

Parker, C. C.
    Information sources in science and technology.—2nd ed.
    1. Technology—Information services
    2. Science—Information services
    I. Title   II. Turley, Raymond V.
    607   T10.5

    ISBN 0-408-01467-9

Library of Congress Cataloging in Publication Data

Parker, C. C. (Christopher Charles)
    Information sources in science and technology.
    Bibliography: p.
    Includes index.
    1. Science—Information services.   2. Technology—
    Information services.   I. Turley, Raymond Victor.
    Q224.P37   1986   507   85-13324
    ISBN 0-408-01467-9

Typeset by Phoenix Photosetting Ltd, Chatham
Printed and bound in Great Britain at the University Press, Cambridge

Preface

For whom is our book written? Initially we tried to write the kind of guide we should have liked to have had available on our own desks. Although this new edition has been extensively revised, we have not departed significantly from our original intentions. We want our book to be used by librarians and information officers, especially those working in scientific or industrial environments, and by practising scientists and engineers. It should prove of value to students associated with all these professions as well. Our aim has been to produce a text which is suitable for beginners, but which will also serve as a handy reference work for the more experienced.

In surveying quite a wide field within a manageable volume, the multidisciplinary approach we have adopted is that demanded by many real-life situations. Furthermore we have paid due attention to both traditional and online methods of information retrieval. We must admit, though, that because of the way in which we have dealt with our subject matter, sometimes we find ourselves addressing mainly library and information *users*, whereas on other occasions we direct our remarks primarily towards library and information *organizers*. We hope readers will make allowances for this, accordingly.

There is nothing new about most of the factual information contained in these pages; however, we do claim that there is a certain novelty about its presentation, and therein lies our justification for producing yet another contribution to the literature explosion. We have tried to make this book easy to use by arranging its contents in a logically structured fashion, with more or less self-contained sections (like any reasonable guide-book). This is no more than an introductory guide; nor is it in any sense 'complete'. When choosing the various examples of information sources and their guides we sought to include many of current importance. Nevertheless we must emphasize that we have produced only a *selection*: comprehensiveness is not to be looked for in a book of this size. When annotating the various publications mentioned as examples, we planned to give sufficient information for someone to decide (say) whether or not it was worth travelling 20 miles to the nearest library holding the item in question.

New editions of more than a few publications mentioned here will appear during the useful life of this book, and may vary in structure and content from those we describe. Similarly, some works to which we have devoted space will be superseded by entirely new titles; also, the recent growth in the number of online databases becoming publicly available is likely to continue. So be prepared to update our examples by seeking out new editions, titles or databases, especially if you need the very latest sources or guides.

When referring to hosts and databases associated with online information retrieval we have followed the convention of setting both in capital letters. This is a widespread, if not quite universal, custom among other authors. Additionally, we have distinguished between hosts and databases by means of roman and italic type respectively, thus DIALOG and *COMPENDEX*. It seems logical to use italic for databases, since italic is also used for their hard-copy equivalents here, and in general elsewhere (for instance, *Engineering Index* in the case of *COMPENDEX*).

This work is concerned only with current scientific/technical information: the historical approach is not provided for. Anyone interested in the latter might care to consult Southampton University Library Occasional Paper No. 4, *The Literature of Science and Technology Approached Historically: a brief guide for reference* (1973), compiled by Dr Raymond V. Turley.

We have received much help and encouragement in the preparation of this book. It is a pleasure to acknowledge, with grateful thanks, all those librarians and information officers, too numerous, alas, to name individually, who have given freely of their time and advice. However we are particularly grateful to the staff of the British Library Lending Division at Boston Spa (especially Mr J. P. Chillag) who, having read the manuscript of the first edition, offered many useful comments. In carrying out the present revision we are indebted to several of our colleagues at Southampton University (among whom we must mention the Librarian, Bernard Naylor, who provided a sympathetic working environment which made the preparation of this new edition possible), our friends in the Hampshire County Library service at Winchester and Southampton, the staff of the British Library, Science Reference Library, and those of the Online Information Centre (Aslib, The Association for Information Management). Notwithstanding this we must, of course, ourselves retain responsibility for all statements (including errors) made herein.

This book is dedicated to our students, past, present and future—because they inspired the original idea, contributed to its development and will, we hope, help us to improve it still further.

C C P
R V T

Publisher's note

News of name changes affecting sections of the British Library arrived just as this book was going to press. The Science Reference Library (SRL) has become the British Library Science Reference and Information Service (SRIS), and the British Library Lending Division (BLLD) is now known as the British Library Document Supply Centre (BLDSC). Unfortunately it was not possible to incorporate the name changes herein, *except* on pp. 259–260 and in the index.

How you can help improve this book

The authors are naturally anxious to improve any future edition of this book. They would therefore appreciate ideas or suggestions from readers.

Contents

How to use this book

This guide is not meant to be read from cover to cover – use it as a reference book.

If you do not know where to start looking for information:

Are you familiar with the various sources and their guides?

Yes Turn to **Choosing sources of information and their guides** (p. 6) for advice on those appropriate to your problem.

No Start with **A routine for finding and using information** (pp. 2–3) followed by **Defining the problem** (p. 4); then read the descriptive material *only* about individual sources/guides (**People, Organizations, The literature, Information services**) and finally consult **Choosing sources of information and their guides** (p. 6).

If you cannot find information via your chosen sources or guides:

Are you completely confident about your 'searching' technique?

Yes Check the lists of guides to appropriate sources and/or **Choosing sources of information and their guides** (especially the charts on pp. 10, 12–13) for help in selecting alternatives.

No Look at **Searching** (p. 211) and, if necessary, **Defining the problem** (p. 4).

If you experience difficulty in using libraries, or would like to know more about the services they provide:

Consult **Obtaining literature in a usable form** (p. 245).

If your problem is not covered by the alternatives above:

The **Contents** pages give our logically structured arrangement of information sources and searching techniques: when choosing types of source, for example, you can go through the list asking in each case 'Is this likely to provide relevant information?' The **Index** has entries for subjects, types of source, names of organizations, authors and titles of publications, and abbreviations: use it when you have a good idea what you are looking for.

If this book fails to help solve your information problem, remember to try the **general guides** and **subject guides** listed on pp. 40 and 42.

A routine for finding and using information

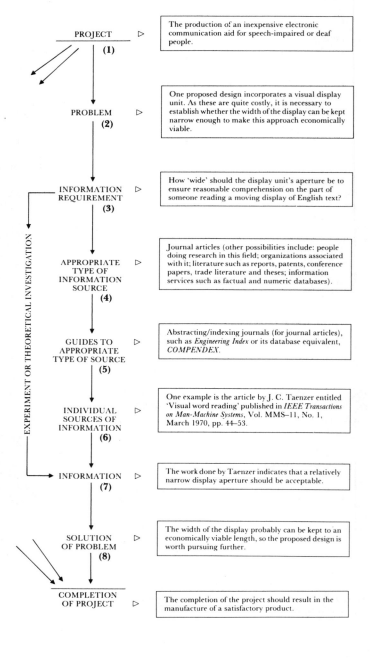

PROJECT **(1)**	▷	The production of an inexpensive electronic communication aid for speech-impaired or deaf people.
PROBLEM **(2)**	▷	One proposed design incorporates a visual display unit. As these are quite costly, it is necessary to establish whether the width of the display can be kept narrow enough to make this approach economically viable.
INFORMATION REQUIREMENT **(3)**	▷	How 'wide' should the display unit's aperture be to ensure reasonable comprehension on the part of someone reading a moving display of English text?
APPROPRIATE TYPE OF INFORMATION SOURCE **(4)**	▷	Journal articles (other possibilities include: people doing research in this field; organizations associated with it; literature such as reports, patents, conference papers, trade literature and theses; information services such as factual and numeric databases).
GUIDES TO APPROPRIATE TYPE OF SOURCE **(5)**	▷	Abstracting/indexing journals (for journal articles), such as *Engineering Index* or its database equivalent, *COMPENDEX*.
INDIVIDUAL SOURCES OF INFORMATION **(6)**	▷	One example is the article by J. C. Taenzer entitled 'Visual word reading' published in *IEEE Transactions on Man-Machine Systems*, Vol. MMS–11, No. 1, March 1970, pp. 44–53.
INFORMATION **(7)**	▷	The work done by Taenzer indicates that a relatively narrow display aperture should be acceptable.
SOLUTION OF PROBLEM **(8)**	▷	The width of the display probably can be kept to an economically viable length, so the proposed design is worth pursuing further.
COMPLETION OF PROJECT	▷	The completion of the project should result in the manufacture of a satisfactory product.

EXPERIMENT OR THEORETICAL INVESTIGATION

Very often, solving one particular problem is only part of a much larger assignment—we shall call this a *project*. Here we illustrate by means of an example how the problem–solution approach to finding information operates within the wider context of project work. In practice the following process becomes cyclic, as it is often necessary to go back to earlier stages as your work progresses.

(1) Examine the project to identify a series of problems which can be tackled individually (this process should continue throughout the project's life).

(2) Analyse each problem in terms of its information requirement; determine what information you need (which is not immediately available) to help solve that problem.

(3) Consider the different types of information source (assuming it is inappropriate to satisfy the information requirement by experiment or calculation) and decide which are likely to be most useful.

(4) Choose and locate guides to appropriate types of information source.

(5) Search these guides using relevant search terms. Select and record details of individual sources to be approached, including literature references to be examined. Locate and obtain the literature concerned.

(6) Examine the literature referred to and extract the required information. Where necessary, contact people, organizations or information services.

(7) Evaluate (and if appropriate report on) the information acquired. Use it to construct solutions to each problem.

(8) Employ these solutions towards completion of the project. Arrange for current awareness where interest in particular problems continues.

Some hints on solving problems and planning an information search

● Make sure you really understand the subject and nature of your project (seek advice from colleagues or whoever is in charge of your work, if in doubt).

● Write down a statement of your overall aims and objective, including any specific limitations.

● Establish a list of priorities with regard to the problems you need to solve and the information that must be found.

● Before you start to solve a problem or look for information, be certain this directly contributes to your overall objective (i.e. do not indulge in these activities for their own sake).

● When tackling a problem it is often helpful to begin by predicting likely solutions.

● Do not look for information until you have a fairly clear idea of what you require (during a search you tend to find only what you are looking for), but be ready to modify your requirement in the light of what is found as the search progresses.

● When planning an information search bear in mind the time you have available, the amount of information you need to acquire (judging when to stop can be as difficult as knowing where to begin), and the risk of stifling your own (original) ideas by becoming overwhelmed with those of other people.

● Remember that most projects are multi-disciplinary in character, so be prepared to draw on the information resources of subject fields other than your own.

Defining the problem

'Surely no man would work so hard to attain such precise information unless he had some definite end in view.' (Dr Watson, on Sherlock Holmes)

'It is of the highest importance, therefore, not to have useless facts elbowing out the useful ones.' (Sherlock Holmes, on the science of deduction)

From *A Study in Scarlet*, by Sir Arthur Conan Doyle (chapter 2)

Your reason for wanting information is always associated with a problem which you hope to solve. It may be a simple problem like satisfying curiosity, or it may be complicated like passing an examination, performing creative research, keeping up to date, designing a piece of equipment, measuring the properties of a new material or even ridding a Pacific island of poisonous starfish. Information must help you solve your problems if it is to be of use to you. (We shall hopefully assume that you have already organized your work in relation to your personal priorities and objectives.) It is a sound practice to write out a 'statement of intent' and keep it before you.

Questions

Do you know what information to look for? Would you recognize useful information if you saw it?

If you think you do and you would, please turn to the next section, which considers the types of information sources (p. 6). If you are not sure what to look for, please read on.

Make sure you fully understand the nature of your problem

It may be necessary to analyse your problem into its constituent parts; considering causes, effects, urgency, importance and even possible solutions. Problems are sometimes solved by comparing them with similar problems where a solution has been found. If your problems are particularly difficult to solve, you may find the following books of help:

The Art of Scientific Investigation, by W. I. B. Beveridge. London, Heinemann, 1968.

How to Solve it: a new aspect of mathematical method, by G. Polya. Princeton, NJ, Princeton University Press, 1971.

Make the Most of Your Mind, by T. Buzan. London, Pan, 1981.

The Use of Lateral Thinking, by E. de Bono. London, Penguin, 1971.

Work out how the information might help you with the problem

To be helpful, information must be applicable to the problem, comprehensible to you, and worth the cost and effort involved in obtaining it.

(a) Decide what 'job' the information must do as far as your *problem* is concerned. This will indicate the subject nature of the information you require. For example:

● If you are trying to identify a flower, the information must be descriptive so that you can make comparisons.

● If you are unable to find the thermodynamic data which are required to set up an experiment, it may be possible to calculate them from other thermodynamic data which you *can* find (using an equation).

● If you are unable to find information on your specific problem (say sedimentation around pipe junctions), you might think of information on similar problems in other subjects (such as deposits in blood vessels).

● If you are trying to be creative, too much detail of other work in your field may bias you before you can develop ideas of your own.

● If you need to know the properties of a particular material, make sure that you can define any variables which may affect them. (Some of these variables may be dictated by the environmental conditions in which the material will be used. For example, the boiling point of a liquid will be higher than normal when it is in pressurized apparatus.)

(b) Decide what 'job' the information must do for *you*. The information you obtain will have to bridge the gap between what you know already and what is necessary for you to know in order to solve your problem.

(c) Decide how important the required information is to you and your problem. What would happen if you could not find it, or even did not bother to look for it? Can you put a value on it in terms of time and money?

Decide what information you are going to look for

At this stage you should have a good idea of what you want, so this will help you to decide where to look and will also help you to recognize helpful information when you see it. Try also to think in terms of what you do *not* want. For example, do you really want all the information on the electrochemistry of aluminium, or just its properties in aqueous systems? Never forget the reason for requiring information and try to think of information *for*, not just information *on*. You will also know whether you require a comprehensive search, some data, the very latest information or a 'state-of-the-art' survey. You will rarely have to do a comprehensive search: the usual procedure is to obtain items of information as and when required by your work.

It is usually impossible to retrieve *all* the information on a topic, owing to limitations of time and money, so a realistic compromise is normally necessary.

Choosing sources of information and their guides

Types of source

We will assume that you now know the nature of the information required. Next, you must choose which type(s) of source will supply this information in a convenient form, within any time limit there may be, to help you with your problem. When we try to obtain information, we are considering several possibilities:

● We may be hoping that there is someone, somewhere, who could supply us with the required information, or at least point us in the right direction. We may think of that 'someone' as an individual, on a personal level, or as an employee of an organization, on an impersonal level. Although we are always dealing with people when we contact an organization, it is convenient to treat organizations as a separate type of source.
● We may hope that the required information already exists in a recorded form. In this case, we would normally turn to the literature, perhaps through a library.
● We may hope that the required information has been processed by an information service or stored in a library. Although information services and libraries are organizations, or parts of organizations, it is convenient to consider them separately.
● We may hope that *no* information exists on a particular subject, especially if we are in the process of applying for a patent, or performing 'original' research for a higher degree. In such cases it is important to conduct a thorough search in spite of any desires or expectations to find nothing.
● We may require the information urgently, and may not have enough time or immediate facilities to search ourselves. In these circumstances, we should normally turn to an information service of some kind.
● We may be reasonably certain that the information does not exist in recorded form, or that its retrieval is not worth the cost. This may cause us to decide to abandon any search and to create our own information, for example by experimentation, or to employ someone else to create the required knowledge, such as a laboratory, research organization or university department.

We have therefore divided the types of information sources into four major groups:

(1) People (including ourselves)
(2) Organizations (such as commercial organizations, professional associations, etc.)
(3) The literature (such as books, reports, standards, etc.)
(4) Information services (such as computerized services, information bureaux and brokers).

Libraries, because of their unique role in information retrieval, are dealt with separately in pp. 245–262.

People includes you, and you should remember that sometimes the best way to obtain new information is via experimentation.

Always remember that you can seek advice at any stage of your search for information from other people, including colleagues, librarians, information officers and staff of other organizations.

Question

Do you already know which type(s) of information source are likely to be of use? For example, books, research associations, journal articles, etc.

If you think you do, please turn to the section which considers guides to sources (p. 9). If you are not sure which type to choose, please read on.

Consider the different types of information source and decide which ones are likely to be most helpful

The different types of source are listed in the Contents and are described in the body of this book. The more you know about these sources, and the more experience you have in using them, the easier it will be for you to decide which ones to use. If you get stuck, you can always consider each type in turn and ask yourself whether it is suitable from the following points of view:

● *Coverage.* Is it likely to provide the particular information you require? Does it cover the right subject area and is it sufficiently comprehensive? For example, bibliographies and reviews are especially helpful in comprehensive searches or state-of-the-art surveys.
● *Level.* Will it provide information which you can understand? For example, articles in learned journals usually require a higher level of understanding than articles on the same subject in encyclopaedias.
● *Emphasis.* There are two aspects of emphasis, namely approach and detail. For example: (1) theses, patents, reports, etc., are all written for different reasons and would approach the same subject in different ways. (2) You could not expect a newspaper article to go into as much detail in describing a new diesel engine as you could an engineering report.
● *Currency.* Will it cover the right period in time? For example, newspapers are more likely to give the latest information on a subject than a textbook.
● *Accessibility.* There are various aspects of accessibility, including urgency and convenience. For example, foreign theses may be difficult to obtain quickly, may involve some expense, and may only be available from the country of origin. (It has been observed that people sometimes tend to rely on inferior sources of information merely because they are readily accessible.)
● *Security.* Will there be problems involving security? For example, in the case of people or organizations, will they wish to give you their information, or will you want them to learn the nature of your work?

• *Cost.* How much money is available? Online services and information bureaux or brokers can be very expensive, although they can be very fast and efficient.

When considering the various types of source, it is also helpful, bearing in mind what was said about information being recorded, used, processed and stored by people and their organizations, to ask yourself these questions:

• What is the likely origin of the information you seek? For example, the properties of a material may well have been investigated and recorded by its manufacturer. This should lead you to commercial organizations or trade literature.
• Who else is likely to have had problems similar to your own, and have already obtained and used the information you are seeking?
• Who is sufficiently concerned with the general subject area of your problem to have stored or processed appropriate information? For example, if you require statistics of particular types of road accident, you might try to find an appropriate official organization or research association.

If you still have difficulty in deciding on a likely type of source, try the chart on p. 13. The **subject guides** mentioned on pp. 42–46 may also be of help.

Access to sources

In the case of people or organizations, you have the choice of visiting, writing, telephoning, or possibly using telex or electronic mail. Visiting usually provides an opportunity to acquire a wider range of information, but is not appropriate if time is short or only straightforward facts are required. When visiting, it is usually helpful to make an appointment, stating your requirements, so that adequate arrangements can be made to help you when you arrive. Writing allows of the accurate communication of information, but frequently involves time delays in the act of replying, or due to the postal service. Telephoning is fast, but can lead to misunderstandings and mistakes. It is a good policy to write down exactly what you want to know and why, before you telephone someone, so that you can present them with a clear idea of your information requirement. Telex is a good compromise between writing and telephoning, providing a fast, accurate way of communicating small amounts of information. It is a neglected facility (mainly, one suspects, through lack of awareness) which is undergoing extensive development (see also p. 185) and which has much to commend it. Electronic mail is becoming increasingly available via British Telecom, Prestel, online hosts etc. It has the advantage over Telex that you may be able to control it from your desk, whereas most organizations have Telex as a central service. The problem is that the person at the other end has to be a user of the same service and must also check for messages as a matter of routine.

In the case of the literature, you normally have to consult it yourself, through a library, colleague or purchase. You may be able to obtain the information direct via an information service or library, especially if your requirement is simple. For example members of professional bodies, which have their own libraries, may be able to obtain limited amounts of information over the telephone. Otherwise, you may pay for information services to do the work for you.

Guides to sources

Question

Do you know which individual sources of information are likely to be of use? For example:

Handbook of Physics and Chemistry
The Institution of Mechanical Engineers
British Patent Specification No. 1 000 301
Electronics, Vol. 5, No. 3, pp. 25–30

If you think you do, and you wish to locate or approach the sources, please turn to the section in the text (via the Contents or Index) which deals with that type of source, and look under *Access*. You may find it necessary to refer to further sections, such as **Library systems and services**.

If you are not sure which particular individual sources will be of use, please read on.

Choose and locate guides to the appropriate type(s) of information source

This is simply done by turning to the sections in the text (via the Contents or Index) which deal with the type(s) of information source you wish to use, and then by referring to the details of their guides. Your choice will be influenced by commonsense, availability (most guides are only available in libraries and some of them may only cover the last decade or two) and ease of use (some guides are difficult to use and others are very time-consuming). You may have to select a subject area at this stage, e.g. chemistry in the case of *Chemical Abstracts*. You should remember that most problems are multidisciplinary and appropriate information may be found in various subject areas and so you should not always rely on just your own subject field. Subject family trees may help you here (see p. 214). For example, you decide that British Standards are a suitable type of information source. You turn to the appropriate place in the text (p. 127), look under **Guides to standards**, select the *BSI Catalogue*, and then check your nearest libraries to see if one has a copy.

Chart 1. Information sources and guides versus coverage

The relationships between the various types of information source and their guides are not always simple. Sometimes you will not be able to find a suitable guide, either because the guides you do find are of no use in your particular case or because the guide you want is simply not available in your locality. To help you with this difficulty, Chart 1 attempts to show all the important relationships between the various types of information source and the different types of guide.

For example, if you wanted to identify recent patents in your field, but you were unable to gain access to the specific **Guides to patents** mentioned in this book, the chart might provide an alternative. In this case the chart shows that abstracting and indexing journals may include patents in their coverage. The size of the circle indicates the importance of each relationship.

Chart 1. Information sources and guides versus coverage

The size of the circle indicates
the importance of the relationship

Use these ⟩⟩

(in addition to any specific guides)

to find out about

these

Use these ↓ to find out about these	Abstracting & indexing journals	Guides to abstracting & indexing journals	Books	Guides to books	Conference proceedings	Commercial organizations	Educational organizations	General guides	Government publications	Information services	Journal articles	Newspapers and articles	Official organizations	Patents	Periodicals	Guides to periodicals	Professional organizations	Reference material	Guides to reference material	Reports	Review serials	Subject bibliographies	Guides to subject bibliographies	Subject guides	Theses and dissertations	Trade literature
Abstracting & indexing journals		●							•	•						●	•		●					●		
Guides to abstracting & indexing journals		●							•	•									●							
Books	•	●						●	•		•			●							•	•		●		
Guides to books			•		●				•															●		
Commercial organizations	•	●			•			•	●		•	●	•	•		•	●	•								●
Guides to commercial organizations		●						•		•							●									
Conference proceedings	●	●						●					•	•				•			•	•				
Guides to conference proceedings								•																		
Conferences								•					●		•			•								
Guides to conferences								•					●													
Educational organizations						•		•	•				•					●								
Guides to educational organizations		●		•		•		•										●								
General guides		●					•	•										●								
Government publications								●		•																
Guides to government publications		●		•		•		•	•				•					●						•		
Information services		•				•	•		●							●								•		
Guides to information services		●						•										●								
Journal articles	●		•	•				●	•				•							●	●●	•		•		
Newspapers and articles					•			●							●	•										
Guides to newspapers and articles		●				•		●										●								
Official organizations						•	•	•			•							●								
Guides to official organizations		●					•	•			•							●								
Patents	●		•		•			●			●	•								•	•					
Guides to patents	•	●		●			•	●															•			
People	•	●					●	•	•		•	•		•	●		•						•			
Guides to people		●			•		●											●						•		
Periodicals	•	●					●							●	•							●				
Guides to periodicals		●			•		●											●								
Products		●					●	●		•	●		•		●										●
Professional organizations		●			•		•						●	•								•				
Guides to professional organizations		●		•			●									●						•				
Reference material		●			●		●										●						●			
Guides to reference material		●					●										●									
Reports	●				●														•	•						
Guides to reports	•	●			•		●									●					•					
Review serials		●			●							●														
Standards		●			●				•		●													●
Guides to standards		●				●								●											
Subject bibliographies	●	●			●												•					●			
Guides to subject bibliographies		●		•		●									•									
Subject guides		●		•		●									●						•		
Theses and dissertations	•		•		•		●										●			•	•			
Guides to theses and dissertations		•	•		•		•									●					●		
Trade literature				●			●																	●

Chart 2. Routes from guides to sources of information

Another problem is that one guide may only take you a short way along the route to information.

Chart 2 shows the main routes from the simplest guides to the ultimate sources of information. It also emphasizes the way in which some guides can help you by giving a broad description of the sources (such as **Guides to periodicals**, which may give descriptions of the periodicals in your subject), or by giving details of specific information available from those sources (e.g. abstracting and indexing journals may give details of the articles in the periodicals).

For example, the chart suggests that if you are unable to identify any source or guide, you move one stage to the left. So if you are not sure which journal article you require, you can go to an abstracting or indexing journal. If you are not sure which of these to consult, you can go to a guide to abstracting and indexing journals. If you are not sure which guide you require, you can go to a general guide or a subject guide. These basic guides are covered on pp. 40 and 42.

Chart 3. Information required versus sources and guides

The information you require may be available from a number of different sources. This is partly because the same work is repeated by different people, and partly because one piece of work may be recorded in more than one place. For example, a boiling point may be measured by two different people and be reported in two different journal articles. The boiling point may be copied from one article into a data-compilation.

Chart 3 shows which sources and guides are likely to be the most useful when searching for particular types of information.

For example, if you wanted a brief introduction to a particular subject, the chart indicates that books, journal articles and reference material (and their appropriate guides) are likely to be useful. The size of the circle indicates the importance of the relationship.

12

Chart 2. Routes from guides to sources of information*

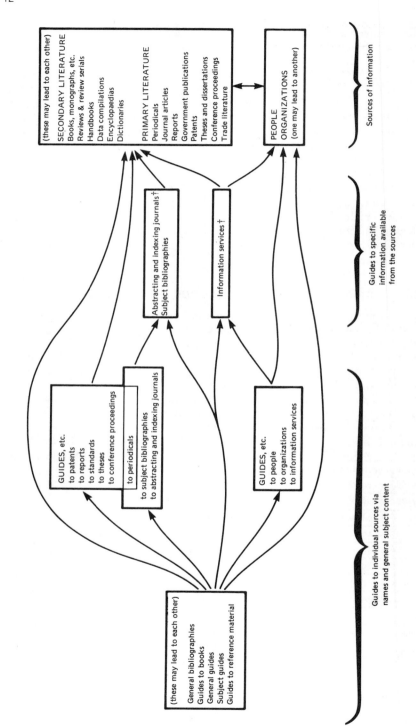

(these may lead to each other)
SECONDARY LITERATURE
Books, monographs, etc.
Reviews & review serials
Handbooks
Data compilations
Encyclopaedias
Dictionaries

PRIMARY LITERATURE
Periodicals
Journal articles
Reports
Government publications
Patents
Theses and dissertations
Conference proceedings
Trade literature

PEOPLE
ORGANIZATIONS
(one may lead to another)

} Sources of information

Abstracting and indexing journals†
Subject bibliographies

Information services†

} Guides to specific
information available
from the sources

GUIDES, etc.
to patents
to reports
to standards
to theses
to conference proceedings

to periodicals
to subject bibliographies
to abstracting and indexing journals

GUIDES, etc.
to people
to organizations
to information services

(these may lead to each other)
General bibliographies
Guides to books
General guides
Subject guides
Guides to reference material

} Guides to individual sources via
names and general subject content

* Libraries and information offices may be involved along all these routes by making literature available or offering advice
† These may also act as sources

Chart 3. Information required versus sources and guides

The size of the circle indicates the importance of the relationship

To find this type of information ▶

use these to guide or supply you

	Properties of materials and numerical data	Methodology and techniques	Product and service data	Brief introduction to a subject	'Some' references on a subject	Comprehensive literature surveys and reviews	Details of current research	The latest information on a subject	Detailed or advanced information on a subject
Abstracting and indexing journals	●	●		●	●	●	•	●	●
Bibliographies		•		•	●	●			●
Books and guides	●	●		●	•	●			●
Commercial organizations and guides	●	•	●		•	•	●	●	•
Conference proceedings and guides		•			•	●	•	•	●
Educational organizations and guides	•	•				•	●	●	
General guides	•		•	●		●	•		
Government publications and guides	•	•	•	•		•			●
Information services and guides	●	●	●	●	●	●	●	●	●
Journal articles	●	●		●	•	●	•	●	●
Newspapers (and articles) and guides			•	•		•	•	●	
Official organizations and guides	•	•	•		•	•	•	●	
Patents and guides		•	●			•			
People and guides	●	●	•	•	•	•	●	●	●
Periodicals and guides	•		•			•	●	●	●
Professional and research organizations and guides	●	•	•			•	●	●	●
Reference material and guides	●	•	●	●	●	•			•
Reports and guides	•	●	•		•	●	●	●	●
Reviews and review serials	•	•			•	●	•		●
Standards and guides	•	●	●			•			
Subject guides	•	•	•	●	•	●	•		
Theses and dissertations and guides	•	●				●		●	●
Trade literature	●	•	●	•	•	•	•		

People

People, ultimately, are the source of all information (i.e. human knowledge—recorded or otherwise) whether you approach them directly, or by way of the organization which employs them, or via something they have written, or through a reference to their work provided by an information service. It is important to bear this in mind, because it follows that all information is subject to the possibility of human error and it is sometimes right to question the accuracy of a piece of information (be it a theoretical statement, an item of numerical data, a reference to someone's work, an address, the title of a journal, or whatever).

You, yourself, are a potential source of information: you have acquired knowledge over the years, perhaps generating new information through your work, and you may have the ability to design/perform experiments which could provide data not available elsewhere. When you have a need for information, the first 'source' to try is your own brain (possibly backed up by its external store—any notes you may have made on the subject in the past). If this fails, you turn to outside sources on the basis that someone, somewhere may have, or may have recorded, the information required. Do not forget, however, that it is sometimes better to perform experiments than to seek information from (say) the literature. For example, if you were a mechanical engineer and needed to know, accurately, the physical properties of a particular sample of metal, it might be better to measure them yourself rather than rely on data obtained from the literature (which, after all, may not be valid for your sample because of some peculiarity in its composition). Approaching people directly can be a quick way of obtaining information, as well as giving access to their unpublished work. Bear in mind, however, that the help you obtain may depend very much on the manner of your approach. When dealing with people you should always be courteous, remembering their time may be of value: your attitude must certainly not be that of demanding assistance as as 'right'. A considerate, tactful approach may make all the difference between a generous exchange of information and no help at all. Do not forget that people like to be thanked when they have tried to assist you: a short note of appreciation takes very little time to write.

Often you will contact people through the organization which employs them: for more details of this see the section on **Organizations** (pp. 19–36). See also the section on **Choosing sources of information and their guides** (pp. 6–13).

Uses

- When the required information cannot be found (or does not exist) in the literature.
- Where speed and convenience are important.
- When advice is required, especially on alternative sources of information.
- For an expert opinion on a topic.
- To obtain further details of a particular piece of work from the person responsible.
- In the kind of situation where immediate two-way communication is a distinct advantage (obtaining instruction in the operation of a complicated machine or process, for instance).

Access

Via personal contact, telephone or letter or, sometimes, through a third party. A librarian or information officer may be able to put you in touch with a suitable person.

Caution

- It is unwise to bother busy or senior people with queries that can easily be answered elsewhere (if you do, you may get a short sharp reply).
- The telephone can be a poor means of requesting and communicating information; if you decide on its use, take extra care to be clear and precise. Cultivate the right way of approaching people; you will find it pays dividends.

Examples

- People already known by you who are active in your subject field.
- People recommended to you.
- People working in a similar field who may have had the same information requirement (how did they solve their problem?).

Guides to people

See also **Guides to organizations**, pp. 20, 25, 29 and 33; **Dictionaries of biography**, p. 107; **Information services**, p. 206

How do you find out who is the right person to consult? There are four effective methods.

(1) Ask someone (a colleague, for example, or a librarian/information officer if you have access to one).

(2) Examine the literature of the subject to see who has written journal articles, reports, books, etc., dealing with your problem area. If somebody writes something, he or she is presumably an 'expert' on it.

(3) Seek advice on who to approach from an appropriate organization (a professional or research association, for example).

(4) Consult one or more of the printed guides mentioned below, or search in relevant online databases.

The guides to people described in this section are basically lists of names, which may include addresses and biographical details. They are more useful for obtaining information about a known person than finding the names of people working in a given subject area or working for a particular organization. In the latter cases **Guides to commercial organizations** (p. 20), **Guides to educational organizations** (p. 25). **Guides to official organizations** (p. 29), and **Guides to professional and research organizations** (p. 33) should be consulted.

Uses

● To find information about a known person: e.g. address(es), telephone number(s), professional qualifications, career details, subject interests, organization(s) with which associated (including employer).
● Occasionally, to find out who is working in a particular subject area.

Access

● By visiting or writing to an appropriate library, or telephoning in the case of a simple query. The City Business Library (see p. 260) maintains a good collection of these guides.
● If you need information about someone living abroad, it may be helpful to contact an embassy or consulate of the country concerned (which may have a selection of guides not readily available elsewhere).

Caution

● Make sure the guide is recent. People change their jobs/addresses from time to time, so guides to people become out of date sooner or later.
● Less well-known people may be difficult to trace using standard guides. If they have published anything recently, it should be possible to contact them through the publisher concerned. Alternatively, the name/address of an employing organization often appears in association with the author's name at the head of a journal article (and is repeated in some abstracting journals).

Examples

Telephone directories.
○ These are the best-known guides of all, and hardly need description here. Remember to make use of the classified sections, or 'yellow pages', when appropriate. For the UK, the alphabetical sections are also available in microfiche form. More than 90 directories are required to cover the UK, so it is worth noting that British Telecom publishes an *Index to Telephone Directories* (consisting of an alphabetical list of places with the directory section number concerned). If you do not have access to a range of directories, use British Telecom's Directory Enquiries service for an inland telephone number, or follow the instructions given in the latest issue of your *Telephone Dialling Codes* booklet if you want to enquire about the number of a foreign telephone subscriber. In the latter case you may find the appropriate embassy or chamber of commerce able to help: the City Business Library (see p. 260) also maintains a collection of foreign telephone directories.

Who's Who: an annual biographical dictionary. London, A. & C. Black, annually.
o Coverage limited to 'top' people. Entries, which are arranged alphabetically by biographee, include full career details, with addresses. A companion publication, *Who was Who* (currently issued at 10-year intervals) contains the entries for those who have died during that period, and a cumulative index covering the period 1897–1980 is also available.

The International Who's Who. London, Europa Publications, annually.
o No connection with the previous example. Entries, which are arranged alphabetically by biographee, include career details and addresses. Coverage is limited to people of international fame.

Details of biographical dictionaries covering individual foreign countries may be obtained from *Walford's Guide to Reference Material* (see p. 115 of this book): Vol. 2, pp. 500–519 of the 4th edition.

Who's Who of British Scientists, 1983/84. Folkestone, Simon Books, 1983.
o Coverage limited to fairly senior people. Entries, which are arranged alphabetically by biographee, include career details and addresses. This publication, which has gone through several editions, was preceded by a two-volume *Directory of British Scientists, 1966–67.* The latter retains some value, as its coverage was very much wider; entries are arranged in the same way, and a subject approach is provided for.

Who's Who of British Engineers, 1982–83. London, Simon Books, 1982.
o Coverage is again limited to fairly senior people. Entries, which are arranged alphabetically by biographee, include career details and addresses. A category index lists names under broad subject headings, and information about some professional institutions is also provided.

Who's Who in Science in Europe. 4th ed. 3 vols. London, Longman, 1984.
o Coverage limited to senior people. Entries, arranged alphabetically by biographee, include brief career details and addresses. A subject listing of individuals by country is provided. Longmans also publish *International Who's Who in Energy and Nuclear Sciences* (1983), *International Medical Who's Who* (2nd ed., 2 vols., 1985) and *Who's Who in World Agriculture* (2nd ed., 2 vols., 1985).

American Men & Women of Science; Physical and Biological Sciences, edited by Jaques Cattell Press. 15th ed. 7 vols. New York, Bowker, 1982.
o A companion set of volumes deals with the social and behavioral sciences. The coverage of this guide is much wider than that of its British equivalents. Entries, which are arranged alphabetically by biographee, include brief career details and addresses. A microfiche version of earlier editions is available, and a hard-copy cumulative index to editions 1–14 has been produced. The database associated with this publication may also be searched online.

Current Bibliographic Directory of the Arts & Sciences: an international directory of scientists and scholars. Philadelphia, Institute for Scientific Information, annually.
o Produced by the publishers of *Science Citation Index* (see p. 52), this is a continuation of two earlier series: *International Directory of Research and Development Scientists* (1967–69) and *ISI's Who is Publishing in Science* (1970–77).

Information is currently drawn from the many periodicals and books whose contents are indexed by ISI every year. This is an up-to-date guide to the first authors of papers in journals covered by the ISI scheme, provided their addresses appear in the articles concerned. Co-authors are not normally listed, unless the publication states that reprint requests should be sent to a co-author rather than to the first author. Previous volumes are also of value, as the work does not cumulate. The author section consists of an alphabetical list of names with the (abbreviated) name/address of corresponding organizations, and bibliographic references for relevant papers published during the year in question. An organization section lists organizations alphabetically, giving their location. A geographic section arranges authors under their organizations in two sequences: by city within state for the USA, and by place within country for the rest of the world. In addition, there is a list of journals and books covered by the system.

Directory of Directors: a list of the directors of the principal public and private companies in the United Kingdom, with the names of the concerns with which they are associated. East Grinstead, Thomas Skinner Directories, annually.
○ This is basically an alphabetical list of directors with names (and sometimes addresses) of associated companies. It is not possible to identify the directors of a particular company using this work: if you need to do so, consult *The Stock Exchange Official Year-Book* (see p. 24 of this book).

Register of Consulting Scientists, Contract Research Organisations and other Scientific and Technical Services, edited by D. J. B. Copp. 6th ed. Bristol, Adam Hilger, 1984.
○ This guide is published for the Council of Science and Technology Institutes (CSTI), and only those practices or contract research organizations which have a corporate member of a CSTI body as principal, partner or director are eligible for inclusion. The first major section is a subject index, with entries arranged under fairly broad headings. These refer to a classified register where there are five categories: independent full-time consultants, analytical and test houses; independent organizations offering contract research in addition to consulting and test facilities; university departments or bureaux offering contract research in addition to consulting and test facilities; part-time consultants, including academics; and useful addresses. A name index is also provided. Entries in the register section give names, addresses, telephone and telex numbers of consultancy services, names of senior personnel, and descriptions of the subject areas covered.

Finally, do not forget that many professional organizations publish a *List of Members* (for relevant guides, see p. 33); individual *University Calendars* cover academic staff; and there are *Research Indexes* in various fields, e.g. *Aerospace Research Index, Agricultural Research Index* (both published by Longman in the UK and distributed by Gale in the USA) – for details of research in your subject area you can consult one of the **Guides to reference material** (see p. 115) or an appropriate subject guide (see pp. 42–46). See also **Guides to professional and research organizations** (pp. 33–36).

Organizations

We take an organization to mean an organized body of people working together for a particular purpose or purposes. Apart from the individual people in the group, there may be associated collections of internal records, literature and equipment. The organization is usually based in buildings and may be contacted via its addresses and telephone numbers, and occasionally telex numbers. Organizations are a very important source of information. You may obtain information via an organization's personnel (see also the previous section on **People**); via its document collection, which may include a specialized library, internal reports, etc.; via its publications; or via its facilities (it may hire out its equipment, or do contractual work on request such as testing or analyzing materials).

When choosing an organization as a potential source of information, you should consider what was said about choosing sources in an earlier section and also what is said under the different types of organization (see below). In general, you will contact an organization which has expertise in a particular field, whether it be a subject (e.g. detergent chemistry) or a product or service (e.g. electric motors). It can sometimes help to approach an organization through one of its personnel whom you know (if only by name), even though you think that your enquiry will be passed on. You will sometimes have to accept that information is either not available to you (it may be confidential to the organization concerned) or that you may even be required to pay for it. Remember that, as with 'people', organizations can frequently offer advice or pass on your enquiry, even though they cannot be of direct help themselves.

It is convenient to divide organizations into four main categories:

(1) Commercial
(2) Educational
(3) Official
(4) Professional and research

Apart from the conceptual differences between these categories, many of the guides to organizations follow similar divisions. Some organizations fall into two or more of these categories, and where there is overlap, be prepared to consult the guides for all the categories concerned. As with many other types of guide, the guides to organizations are becoming available as databases in online information services, so be prepared to check in the appropriate directories (p. 186) if you wish to use this type of service.

When you contact an organization, you do so through people, and in all communication with people you should remember that a polite and considerate approach can make all the difference to the quantity and quality of the information that you obtain.

Commercial organizations

These are organizations which offer products or services, usually on a purely profit-making basis: manufacturers, suppliers, consultants, etc.

Uses

• To obtain specific details of products or services, including after-sales services.
• For information associated with these products; research, development, exploitation, etc.
• Occasionally to do work for you, usually under contract, in order to provide special information.

Access

Write, telephone or visit the sales or enquiry office, or library/information service, unless you already have a personal contact. Representatives will usually visit you if required, sometimes on a routine basis.

Caution

• They may be naturally biased towards their own products, and may be reluctant to advise on the products of other organizations.
• They may expect you to buy something (or pay) if they offer substantial help.

Guides to commercial organizations
See also Trade literature, p. 136

These contain lists of organizations giving various details such as address, telephone number, product or service, etc. Their names may usually be found alphabetically, or via the product, service or subject activity of each organization. Some of the guides mentioned below cover foreign organizations, or their UK associates, but in cases of difficulty the Science Reference Library (p. 259) or the larger public reference libraries are usually well equipped to help you; alternatively, you may contact the appropriate foreign embassy or chamber of commerce. The Companies Registration Office in London is especially useful for detailed financial information, structure and affiliations of known British registered companies.

The number of online databases effectively duplicating hard-copy guides is growing, and includes *EUROPEAN KOMPASS ON-LINE (EKOL)* which is almost identical to the printed data. Furthermore there are completely new databases with no hard-copy equivalent, such as *ICC DATABASE* which consists of a directory of some 900 000 companies and a financial data file for some 60 000 companies.

Uses

• Knowing the organization's name you can find its address, telephone number, what it produces or supplies, its connections with other organizations, the names of its directors, its research interests, etc.
• If you want a particular product or service, you can find out which organization manufactures or supplies it.
• If you are interested in a particular field of research, you can find out which organizations are active in that field.
• To find which organizations have libraries covering your subject interests.

Access

• By visiting or writing to an appropriate library, or telephoning in the case of simple enquiry (such as the Science Reference Library or the City Business Library in London).
• Use of an online information service or broker.

Caution

Make sure the guides are recent. Some guides are surprisingly inaccurate, and it is always advisable to check the address in a second guide, if you are writing to an organization, or the telephone number with Directory Enquiries, if you are telephoning the organization.

Examples

Trade directories and buyer's guides.
○ These exist in most product or service areas, such as plastics, chemicals, engineering, instrumentation etc., including *Machinery Buyer's Guide, Engineer Buyer's Guide, Electrical and Electronics Trades Directory*, and *Laboratory Equipment Directory*. Details of these and other guides may be found in **General guides**, **Subject guides** and **Guides to reference material**. The following examples are more general in coverage.

KOMPASS: United Kingdom. 2 vols. East Grinstead, Kompass Publishers, annually.
○ Volume 1 (Products and Sources) enables you to find details of suppliers of a product or service. There is an alphabetical product and service index which indicates the corresponding classification number. Using the classification numbers in the 'classified section', you can find the names and addresses of the appropriate suppliers. This section includes broad subject tables in English, German and French, and a series of grids which enables a large amount of information to be contained in a relatively small space. Volume 2 (Company Information) helps you to find out more about a company when you know its name. There is an alphabetical index of companies indicating which companies are covered, their general location, whether or not they are members of the Confederation of British Industry, and a brief alphabetical list of trade names and corresponding manufacturers. In the main section the companies are arranged alphabetically under their locations, county then town, and then under their own names. Under each entry you may find such details as full address and telephone numbers, names of directors, office hours,

bankers, executives, share capital, number of employees, and an indication of the nature of business of each company or a list of product groups. (*KOMPASS: UK* is also available online as part of *EKOL*.)

There are separate sets of volumes for organizations in the following countries: Australia, Belgium, Denmark, France, Holland, Indonesia, Italy, Luxembourg, Morocco, Norway, Singapore, Spain, Sweden, Switzerland and West Germany. There are also UK Kompass Management Registers covering the main regions of the UK, such as Greater London, South England, North West England etc. For online availability, see KODA ONLINE (p. 157).

British Exports. 2 vols. 16th ed. East Grinstead, Kompass Publishers, 1984.
○ This is primarily intended for the overseas buyer as a guide to sources of British products and services. In volume 1, the main section is an alphabetical list of some 16500 products which are available for export from the UK, together with names and addresses of exporters and a code that indicates the parts of the world to which they are interested in exporting. There are separate indexes to products in French, German and Spanish. Volume 2 contains two main sections covering certain technical data and illustrations for some of the products in volume 1, under company name; and an alphabetical listing of exporters giving names, addresses, and overseas agents. There are two small sections covering trade names and the British property market.

Kelly's Manufacturers and Merchants Directory. East Grinstead, Kelly's Directories, annually.
○ This directory contains five sections: classified listings of manufacturers, merchants, wholesalers and firms offering an industrial service, under separate subject headings; a small classified section for the offshore oil trades and industry; a short alphabetical list of brand and trade names giving name of the manufacturer or UK distributor for foreign products; a large section on company information with organizations listed alphabetically and giving details of addresses, telephone number, telex, and their trade; and an international exporters and services section under separate trade headings.

Key British Enterprises. The top 20000 British companies. 2 vols. London. Dun and Bradstreet, annually.
○ The main section in both volumes consists of an alphabetical list of companies giving details of address, telephone and telex numbers, trades, trade names, directors, financial data, number of employees and standard industrial classification (SIC) codes. Volume 2 also contains an index to the SIC codes (subject and numeric), a SIC classified listing of companies and a geographical (country) list of companies under SIC code headings. Dun and Bradstreet publish similar guides for Austria, Belgium and Luxembourg, Denmark, Holland, Finland, France, Italy, Norway and Sweden. There are several directories covering the US and Canada, too. This guide is also available as an online database.

Sell's Directory: products and services. Epsom, Sell's Publications, annually.
○ The main section consists of an alphabetical list of organizations giving addresses, telephone and telex numbers. There is a products and services subject index and an alphabetical listing of trade names, both using a simple coding system to link up with the main section.

Who Owns Whom: United Kingdom and Republic of Ireland. 2 vols. London, Dun and Bradstreet, annually with quarterly supplements.

○ Volume 1 consists of four sections: UK parents showing subsidiaries and associates registered in the UK; Ireland parents showing subsidiaries and associates registered in the Republic of Ireland; foreign parents showing British and Irish subsidiaries and associates registered abroad; and a list of consortia groups showing member companies where appropriate and an alphabetical index of member companies showing the parent consortium. Volume 2 provides a complete index of subsidiaries and associates listed in volume 1 showing their parent companies. Dun and Bradstreet publish similar guides for continental Europe, North America, and Australasia and the Far East. This guide is also available as an online database.

Major Companies of Europe 1983, edited by R. M. Whiteside. 2 vols. London, Graham and Trotman, 1982.

○ Volume 1 covers companies in the European Economic Community, while volume 2 covers the rest of Western Europe. Entries are listed alphabetically within countries, and include name, address, telephone and telex numbers, directors, principal activities, financial information, affiliations, trade names, and number of employees. There are volume indexes to companies, companies by country, and companies by business activity subdivided by country.

Europe's 10 000 Largest Companies 1984. London, Dun and Bradstreet, 1984.

○ This guide is a source of financial and statistical information, presented in a tabular form. There are various listings, with the main sections covering the largest industrial companies and trading companies. Separate lists cover banking, transport, insurance, advertising, and hotel/restaurant industries. There is an index to companies.

Principal International Businesses. New York, Dun and Bradstreet International, annually.

○ One large volume covers about 50 000 leading businesses in 133 countries. The first main section is divided alphabetically by country and then by business, and includes for each organization details of address, cable/telex number, lines of business, number of employees, and sales figures. The second main section consists of businesses classified by Standard Industrial Classification (SIC) business code numbers. The third section is an alphabetical list of organizations. The directory includes numerical and subject indexes for the SIC numbers.

Anglo American Trade Directory. London, American Chamber of Commerce (United Kingdom), annually.

○ This directory includes details and background information of American activities in the United Kingdom, e.g. American Chamber of Commerce, American Embassy in the United Kingdom, Anglo activities in America, Anglo American organizations. The main section is the trade register, which lists British and American businesses having trade and/or investment relations with each other. This register is particularly useful for finding the name, address and telephone number of the British representative or distributor of a known American company. It may also be used to find the addresses of American companies, if they have UK affiliations. Industrial subject and regional (US and UK) classified lists are provided as well.

Thomas Register of American Manufacturers and Thomas Register Catalog File. New York, Thomas Publishing Company, annually.
○ This is a very large multivolume (currently 18 volumes) work covering some 120 000 US companies. The first group of volumes lists products and services alphabetically; the next two volumes give company profiles and include addresses, telephone numbers and other company information; and most of the remaining volumes contain catalogue data for nearly 1200 companies. There is also a brand names index, a subject index to the classified section, and a section dealing with transportation. An online database version is available too.

Standard and Poor's Register of Corporations, Directors and Executives. 3 vols. New York, Standard and Poor's Corporation, annually with cumulative quarterly supplements.
○ This covers US, Canadian and major international corporations. Volume 1 consists of an alphabetical listing of corporations giving such details as address, telephone number, senior officers, annual sales and number of employees. Volume 2 is an alphabetical listing of directors giving brief biographical details. Volume 3 contains various indexes covering corporations classified by standard industrial classification codes; geographical locations; cross-references between subsidiaries, divisions and parents; and individuals and companies new to the latest edition of the guide.

The Stock Exchange Official Year-Book. London, Macmillan, annually.
○ Covers companies and public corporations, utilities and other organizations throughout the world, whose securities are handled by British stock exchanges. The entries for each company are arranged in alphabetical order, and usually include details of address, telephone and telex numbers, directors, history, principal subsidiaries, capital and other financial items. The *Year-Book* includes details of other securities, general information and statistics.

CRO Directory of Companies. Cardiff, Companies Registration Office, three-monthly with weekly cumulative supplements.
○ A subscription service produced on magnetic tape, microfilm or microfiche, and containing information on all UK and foreign companies registered in Great Britain (England, Wales and Scotland). Information includes company name, registered office address, annual accounts and returns.

Extel Cards: The British Company Service. London, Extel Statistical Services, daily.
○ A subscription service with daily updates, covering almost every company listed on the British and Irish Stock Exchanges, giving such details as activities, balance sheets, board members, capital, dividend record, yields, earnings etc. Extel also offer services for unlisted securities, unquoted companies, overseas companies, and a special service for analysts and also for enquirers wishing to know about companies not covered by Extel cards. For online availability see FINSBURY DATA SERVICES, p. 156.

See also:
Guides to people (p. 15)
Guides to official organizations (p. 29)

Guides to professional and reseach organizations (p. 33)
Guides to libraries (p. 260)
Guides to information services and suppliers (p. 186)

Educational organizations

These are mainly the universities, polytechnics, colleges and institutions where research and development work is taking place in addition to teaching.

Uses

- Expert advice especially on techniques, use of equipment, etc.
- Information on very advanced and narrow subject fields.
- Details of the very latest academic research in a particular field.
- Information via contractual work or available facilities.
- Library services.
- Educational courses, summer schools and special lectures.

Access

It is usually preferable to contact a known individual; otherwise try one of the following: enquiries, reception, public relations, industrial liaison, departmental administrators or secretaries, library or information office.

Caution

- Universities, in particular, cover a great range of subjects, and 'official' channels are frequently out of touch with a lot of the work that is going on. It is sometimes advisable to treat individual university departments as separate organizations!
- Individuals may be difficult to contact, especially over the telephone. This is due to the nature of their work, which may result in unusual hours, travel, meetings and lectures. Vacations are particularly troublesome in this respect.

Guides to educational organizations
See also **Theses and dissertations**, p. 133

These range from guides dealing with individual organizations to those with international coverage, which list the various organizations, giving appropriate information about each one, and often including a subject index. Although most of these guides are currently available only as printed versions, it is to be expected that they will become accessible as online databases in the future.

Uses

- Finding information about a particular organization such as address, structure, staff, research, etc.
- Identifying organizations working in a particular subject field.

Access

By visiting or writing to an appropriate library, or telephoning in the case of a simple enquiry.

Caution

Bear in mind that even the latest guides refer to the previous academic year. Subject indexes are frequently inadequate.

Examples

Individual educational organizations usually issue their own guides, such as a university 'Calendar' or 'Prospectus' which may give details of courses, staff, committees, departmental interests, etc. The following guides offer wider coverage.

Research in British Universities, Polytechnics and Colleges (RBUPC). (Also covers government departments and some other institutions.) 3 vols. London, British Library, annually. Volume 1, Physical Sciences (includes engineering); Volume 2, Biological Sciences; Volume 3, Social Sciences.
○ Brief details of active research topics are given for the academic year. Each volume contains an alphabetical list of departmental names of all contributing institutions, giving addresses and telephone numbers. The main section consists of institutions and departments (arranged under broad subject headings), listing members of staff, subject areas of research, sponsoring bodies and dates of work. There is also a name index of research workers and a subject index. The lists of research topics and the indexes are related by subject and organization code numbers.

From 1985 *RBUPC* will be renamed *Current Research in Britain*, and may include an additional volume to cover humanities.

Chemical Research Faculties: an international directory. London, The Royal Society of Chemistry, 1984.
○ This new guide gives details of more than 8900 faculty members and 737 departments in 62 countries, including current research and recent publications. There are indexes covering research subjects, faculties, and institutions.

Graduate Studies: a guide to postgraduate study in the UK. Cambridge, Hobsons Press, annually.
○ The main section comprises details of the research facilities and courses, and basic information about the institutions in which postgraduate study is available. There are four sections: humanities and social sciences; biological, health and agricultural sciences; physical sciences; and engineering and applied sciences. Each section is divided into subject areas. There is a list of institutions giving address and telephone number, and there are indexes to course titles, and broad subject headings.

Graduate Programs in the Physical Sciences and Mathematics. Book 4, New Jersey, Peterson's Guides, annually.
○ This is one of a set of five guides covering the US, with Book 1 comprising a general overview and directory; Book 2 covering humanities and social

sciences; Book 3: biological, agricultural and health sciences; and Book 5: engineering and applied sciences. The main part of the subject books is divided into broad subject sections, each containing brief entries for universities and colleges in alphabetical order, including profiles of the organizations and their programs. Sometimes these profiles are enhanced by 'announcements' which give details of specific research interests and facilities. In other cases, additional two-page descriptions are given at the end of each subject section. Book 1 contains profiles on all the universities and accredited colleges, some with announcements, and some with two-page descriptions. It also contains a directory that lists 251 subject fields, showing which organizations offer graduate studies, and another directory which lists the institutions, showing which subject fields they offer and in which Book and in what detail they are covered.

Directory of Technical and Further Education. 19th ed. London, George Godwin, 1983.

○ This book gives a general survey of the framework of further education in the UK and provides limited information on all technical and further education colleges, polytechnics, universities, and professional organizations. There is an index to the teaching bodies, but no subject index. This is a good starting point for identifying facilities in particular areas or for obtaining brief details of specific institutions.

The World of Learning. London, Europa Publications, annually.

○ The organizations covered include learned societies, research institutes, libraries, museums, universities, polytechnics and colleges. There is one section for each country and one for international organizations. The amount of information given varies considerably according to the type of organization. For example, in the case of universities all senior academics and officers are named, but in the case of learned societies only a brief paragraph is given. An index of institutions is provided.

International Handbook of Universities: and other institutions of higher education. 9th ed. Paris, International Association of Universities, 1983.

○ Published every 3 years, this is a companion volume to the two guides below, and supplements their coverage. The contents are arranged in order of country, and extensive information about each organization is given in many cases. There is an index of institutions.

Commonwealth Universities Yearbook: a directory to the universities of the Commonwealth and the handbook of their Association. London, Association of Commonwealth Universities, annually.

○ Each entry usually contains details of staff and a section on general information. There is a general subject and institution index, also a names index.

American Universities and Colleges. 12th ed. New York, American Council on Education, 1983.

○ There are two general sections giving background information on higher education and professional education in the USA. The main section, which is arranged alphabetically by state, lists the local universities and colleges, giving brief information about each. There is a general index and an institutional index.

British Qualifications: a comprehensive guide to educational, technical, professional and academic qualifications in Britain. 14th ed. London, Kogan Page, 1983.
○ This guide includes details of qualifications obtained in secondary schools; in further education establishments, universities and polytechnics; and in connection with professional associations. There is an index of abbreviations and designatory letters, a list of qualifications arranged by trades and professions, and a general index covering qualifications, subject fields, institutions or examining bodies.

See also:
Guides to people (p. 15)
Guides to official organizations (p. 29)
Guides to professional and research organizations (p. 33)
Guides to libraries (p. 260)
Guides to information services and suppliers (p. 186)

Official organizations
See also Patents, etc., p. 78; Standards, p. 127

These are the organizations associated with government, administration and control, including national and local government departments, offices, library and information services. Some are international in character.

Uses

● Official information of any kind, including statistics and legislation.
● Information connected with all branches of government work (this covers most subject fields to some extent), including associated library and information services.

Access

● If you do not have a specific department or individual to contact, it is normally preferable to write a letter unless you are in a hurry for your information. If you do telephone, without a specific contact in mind, you may have to rely on a switchboard operator. In this case, ask the operator to connect you with an enquiry office, or explain your problem in simple terms and ask to be connected to a likely department.
● In some cases, particularly those where it is necessary to search files or collections of documents, it is often fairer and preferable to visit the organization and conduct the search yourself, under guidance if appropriate.

Caution

With really large organizations it is often difficult to locate the person who can help, especially over the telephone. You may have to say which library you require, or which Mr T. Smith. The various departments may operate independently and be unaware of what is done outside their section. At the worst, you may find yourself shunted from one telephone extension to another, and this can be frustrating, expensive and time-consuming (see *Access*, above).

Guides to official organizations
See also Government Publications, p. 69; General guides, p. 40; Subject guides, p. 42

These are usually lists of organizations giving appropriate details about each one and often including a subject index. It is likely that these will gradually become available via online information services, and in future it may be necessary to consult the appropriate directories (p. 186).

Uses

● Finding information about a particular organization such as address, telephone number, names of staff, structure, relationships with other organizations, and subject activities.
● Identifying organizations concerned with particular subject areas.

Access

By visiting or writing to an appropriate library, or telephoning in the case of a simple enquiry.

Caution

Name changes sometimes occur in this type of organization (e.g. government departments). Subject indexes are frequently inadequate.

Examples

Britain: an official handbook. London, HMSO, annually.
○ This is a general reference work covering administration and national economy, including the activities of many national institutions and the part played by government in the life of the community. There are 24 chapters, which include industry; agriculture; energy and natural resources; fisheries and forestry; transport and communications; and promotion of science and technology. There is a subject index.

Technical Services for Industry: technical information and other services available from government departments and associated organizations. London, Department of Trade and Industry, 1981.
○ The organizations covered include the Departments of Trade and Industry, Education and Science, Employment, and Health and Social Security; the Ministries of Agriculture, Fisheries and Food, and Defence; and the various associated services, centres, research associations, authorities, establishments, laboratories, etc. There is an index covering subjects, organizations, publications and acronyms.

CODATA Bulletin. Oxford, Pergamon, bi-monthly.
○ Although this is basically an interdisciplinary journal which publishes conference proceedings on the subject of data in science and technology, research papers, and CODATA reports on new developments in data handling and presentation, it also incorporates the *Directory of Data Sources for Science and Technology*. Certain issues of the *Bulletin* are essentially concise directories of data centres and other formal projects which serve as sources of

quantitative numerical data on properties of well-defined physical and chemical systems, and with each issue devoted to a subject area, such as corrosion or geomagnetism.

The Municipal Yearbook, and public services directory. London, Municipal Publishers, annually.
○ A general reference work for people concerned with government and public affairs. The book contains 30 sections for general information on public services, such as civil aviation, education, health, police, power, roads and transport, town and country planning, and water. Each section gives appropriate descriptions, names, addresses and telephone numbers. There are also separate sections on such topics as government departments, county councils, the GLC, district and local councils, and government in Scotland and Northern Ireland. There is a general index, indexes to existing and former local authorities, and a list of principal officers of local authorities.

The Civil Service Yearbook. London, HMSO, annually.
○ This includes coverage of the Royal households and offices; parliamentary offices; ministers and departments (the main section); libraries, museums, galleries, and research councils; and the departments and other organizations for Northern Ireland, Scotland and Wales. Details include structure, staff, addresses and telephone numbers. There is an index to departments and other organizations, and an alphabetical list of officers.

The Statesman's Year-Book: statistical and historical annual of the states of the world, edited by J. Paxton. London, Macmillan, annually.
○ A general reference work covering international organizations and the different countries of the world. There is an index to place names and international organizations.

Dod's Parliamentary Companion. Hailsham, East Sussex, Dod's Parliamentary Companion Ltd, annually.
○ This guide includes a large biography of the members of the House of Lords and the House of Commons, details of the select committees, principal officers and clerks, government and public offices, international, European and parliamentary organizations. There are various explanatory sections and a short general index.

Councils, Committees and Boards: a handbook of advisory consultative, executive and similar bodies in British public life, edited by L. Sellar. 6th ed. Beckenham, CBD Research, 1984.
○ The book provides information on bodies formed to advise the government or public authorities, or to exercise regulatory, executive, administrative or investigatory functions in the public interest. It is arranged in four sections: (1) alphabetical register of bodies, giving full name, abbreviation, address, telephone and telex number, date and authority of establishment, constitution, names of chairman and secretary, terms of reference, responsibilities, duties, activities and publications; (2) index to names of chairmen; (3) index to acronyms and abbreviated names; (4) subject index to fields of interest and activity.

Exporters' Encyclopaedia—the world marketing guide. London, Dun and Bradstreet, 1984, with fortnightly supplement.

○ This publication contains comprehensive data on export regulations around the world. The *Export Documentation Handbook* is included in the price, and gives export documentation requirements in chart form for over 220 world markets.

Export Handbook, prepared by the British Overseas Trade Board. London, HMSO, 1983.
○ A useful guide to export services, regulations, insurance, finance, packing, transportation, markets, and education and training for export. There is an index and various helpful appendices giving names, addresses, telephone and telex numbers of organizations and departments.

Selling to Western Europe, prepared by the Exports to Europe Branch of the British Overseas Trade Board. 2nd ed. London, HMSO, 1982.
○ A guide aimed at those with little or no experience, and intended to help with the basic procedures. There are useful appendices similar to those in the previous guide, and a brief index.

Directory of Grant-Making Trusts. 8th compilation. London, Charities Aid Foundation, 1983.
○ The book is divided into four parts: a subject classification scheme; trusts listed under classifications; register of grant-making charitable trusts, in alphabetical order, giving details of the establishment and operation of each trust, including finances and name and address for correspondence; geographical index of trusts, alphabetical index of subjects and alphabetical index of trusts.

The Grants Register 1983–1985, edited by C. A. Lerner and R. Turner. London, Macmillan, 1982; published every 2 years.
○ Primarily intended for students, this guide has international coverage. The main section is an alphabetical list of the awarding bodies, which gives details of the awards such as purpose, subject, value, tenure, eligibility and address for further information. There is a separate subject index giving nationality codes for students, and an index of awards and awarding bodies.

The International Foundation Directory, edited by H. V. Hodson. 3rd ed. London, Europa Publications, 1983.
○ A directory covering international foundations, trusts and other similar non-profit institutions, listing them alphabetically by country and giving brief details such as history, activities, finance, administration, address and telephone numbers. There are indexes to foundations and to main activities.

The United States Government Manual. Washington DC, Office of the Federal Register, annually.
○ This is the official handbook of the Federal Government providing comprehensive information on the agencies of the legislative, judicial and executive branches, and on quasi-official agencies, international organizations in which the United States participates, and boards, committees and commissions. There is a list of acronyms and abbreviations, and indexes to personal names and to subjects/agencies.

Congressional Directory 98th Congress 1983–1984. Washington DC, Government Printing Office, annually.

○ Similar in much of its coverage to the previous publication, but including a large biographical section on senators and representatives.

The Europa Year Book: a world survey. 2 vols. London, Europa Publications, annually.
○ An authoritative reference work covering political and economic institutions as well as commercial organizations. It is divided into three sections: international organizations; European countries; and countries of the rest of the world. There is an index to international organizations and another to territories. For each territory there are introductory and statistical surveys and a directory covering all kinds of national institutions, including broadcasting, finance, publishing, insurance, trade and industry, and transport.

Yearbook of International Organizations, edited by the Union of International Associations. 3 vols. Munich, K. G. Saur, bi-annually.
○ Volume 1 consists of a general index including names, abbreviations, subject keywords, former names, and names of executive officers. The body of entries is accessed via a code made up from a section/entry number. Information given for each entry, apart from address and telephone number, may include aims, members, structure, officers, finance, relations with other organizations, activities and publications. Volume 2 lists organizations classified by country of secretariat location. Volume 3 contains organizations classified by subject, organization interlinks, and cross-references, and publications.

See also:
Guides to people (p. 15)
Guides to professional and research organizations (p. 33)
Guides to libraries (p. 260)
Guides to information services and suppliers (p. 186)

Professional and research organizations

These organizations exist to assist members in their individual professions, trades or pastimes, and also to promote knowledge in particular subject fields. They include most associations, societies, institutes and institutions.

Examples are: The Royal Society (a learned body), The Institution of Mechanical Engineers (a professional association), The National Union of Railwaymen (a trade union), British Rubber Manufacturers' Association Ltd (a trade association), British Beekeepers Association (an interest group), The Electrical Research Association (a research association).

They do not include professional firms, such as consultants (which are treated as commercial organizations), or the research sections of educational or commercial organizations.

Uses

● Expert sources of information on all aspects of a particular subject field, profession, trade, etc.
● Specialist library and information resources, and, in some cases, research facilities.

Access

By visiting, writing, or telephoning in the case of simple enquiries. Contact may often be made via their library or information office.

Caution

They may offer a limited service or even refuse help to non-members.

Guides to professional and research organizations
See also **General guides, p. 40; Subject guides, p. 42**

These usually contain lists of organizations and give various details such as addresses, telephone numbers, names of staff, subjects covered, and services offered. There may also be a subject index. It is likely that these guides will eventually become available as databases in online information services.

Uses

• To find information on a given organization.
• To find out which organizations can give information on a particular subject, class of product, service, etc.

Access

By visiting or writing to an appropriate library or telephoning in the case of a simple enquiry.

Caution

Subject indexes often only offer a rough guide, resulting in several organizations being listed under one broad heading. This usually means contacting them one after another until one that can help is found.

Note that many professional associations regularly issue lists of members, which may include addresses, telephone numbers, etc.

Examples

Trade Associations & Professional Bodies of the United Kingdom, by Patricia Millard. 6th ed. Oxford, Pergamon, 1979.
○ This guide consists of 3 parts: Part 1 is an alphabetical directory of organizations, giving address and telephone number, with an alphabetical subject index to these organizations; Part 2 is a geographical index by town (excluding London); Part 3 is a list of chambers of commerce, trade, industry and shipping, and a list of UK offices of overseas chambers of commerce.

Directory of British Associations: interests, activities and publications of trade associations, scientific and technical societies, professional institutes, learned societies, research organizations, chambers of trade and commerce, agricultural societies, trade unions, cultural, sports and welfare organizations in the United Kingdom and in the Republic of Ireland, edited by G. P. Henderson and S. P. A. Henderson. 7th ed. Beckenham, Kent, CBD Research, 1982 (8th ed. due in 1986).

○ This guide consists of an alphabetical directory of organizations, giving information about each, using a system of abbreviations and symbols. There is an index to abbreviations (acronyms of organizations) and a subject index. An 8th edition is due in 1986.

National Trade and Professional Associations of the United States. Washington DC, Columbia Books, annually.
○ The body of this directory comprises an alphabetical listing of associations, giving brief details such as address, officer for contact, numbers of members and staff, annual budget, historical notes, publications and annual meetings. There are subject, geographic and budget indexes.

Associations' Publications in Print. 2 vols. New York, Bowker, 1982.
○ A comprehensive listing of publications and audio-visual materials produced and sold by national, state, regional, local and trade associations in Canada and the United States. Volume 1 consists of a subject index, and volume 2 contains title, publisher (giving titles), association and acronym indexes.

Directory of European Associations. Part 1: National industrial, trade & professional associations, edited by I. G. Anderson. 3rd ed., 1981. *Part 2: National learned, scientific & technical societies*, edited by R. W. Adams. 3rd ed., 1984. Beckenham, Kent, CBD Research.
○ These directories do not cover international, pan-European or British associations. There are subject indexes in English, French and German; a subject-classified directory of organizations, giving a wide range of information about each, using a system of abbreviations and symbols; an abbreviations index and an index of organizations.

Pan-European Associations: a directory of multi-national organisations in Europe, edited by C. A. P. Henderson. Beckenham, Kent, CBD Research, 1983.
○ This new guide covers most European voluntary organizations and international voluntary organizations with memberships limited almost entirely to Europe, but excluding ordinary national organizations or those only covering specific regions such as Scandinavia. Entries are arranged alphabetically with extensive cross-referencing. Main entries are under English titles, giving such details as address, telephone and telex numbers, activities, membership data, affiliations and publications. There are separate indexes to abbreviations and subjects.

World Guide to Scientific Associations and Learned Societies, edited by B. Verrel and H. Opitz. 4th ed. Munich, K. G. Saur, 1984.
○ This guide lists about 22 000 associations from 150 countries. Regional, national and international associations in every area of science, culture and technology are included. Entries are in alphabetical order under country, giving name, telephone number, membership, and brief details of publications. There is a subject index.

Encyclopedia of Associations, edited by D. S. Akey. 4 vols. Detroit, Gale Research, annually.
○ Volume 1 is the basic work in 2 parts, providing detailed entries for over 18 000 active associations, organizations, clubs and other non-profit membership groups. Entries include name, address, officers, membership

size, areas of concern, publications, conventions and meetings. There is an alphabetical and keyword index for names and subjects. Volume 2 contains geographical and executive name indexes; volume 3 contains new associations and projects with indexes; and volume 4 covers international organizations whose headquarters are located outside the US. Gale has also published *World Guide to Scientific Associations* in 1982 and *World Guide to Trade Associations* in 1980.

Industrial Research in the United Kingdom: a guide to organizations and programs, edited by T. I. Williams. 10th ed. Harlow, Longman, 1983, with supplements.
○ The guide covers industrial firms, public corporations, research associations and consultancies, government departments, universities, polytechnics, trade and development associations, and learned and professional associations. Entries for each organization may include address, telephone and telex numbers, product range, activities, publications, number of graduate research staff, annual expenditure, and names of some senior staff. There are three indexes covering personal names, establishments, and subjects. An 11th edition is due in 1985.

European Research Centres: a directory of organizations in science, technology, agriculture and medicine, edited by T. I. Williams. 5th ed. Harlow, Longman, 1982.
○ This directory covers major industrial research laboratories in private and public corporations, government laboratories, research funding organizations, universities and their research institutes. Information given for each centre includes title of organization/English translation, address, telephone and telex numbers, affiliation, senior officers, numbers of graduate staff, annual expenditure, activities and major projects, contract work and publications. There are indexes to organizations and to subjects. A 6th edition is due in 1985. Longman have published or expect to publish the following world directories of organizations and programmes:
  Engineering Research Centres
  Earth and Astronomical Sciences Research Centres
  Medical Research Centres
  Agricultural Research Centres
  Materials Research Centres
  World Nuclear Directory
  World Energy Directory
  Pollution Research Index
  Aerospace Research Index
  Science and Technology in China
  Science and Technology in Japan

International Research Centers Directory, edited by K. Gill and A. T. Kruzas. 2nd ed. Detroit, Gale Research, 1984, with supplements.
○ This work covers research facilities of all types, including government, university, and private research firms throughout the world, but excluding the US. Entries are arranged by country, and include: name, address, telephone number, director, date established, affiliations, number of staff, activities, fields of research, library holdings, publications and seminar programs. There are name and keyword, and country indexes. Gale Research publish two other directories which cover the US, both with regular supplements:
  Research Centers Directory
  Government Research Centers Directory

Research in the United States is also covered by two guides published by Bowker:

Industrial Research Laboratories of the United States
Energy Research Programs Directory

See also:
Guides to people (p. 15)
Guides to commercial organizations (p. 20)
Guides to educational organizations (p. 25)
Guides to official organizations (p. 29)
Guides to information services and suppliers (p. 186)
Guides to libraries (p. 260)

The literature

Much of this book is concerned with the literature of science and technology and how it should be used. This reflects the fact that, sooner or later, most scientific/technical knowledge (or information) becomes available in one or other of the various printed forms. While people and organizations are especially valuable sources for information about current research and development work, it is the literature that provides a permanent record of what has been done in the past, usually with sufficient detail to avoid repetition (unless this seems desirable for some special reason).

The great difficulty most people experience with the literature stems simply from the fact that there is so much of it! A high proportion of the information you require will be found somewhere in the literature: the problem is to know where and how to look. In this context we think it is helpful to be aware that the literature can be structured: you may use it more efficiently if you know about the different components and the way in which they fit together. We must emphasize, however, that there are several ways of defining a structure: our method is simple and (we believe) practical – but, of course, the true situation is more complex than our scheme suggests, so it is not offered as a model of academic perfection, but as a rough-and-ready guide to help those who intend to use the literature.

Uses

- For finding information of almost any kind (e.g. data, names and addresses, details of specific research and development projects, experimental methods and techniques, and many, many, others).
- To see whether anyone has carried out a particular piece of work (you may save time by not having to repeat something that has been done already).
- To see who is working/has worked on a given topic, or where the work was done.
- To encourage inspiration (by browsing).
- For keeping up to date in your own, or a related, subject field.

Access

- Via libraries, or purchase.
- Your own organization may have a library; if not you will almost certainly be within reach of one of the many public libraries (or perhaps there may also

be a technical college or university library to which you can obtain access). Libraries and their services are dealt with in more detail on pp. 245–260; it is worth noting here, however, that even if your local library does not stock everything you require, it should be able to borrow items from other libraries, using the inter-library loans system.
● If you wish to buy any literature yourself, this may be done through bookshops or, in some cases, direct from the publisher. Before ordering literature which is published abroad, it is wise to enquire how long the item will take to come and what it is likely to cost – do not assume the price will necessarily be the sterling or dollar equivalent of that in the country of origin. There is a section on **Guides to the book trade** on pp. 263–265.

Caution

● Try to keep the correct balance between literature searching and your own theoretical/practical work.
● When using the literature, be on the lookout for personal or national bias as well as out-of-date and inaccurate material: just because something gets into print, it isn't necessarily true!
● Remember that it is sometimes quicker to approach a person, or organization, for information (especially by telephone) than to spend hours slogging through quantities of literature.

Examples

The various types of literature are listed below according to a simple structure: this is based on a chronological division into primary and secondary literature, followed by guides to the literature (referred to as tertiary literature by some commentators).

Structure of scientific and technical literature

● *Primary literature*. This contains the first, and sometimes only, publication of specific information: here one must look for reports of the most recent advances in knowledge (i.e. the 'latest' information).
   Periodicals
   Reports
   Conference proceedings
   Trade/commercial literature
   Patents
   Standards
   Theses/dissertations
   Government publications

● *Secondary literature*. This consists of republication of material distilled from primary sources, generally after a substantial period of time has elapsed.
   Books
   Reference material (data compilations, dictionaries, encyclopaedias, handbooks)
   Review serials

- *Guides to the literature.* These have been developed to assist the retrieval of information from both primary and secondary sources.
  Abstracting and indexing journals
    Guides to abstracting and indexing journals
  Bibliographies
    Bibliographies of bibliographies
  Subject guides
  General guides

Notes

- Within each category items have been arranged in the approximate order of importance to the average person (this is a matter of opinion and must not be regarded as definitive).
- The divisions between different types of literature are not entirely watertight. Take, for instance, the case of papers presented during a symposium, at first glance an example of *conference proceedings*. However, these might be published as journal articles in a *periodical*, or they could appear individually with serial code numbers for identification purposes like *reports*. Hence documents exist with at least dual status, as demonstrated by the conference paper/journal article and conference paper/report possibilities just mentioned. This reinforces the point that our structure of the literature is only a simplified, though useful, model of a more complicated system.
- While it is not possible to say that one class of guide covers the primary and another the secondary literature, it is fair to comment that (a) abstracting/indexing journals cover more primary than secondary literature, and (b) certain kinds of bibliography provide the principal guides to secondary literature. Beyond this it is unwise to generalize.
- Maps, especially the geological variety, are one type of literature not dealt with in this book. Anyone interested in these should consult *Information Sources in the Earth Sciences*, listed under **subject guides** for geology (p. 44).
- Online databases can act as guides to the literature but are not themselves 'literature', hence we discuss them separately.

Use of the literature (summary)

The most recent information must be obtained from the primary literature (e.g. periodicals or reports) using abstracting/indexing journals (or bibliographies) as guides. When entering a new field begin, at a lower level of information, with the secondary literature (e.g. books), perhaps working up to the primary sources through review serials, again using bibliographies as guides. For a brief account of an unfamiliar topic, an encyclopaedia (or dictionary) is often sufficient; while facts, figures and formulae are best obtained from a data compilation (or handbook).

Guides to the literature

A guide to the literature we define as any publication whose principal purpose is to help you retrieve information published elsewhere in the literature: it may do this by directing you to a specific reference (as in the case of an abstracting/indexing journal, or subject bibliography); or by leading you to a particular type of source (e.g. a journal/book, indicated by guides to periodicals/books); or by offering general advice about the literature of a given subject or wide range of subjects.

This definition covers many different kinds of publication, of which examples will be found distributed throughout this book. Since many guides deal with a particular type of literature, it is our practice to deal with them under the type concerned. The only exceptions to this are the general sections on **Abstracting and indexing journals** and **Bibliographies** (which are sufficiently important to deserve individual treatment) and the 'advisory' guides dealt with below. The latter may conveniently be divided into two categories: general guides, and subject guides.

General guides

Under this heading we include guides to the literature of science and technology in general, some of which may cover other sources of information (e.g. organizations, information services).

Uses

● As a starting point, especially if you have to move into an unfamiliar subject field.
● For general reference in connection with the problem of finding information.

Access

Via a library, or by purchase.

Caution

● Some cover a wide field with insufficient depth, perhaps because they are intended for library school students rather than users of scientific information.
● They are apt to go out of date very quickly, as the material covered is liable to constant change.
● They are not always arranged in a way that makes them easy to use.

Examples

This book is, we hope, a good example of a general guide. Others are given below.

How to Find Out: printed and on line sources, by G. Chandler. 5th ed. Oxford, Pergamon Press, 1982.
○ Covers a very wide field, but has chapters devoted to the sciences and engineering. Useful as a general introduction to the subject.

Understanding the Structure of Scientific and Technical Literature: a case-study approach, by R. V. Turley. London, Clive Bingley, 1983.
○ Novel pictorial treatment in which the literature's structure emerges from consideration of one real-life research project, a 'talking brooch' communication aid for the dumb.

Science and Technology: an introduction to the literature, by D. Grogan. 4th ed. London, Clive Bingley, 1982.
○ Although it covers only the literature, and the arrangement is not convenient for the rapid extraction of specific information, this provides a fairly thorough (if rather academic) account of its field.

How to Find Out about Engineering, by S. A. J. Parsons. Oxford, Pergamon Press, 1972.
○ Sections on careers, finding information, libraries, particular kinds of literature, organizations and education are followed by chapters devoted to information sources in the different branches of engineering (mechanical, electrical, etc.).

Technical Information Sources: a guide to patent specifications, standards, and technical reports literature, by B. Houghton. 2nd ed. London, Clive Bingley, 1972.
○ Provides detailed coverage of the types of literature mentioned, especially patents.

Information Sources in Engineering, edited by L. J. Anthony. 2nd ed. London, Butterworths, 1985.
○ Previously published as *Use of Engineering Literature*, edited by K. W. Mildren. The opening sections deal with various types of literature and online information services. The remaining two-thirds of the volume describes the sources and guides in particular branches of engineering.

A Brief Guide to Sources of Scientific and Technical Information, by S. Herner. 2nd ed. Arlington, VA, Information Resources Press, 1980.
Exhibits of Sources of Scientific and Technical Information, edited by S. Herner and J. C. Moody. Arlington, VA, Information Resources Press, 1971.
○ The first volume covers a wide range of sources and guides to them, with a chapter on personal index files. The second volume consists entirely of facsimiles of pages from various guides.

Scientific and Technical Information Sources, by C. Chen. Cambridge, MA, MIT Press, 1977.
○ The material is largely arranged by type of source (or guide) sub-divided according to subject. Thus the section on bibliographies has sub-headings for general science, astronomy, mathematics, physics, chemistry, etc. Easy to use if you want, say, a handbook for mechanical engineering, but less convenient should you require a list of all sources and guides in that field.

Library Searching: resources and strategies, with examples from the environmental sciences, by J. M. Morris and E. A. Elkins. New York, Jeffrey Norton Publishers, 1978.
○ Far more than just an annotated list of sources and guides, this book contains much practical advice on using libraries and searching literature.

Research: a practical guide to finding information, by P. Fenner and M. C. Armstrong. Los Altos, William Kaufmann, 1981.
○ Despite its rather broad title, most of this work is concerned with science and technology. As in the preceding example, there is plenty of advice on using libraries and literature searching; some practical exercises are also provided.

Scientific and Technical Information Resources, by K. Subramanyam. New York, Dekker, 1981.

○ This is a mainly descriptive guide, with the emphasis on printed sources and only a brief mention of online services at the end. Select lists of examples are provided for most categories of secondary literature.

Subject guides

These are the guides to the literature relating to particular subjects; again many of them cover other sources of information as well (e.g. organizations, information services).

Uses

● To obtain a good idea of the information sources relating to a specific subject (which may be your own, or one with which you are not too familiar).
● To help you with a specific information problem involving the literature of a subject other than your own.

Access

Via a library, or by purchase.

Caution

● Some cover only the literature, so may not provide a complete answer to every problem.
● They are not always up to date: make sure you have the latest edition in any case.

Examples

Contrary to our usual practice, the examples are given without annotation (in order to increase the number that can be included). They are listed under broad subject headings and it is hoped that most fields of immediate interest to scientists and engineers have been covered under one heading or another.

AERONAUTICAL ENGINEERING AND ASTRONAUTICS

A Guide to Information Sources in Space Science and Technology, edited by B. M. Fry and F. E. Mohrhardt. New York, Interscience, 1963.

AGRICULTURE

Information Sources in Agriculture and Food Science, edited by G. P. Lilley. London, Butterworths, 1981.

ASTRONOMY

A Guide to the Literature of Astronomy, by R. A. Seal. Littleton, Colorado, Libraries Unlimited, 1977.

BIOLOGICAL SCIENCES

Information Sources in the Biological Sciences, edited by H. V. Wyatt. 3rd ed. London, Butterworths, announced for 1986.

Guide to the Literature of the Life Sciences, by R. C. Smith and W. M. Reid. 8th ed. Minneapolis, Burgess, 1972. Previously published as *Guide to the Literature of the Zoological Sciences*.

Biological and Biomedical Resource Literature, by A. E. Kerker and H. T. Murphy. Lafayette, Purdue University, 1968.

Entomology: a guide to information sources, by P. Gilbert and C. J. Hamilton. London, Mansell, 1983.

CHEMISTRY AND CHEMICAL ENGINEERING

The Use of Chemical Literature, edited by R. T. Bottle. 3rd ed. London, Butterworths, 1979.

How to Find Out in Chemistry, by C. R. Burman. 2nd ed. Oxford, Pergamon, 1966.

Chemical Publications; their Nature and Use, by M. G. Mellon. 5th ed. New York, McGraw-Hill, 1982.

How to Find Out about the Chemical Industry, by R. Brown and G. A. Campbell. Oxford, Pergamon, 1969.

Plastics and Rubber: world sources of information, by E. R. Yescombe. London, Applied Science Publishers, 1976.

Guide to Basic Information Sources in Chemistry, by A. Antony. New York, Halsted Press, 1979.

How to Find Chemical Information, by R. E. Maizell. New York, Wiley–Interscience, 1979.

Chemical Information: a practical guide to utilization, by Y. Wolman. New York, Wiley, 1983.

CIVIL ENGINEERING AND CONSTRUCTION INDUSTRIES

How to Find Out in Architecture and Building: a guide to sources of information, by D. L. Smith. Oxford, Pergamon, 1967.

Building Construction Information Sources, by H. B. Bentley. Detroit, Gale, 1964.

Information Sources in Architecture, edited by V. J. Bradfield. London, Butterworths, 1983.

COMPUTER SCIENCES

A Guide to Computer Literature, by A. Pritchard. 2nd ed. London, Clive Bingley, 1972.

Guide to Reference Sources in the Computer Sciences, by C. Carter. New York, Macmillan Information, 1974.

Computer Science Resources: a guide to professional literature, by D. Myers. White Plains, NY, Knowledge Industry Publications, 1981.

ECONOMICS AND STATISTICS

Information Sources in Economics, edited by J. Fletcher. 2nd ed. London, Butterworths, 1984. Previously published as *Use of Economics Literature*, edited by J. Fletcher.

How to Find Out about Economics, by S. A. J. Parsons. Oxford, Pergamon, 1972.

Sources of Statistics, by J. M. Harvey. 2nd ed. London, Clive Bingley, 1971.

How to Find Out about Statistics, by G. A. Burrington. Oxford, Pergamon, 1972.

ELECTRICAL AND ELECTRONIC ENGINEERING

How to Find Out in Electrical Engineering, by J. Burkett and P. Plumb. Oxford, Pergamon, 1967.
Electronic Industries: information sources, edited by G. R. Randle. Detroit, Gale, 1968.
Handlist of Basic Reference Material in Electrical and Electronic Engineering, edited by E. M. Codlin. 6th ed. London, Aslib Electronics Group, 1973.

GEOGRAPHY

How to Find Out in Geography, by C. S. Minto. Oxford, Pergamon, 1966.
Geography and Cartography: a reference handbook, by C. B. M. Lock. 3rd ed. London, Clive Bingley, 1976.
The Literature of Geography: a guide to its organisation and use, by J. G. Brewer. 2nd ed. London, Clive Bingley, 1978.

GEOLOGY

Guide to Geologic Literature, by R. M. Pearl. New York, McGraw-Hill, 1951.
Information Sources in the Earth Sciences, edited by J. Hardy, D. N. Wood and A. Harvey. 2nd ed. London, Butterworths, announced for 1986.
Sources of Information for the Literature of Geology, by J. W. Mackay. London, Geological Society, 1973.

MANAGEMENT AND COMMERCE

How to Find Out: management and productivity, by K. G. B. Bakewell. 2nd ed. Oxford, Pergamon, 1970.
Commercial Information: a source handbook, by D. E. Davinson. Oxford, Pergamon, 1965.
Sources of Business Information, by E. T. Coman. Revised ed. Berkeley, University of California Press, 1964.
Information Sources in Management and Business, edited by K. D. C. Vernon. 2nd ed. London, Butterworths, 1984. Previously published as *Use of Management and Business Literature*, edited by K. D. C. Vernon.

MATHEMATICS

Guide to the Literature of Mathematics and Physics, by N. G. Parke. 2nd ed. New York, Dover, 1958.
How to Find Out in Mathematics, by J. E. Pemberton. 2nd ed. Oxford, Pergamon, 1969.
Use of Mathematical Literature, edited by A. R. Dorling. London, Butterworths, 1977.
Using the Mathematical Literature: a practical guide, by B. K. Schaefer. New York, Marcel Dekker, 1979.

MECHANICAL AND POWER ENGINEERING

Mechanical Engineering: the sources of information, by B. Houghton. London, Clive Bingley, 1970.
Information Sources in Power Engineering: a guide to energy resources and technology, by K. S. Metz. Westport, CT, Greenwood Press, 1976.

MEDICINE

Medical Reference Works, 1679–1966: a selected bibliography, edited by J. B. Blake and C. Roos. Chicago, Medical Library Association, 1967. Supplements have also been issued.
Sources of Medical Information: a guide to organisations and government agencies . . . and selected scientific papers, edited by R. Alexander. New York, Exceptional Books, 1969.
Information Sources in the Medical Sciences, edited by L. T. Morton and S. Godbolt. 3rd ed. London, Butterworths, 1983. Previously published as *Use of Medical Literature*, edited by L. T. Morton.
How to Find Out in Pharmacy, by A. L. Brunn. Oxford, Pergamon, 1969.
Comparative and Veterinary Medicine: a guide to the resource literature, by A. E. Kerker and H. T. Murphy. Madison, University of Wisconsin Press, 1973.
Health Sciences Information Sources, by C. Chen. Cambridge, MA, MIT Press, 1981.
Searching the Medical Literature: a guide to printed and online sources, by J. Welch and T. A. King. London, Chapman and Hall, 1985.

METALLURGY AND MINING

Guide to Metallurgical Information, edited by E. B. Gibson and E. W. Tapia. 2nd ed. New York, Special Libraries Association, 1965.
How to Find Out in Iron and Steel, by D. White. Oxford, Pergamon, 1970.
A Guide to Information Sources in Mining, Minerals and Geosciences, edited by S. R. Kaplan. New York, Interscience, 1965.

OCEANOGRAPHY

Oceanography Information Sources, compiled by R. C. Vetter. Washington, DC, National Academy of Sciences, 1970.

PHYSICS

How to Find Out about Physics, by B. Yates. Oxford, Pergamon, 1965.
Physics Literature: a reference manual, by R. H. Whitford. 2nd ed. Metuchen, NJ, Scarecrow Press, 1968.
Sources of Information on Atomic Energy, by L. J. Anthony. Oxford, Pergamon, 1966.
Information Sources in Physics, edited by D. F. Shaw. 2nd ed. London, Butterworths, 1985. Previously published as *Use of Physics Literature*, edited by H. Coblans.

PSYCHOLOGY

How to Find Out in Philosophy and Psychology, by D. H. Borchardt. Oxford, Pergamon, 1968.

TEXTILE INDUSTRY

How to Find Out about the United Kingdom Cotton Industry, by B. Yates. Oxford, Pergamon, 1967.

How to Find Out about the Wool Textile Industry, by H. Lemon. Oxford, Pergamon, 1968.
Textile Industry: information sources, by J. V. Kopycinski. Detroit, Gale, 1964.

TRANSPORTATION AND PLANNING

Transportation: information sources, by K. N. Metcalf. Detroit, Gale, 1966.
Sources of Information in Transportation, compiled by R. F. Blaisdell and others. Evanston, Illinois, Northwestern University Press, 1964.
Sourcebook of Planning Information: a discussion of sources of information for use in urban and regional planning, by B. White. London, Clive Bingley, 1971.

Abstracting and indexing journals

These are guides to the contents of periodicals (especially) but may also cover many other forms of literature, such as reports, patents, theses, conference papers, reviews and even, on occasion, books.

Every periodical should have its own index, published at regular intervals. An indexing journal is the natural extension of this: it lists, usually under various subject headings, bibliographic information (author, title, publication details) concerning articles, etc., drawn from a large number of primary sources, generally in closely related fields. Author and other appropriate indexes are normally provided as well. A good example is *Index Medicus*, which lists the contents of some 3200 different journals each year.

Although an indexing journal may provide additional subject headings, the basic subject information about an entry derives from its title. It is a sad fact of scientific life that many authors do not give their articles good, descriptive titles; hence, an indexing journal is of less value than an abstracting journal (which does everything done by the former – and more).

An abstracting journal lists, usually under various subject headings, brief summaries or abstracts of articles, etc., drawn from a large number of primary sources (generally in closely related fields) together with the relevant bibliographic information for the item concerned. Again, author and other appropriate indexes are normally provided. A good example is *Chemical Abstracts*, which, among other things, scans some 14 000 different periodical titles each year.

Abstracting journals are the means by which the vast amount of information contained in journals, reports, patents, theses, conference proceedings, and so on, can be both indexed and summarized.

Abstracts vary a great deal from a short statement (perhaps a sentence or two) expanding the title and indicating the general nature of a paper, to several paragraphs outlining the principal results obtained in the work and the conclusions drawn. In the latter case, it may sometimes be possible to obtain useful information from the abstract without consulting the original paper, but, in general, do not expect this to be the case. An English-language abstract may be of particular value in this context when the original item is in a 'difficult' foreign language, for instance Russian, Japanese or Chinese.

Do not forget that abstracts may also be a regular feature of periodicals which are not themselves abstracting journals.

The value of an abstracting journal depends on the following factors:

(1) The number of periodicals, etc., scanned: obviously this should be as large as possible.

(2) The way in which material is selected from the sources covered: even though a particular periodical is covered by an abstracting journal, it is often the case that only a proportion of its articles (perhaps in a given subject area) are abstracted.

(3) The time taken between the appearance of a paper and the subsequent publication of its abstract: obviously this should be as short as possible.

There is a remarkable non-uniformity in the specifications of different abstracting journals: some, such as *Chemical Abstracts* (currently occupying about 8 ft of library shelving per year) are large services covering a very wide subject area; others, *Lead Abstracts* and *Zinc Abstracts*, for example, are much smaller in scope, covering only a narrow field (the two examples quoted each being contained in one small volume per year).

Owing to this general lack of overall cohesion, some periodicals are never abstracted at all, while others may be covered several times by different services.

The main use of abstracting/indexing journals is to assist retrospective searching of the literature – keeping up to date requires something more.

Current-awareness services have been developed to overcome the time-lag between publication of an article and its abstract: ensuring that some information appears within a few weeks rather than months. Some current-awareness publications are intended to be ephemeral (kept for a short time and then discarded) – these are dealt with under **Current-awareness publications** on p. 285. Others are provided with cumulative indexes, and can be used for retrospective searches as well: one example is mentioned below. Current-awareness publications usually take the form of an indexing (rather than an abstracting) journal.

Certain abstracting and indexing journals cover a particular type of literature as opposed to a specific subject area: in this book these are dealt with under the heading of guides to the type of literature concerned. Major examples will be found under **Guides to reports** (p. 119) and **Guides to theses and dissertations** (p. 134). Most abstracting/indexing journals and current-awareness publications have online database equivalents: for details of these see pp. 142–149.

Uses

● As guides to the immense amount of information appearing in journal articles, reports, etc.

● To perform retrospective searches, or keep up to date, in a given subject field.

● To keep in touch with the work of individual scientists/engineers.

● As a first step when looking for information on a topic which is not well covered by books, encyclopaedias, or other forms of secondary literature.

● To obtain correct details in respect of inaccurate references.

Access

Via libraries. Subscriptions to indexing/abstracting journals can be extremely expensive, so a wide selection of them is to be looked for only in major libraries, for instance the Science Reference Library (formerly the National Reference Library for Science and Invention—the old Patent Office Library), university and polytechnic libraries (where it may be necessary to apply for reading facilities) and the larger public libraries. A very wide range is also available for consultation in the reading room of the BLLD at Boston Spa, Yorkshire (see p. 259).

Caution

• Always read the 'how-to-use' section before proceeding to the indexes.
• Be prepared to look under alternative subject headings, especially if you cannot find what you want at the first attempt.
• There is a considerable time delay (occasionally 2 years, or more) before some material is mentioned in certain abstracting journals, so do not expect them to be particularly up to date.
• Never rely on a single abstracting/indexing journal to provide coverage of all literature relevant to your interests. Consulting a range of these guides is also advised because relevant information may be much easier to find in some than in others, on account of differing indexing policies.
• It is sometimes wise to find out which sources are covered by a specific abstracting/indexing journal before making use of it.
• Indexing journals are often of limited use, as titles can be poor guides to the contents of articles, etc.
• Some abstracting journals have misleading titles themselves: e.g. *Engineering Index* and *Mathematical Reviews*. They may also cover a much wider range of subjects than their titles suggest.
• Check whether multi-annual indexes are available, as these can save considerable time when searching.
• Be prepared for index terms and/or subject headings to change from one volume of a given indexing/abstracting journal to another.
• References to journal articles will often cite an abbreviated form of the periodical title: as a first step in expanding this, see whether the indexing/ abstracting journal publishes its own list of journal abbreviations (often in association with the list of titles covered). If it does not, note the reference *exactly* as it is given.

Examples

A number of examples are dealt with as guides to specific types of literature, such as reports (p. 119) and theses (p. 134). Apart from this, all that can be done here is to give details of a dozen of the more important examples, selecting them so that most areas of science/technology will be partially covered by one or other of them.

Bibliography and Index of Geology. Falls Church, VA, American Geological Institute, 1969–   .
○ Originally published as *Bibliography and Index of Geology exclusive of North America (1934–68)*, which was an abstracting journal, this is now an indexing journal, with entries arranged under fairly broad subject headings. Each monthly issue contains subject and author indexes, cumulative bibliography and index volumes being produced annually. *GEOREF*, a computerized database associated with this publication, may be searched online.

The needs of geologists were also catered for by *Geophysical Abstracts* (1929–71), *Abstracts of North American Geology* (1966–71) and *Bibliography of North American Geology* (1923–73).

Another indexing journal of interest here is *GeoSciTech Citation Index*, published in Philadelphia by the Institute for Scientific Information. Like *Science Citation Index* (see p. 52) this has citation, source and permuterm subject indexes. A five-year cumulation (1976–80) has been produced, with annual cumulations from 1981–   and two interim issues each year.

Biological Abstracts. Philadelphia, BioSciences Information Service (BIOSIS), 1927–   .

○ Entries are arranged under narrow subject headings: each issue contains author, biosystematic, generic, concept and subject indexes. The last is a computer-generated KWIC (KeyWord In Context) index based on words appearing in the titles of articles included, or additional terms specified by BIOSIS indexers. *Biological Abstracts* is issued semi-monthly, and cumulated indexes are published semi-annually.

Biological Abstracts/RRM (Reports, Reviews, Meetings), successor to *Bioresearch Index*, is an indexing journal which supplements *Biological Abstracts* by covering conference proceedings, reports, reviews, and some secondary literature. It, too, has author, biosystematic, generic and subject indexes.

A computerized information retrieval service, BIOSIS (see p. 176), is associated with the *Biological Abstracts* database. See, too, the *Bibliographic Guide* mentioned on p. 101. Material of interest to biologists also occurs in *Zoological Record* (1864–   ), and the Commonwealth Agricultural Bureau is responsible for a number of specialized abstracting journals, such as *Forestry Abstracts* and *Review of Plant Pathology*.

Chemical Abstracts. Columbus, Ohio, American Chemical Society, 1907–   .

○ This is the world's largest abstracting service covering a very wide range of subjects: it is useful to the physical and biological scientist, as well as the engineer. Entries are arranged under fairly narrow subject headings, keyword (subject), patent and author indexes being provided in each weekly issue. Cumulative indexes are produced for each completed volume semi-annually (including author, general subject, chemical substance, formula and patent indexes), but one of the most valuable features of the service is the multi-annual cumulative indexes covering 5- (or 10-) year periods. Computerized information retrieval services are associated with the *Chemical Abstracts* database (see page 146). See, too, the *Bibliographic Guide* mentioned on p. 101. Also of interest to chemists is *Current Abstracts of Chemistry and Index Chemicus* (1970–   , *Index Chemicus* 1960–69): a weekly current-awareness publication which can be used for retrospective searches as cumulative indexes are provided. This, together with the associated *Current Chemical Reactions* and *Chemical Substructure Index*, is produced by the Institute for Scientific Information in Philadelphia. *INDEX CHEMICUS ONLINE* is available through TELESYSTEMES using QUESTEL/DARC.

See also **Current-awareness publications** (pp. 285–287).

Current Technology Index. London, Library Association Publishing, 1981–   .

○ This indexing journal replaced *British Technology Index* (1962–80), which was itself preceded by *The Subject Index to Periodicals* (1915–61). Entries are arranged under narrow subject headings, and each monthly issue contains an author index, these individual parts being superseded by a cumulated annual volume. Coverage is limited to articles in British technical journals.

Applied Science & Technology Index (1958–   ) is a somewhat similar publication, produced in New York by the H. W. Wilson Company, but not solely concerned with material of American origin. Online access is possible via WILSONLINE, which is available in the USA but not in the UK at the time of writing.

If you are interested in the contents of popular magazines (for such things as test reports of cameras, hi-fi equipment or cars), reference should be made to *Clover Index* (c. 1975–   ), an indexing journal produced by Clover Publications of Biggleswade, Bedfordshire, England.

Electrical & Electronics Abstracts; Science Abstracts series B. London, INSPEC, 1966–   .
○ This publication first appeared separately as *Science Abstracts, Section B; Electrical Engineering* (1903–40) and was afterwards known as *Electrical Engineering Abstracts* (1941–65). Entries are arranged under narrow subject headings: each monthly issue contains a subject guide to the classification used, an author index, and separate indexes for bibliographies, books, conferences and corporate authors. Semi-annual cumulative indexes are produced, and indexes covering 4- or 5-year periods are available from 1950. *Computer & Control Abstracts; Science Abstracts series C* (1969–   ) is a companion publication which began as *Control Abstracts* (1966–68). Online versions of *Electrical & Electronics Abstracts* and *Computer & Control Abstracts* are available as part of the *INSPEC* database (see p. 146). See also *INSPEC Thesaurus* (p. 114), *Physics Abstracts* below, and **Current-awareness publications** (p. 286). Other titles of interest in this field include: *Electronics and Communications Abstracts Journal* (1967–   ) and *Computer and Information Systems Abstract Journal* (1962–   ), both published in Bethesda, Maryland, by Cambridge Scientific Abstracts; also *Computing Reviews* (1960–   ), an abstracting journal produced in New York by the Association for Computing Machinery.

The Engineering Index. New York, Engineering Information Inc, 1884–   .
○ Despite its title, this is an abstracting journal, with entries arranged under narrow subject headings and an author index in each monthly issue. The individual parts are superseded by a set of annual volumes. Cumulated author and subject indexes are available for the years 1973–77 and 1978–81 (those for 1982–84 are in preparation), otherwise there is no separate subject index. Computerized information retrieval services are associated with the *Engineering Index* database (see p. 146). See, too, the *Bibliographic Guide* mentioned on p. 101, and *Ei Engineering Conference Index* (p. 68). Also of interest to engineers are: *Applied Mechanics Reviews* (1948–   ), a monthly abstracting journal published in New York by the American Society of Mechanical Engineers; *ISMEC Bulletin* (1973–   ), since 1982 an abstracting journal produced by Cambridge Scientific Abstracts of Bethesda, Maryland (before which it was an indexing journal only)—this may be searched online via the *ISMEC* database; and a series of specialized abstracting journals issued by the British Hydromechanics Research Association (BHRA) of Cranfield, Bedfordshire, England, which includes titles such as *Fluid Sealing Abstracts* and

Tribos (Tribology Abstracts)—these are available for online searching through BHRA's *FLUIDEX* database.

Index Medicus. Bethesda, Maryland, National Library of Medicine, 1960–   .
○ This indexing journal, which has been published under various titles since 1879, has entries arranged under narrow subject headings. An author index is provided in each monthly issue, and there is a separate section entitled 'Bibliography of Medical Reviews'. The individual parts are superseded by a cumulated annual issue. A computerized information retrieval service, generally known as *MEDLINE*, is associated with *Index Medicus* (see p. 147). Also relevant in the present context are: a series of more than 40 abstracting journals produced by Excerpta Medica of Amsterdam, including *Excerpta Medica; Physiology, Excerpta Medica; Surgery* and *Excerpta Medica; Public Health, Social Medicine and Hygiene*; and *Psychological Abstracts* (1927–   ) published in Arlington, VA, by the American Psychological Association.

International Aerospace Abstracts. New York, American Institute of Aeronautics and Astronautics, 1961–   .
○ Entries are arranged under broad subject headings: each semi-monthly issue contains subject and author indexes, which are cumulated annually. Coverage includes periodicals, conference papers and translations; a similar publication called *Scientific and Technical Aerospace Reports* (*STAR*) deals with the report literature (see **Guides to reports**, p. 122). Computerized information retrieval services are associated with the *International Aerospace Abstracts* database (see p. 147). See also *NASA Thesaurus* (p. 114).

Mathematical Reviews. Providence, RI, American Mathematical Society, 1940–   .
○ This is an abstracting journal with entries arranged under narrow subject headings; an author index is provided in each monthly issue. Cumulative author and subject indexes are available for the periods 1940–79 and 1973–79 respectively, cumulative subject indexes for 1940–72 being in preparation. *MATHFILE*, a computerized database associated with this publication, may be searched online. See also **Current-awareness publications** (p. 287). One indexing journal of particular interest to mathematicians is *CompuMath Citation Index*, produced in Philadelphia by the Institute for Scientific Information. Like *Science Citation Index* (see p. 52) this has citation, source and permuterm subject indexes. A five-year cumulation (1976–80) has been published, with annual cumulations from 1981–   and two interim issues per year.

Metals Abstracts. London and Ohio, Metals Information (The Metals Society and the American Society for Metals), 1968–   .
○ This was preceded by *Metallurgical Abstracts* (published separately from 1934, previously contained in *Journal of the Institute of Metals*, 1909–   ). Entries are arranged under fairly broad subject headings, and each monthly issue contains an author index.
A companion publication, *Metals Abstracts Index* (1968–   ), which is also issued monthly, consists almost entirely of a detailed subject index to the contents of the former. A cumulated version of the index is produced annually. *Metals Abstracts* may be searched online via the *METADEX* database.

Physics Abstracts; Science Abstracts series A. London, INSPEC, 1941– .
○ This publication began as *Science Abstracts; Physics and Electrical Engineering* (1898–1902) and then became *Science Abstracts, Section A; Physics* (1903–40). Entries are arranged under narrow subject headings; each fortnightly issue contains a subject guide to the classification used, an author index, and separate indexes for bibliographies, books, conferences and corporate authors. Semi-annual cumulative indexes are produced, and indexes covering 4- or 5-year periods are available from 1950. An online version of *Physics Abstracts* is available as part of the *INSPEC* database (see p. 146). For details of companion publications, see *Electrical & Electronics Abstracts* above. See also *INSPEC Thesaurus* (p. 114) and **Current-awareness publications** (p. 285). Other examples of interest to some physicists are: *Meteorological and Geoastrophysical Abstracts* (1960–), produced in Boston by the American Meteorological Society and previously published as *Meteorological Abstracts and Bibliography* (1950–59); *INIS Atomindex* and *Nuclear Science Abstracts* (see **Guides to reports**, p. 122).

Science Citation Index. Philadelphia, Institute for Scientific Information, 1955– .
○ This indexing journal is rather different from the others mentioned here, as it enables a literature search to come 'forward in time' from a given starting point, which is achieved by indexing the citations to earlier work contained in the articles covered. *Science Citation Index* is published in three separate sections:

(1) Source Index, with entries arranged by author, indexes the articles, etc., from over 3000 periodicals.

(2) Permuterm Subject Index, with entries based on pairs of significant words occurring in the titles covered by the Source Index—permutations of significant words (in pairs) are employed, so it does not matter which word you choose to look under first. The name of an author generally appears in connection with each word pair, and the full title of the article concerned may be obtained from the Source Index. For example, if you look in the 1971 Permuterm Subject Index under ELECTRONIC coupled with GLASS you will find the author TRAP H J L. The Source Index reveals that this author's article had the title 'Electronic conduction in glass' and the bibliographic reference is given. Entries in the Permuterm Subject Index for the same article will also be found under CONDUCTION coupled with ELECTRONIC or GLASS; GLASS coupled with CONDUCTION, and so on. Remember that, because this is a computer-produced index based solely on titles, it may be necessary to look under a variety of similar headings; for example, if you were interested in 'conduction' you might find relevant material under CONDUCTANCE, CONDUCTING, CONDUCTION, CONDUC-TIVITY, CONDUCTOR, to name but a few. An index of this kind is especially valuable when you are seeking new terminology or jargon which is only just beginning to appear in the literature. It is often less convenient when you need to search under long-established, much-used terms like laser or lasers, though attempts are being made to reduce some of the difficulty by introducing key phrases such as laser-beam and laser-induced as main headings.

(3) Citation Index, with entries arranged under the authors of *cited* articles, giving the name of the *citing* author (whose title can then be found from the Source Index).

Returning to the example in (2) above, if you were interested in 'conduction of glass' you might be aware of a paper on 'The effect of high fields on the conduction of glasses containing iron' by J. L. Barton, published in *Journal of Non-crystalline Solids*, Vol. 4, pp. 220–30, 1970. Assume that, so far as your work is concerned, this paper by Barton is very significant and you want to know if anyone subsequently cites it (presumably if they do, they must be working on the same topic). If you look in the 1971 Citation Index under BARTON J L, you will find his paper listed in the form 70 J NONCRYSTALLINE SOL 4 220 (note that only the reference is given, no title); underneath this appears the name TRAP J J L, which means that Trap has cited Barton's paper in an article, the title of which you can obtain from the Source Index (it is, of course, the article mentioned in (2) above). Thus are you brought forward in time from Barton (1970) to Trap (1971).

To summarise, you could find Trap's article in three ways:

(1) by knowing that he worked on this topic, and looking him up in the Source Index;
(2) by looking in the Permuterm Subject Index under CONDUCTION and GLASS, etc.;
(3) by knowing of Barton's 1970 paper, and using the Citation Index to see whether it had been quoted since.

It should perhaps be mentioned that Trap's article is in German (it was published in *Nachrichtentechnische Zeitschrift*), so the title given in this example was, in fact, a translation: the Source Index includes the information that the original was in German, also that it cited 13 references—a miniature subject bibliography!

There have been further developments since this illustration was prepared for our first edition. For instance, the 1982 Citation Index reveals four authors who have cited your 'classic' paper, Barton (1970), including M. Chybicki whose article entitled 'Non-ohmic conductivity in iron-phosphate glasses' appeared in *Physica Status Solidi, A*, Vol. 68, pp. K129–34, 1981. Indeed Barton (1970) has been cited 16 times during the period 1971–82 according to *Science Citation Index*, whereas the German article, Trap (1971), has not been quoted at all.

Science Citation Index is published in bi-monthly issues superseded by an annual cumulation. Three 5-year cumulations span the period 1965–79, and a 10-year cumulation of the Source and Citation Indexes (1955–64) extends their coverage beyond that of the annual volumes, which began in 1961. All cumulations include additional material which never featured in the corresponding *Science Citation Index* annuals.

From 1974 onwards, review articles covered by the system can be more easily retrieved using a separate *Index to Scientific Reviews* (see p. 127).

As well as *Science Citation Index*, the Institute for Scientific Information of Philadelphia also produces *Social Sciences Citation Index* (1966–) and *Arts and Humanities Citation Index* (1976–). Computerized information-retrieval services are associated with all these publications (see pp. 147, 208). See also **Current-awareness publications** (p. 285).

Guides to abstracting and indexing journals

The growth in the number of abstracting and indexing journals has mirrored that of primary journals. It is not easy to give a precise figure for the former, but it may be in the region of 2000–3000 titles worldwide; hence the introduction of guides to abstracting and indexing journals.

It should be remembered that abstracting/indexing journals are also periodicals, and covered by guides to periodicals (though they may not be listed separately). In particular, *Ulrich's International Periodicals Directory* (see p. 99) does have a section entitled 'Abstracting and Indexing Services', and this guide is also useful because, as part of the annotation for many journals, it lists coverage by specific indexing/abstracting services.

The guides listed below cover only abstracting and indexing services: they enable a particular item to be located by title, subject and (in some cases) country of origin.

Uses

● To find out what abstracting/indexing journals there are in a given subject area (including those that are less well-known, foreign or apparently exclusive to a different subject).
● To see which titles are held in certain libraries.

Access

Via libraries.

Caution

There are few examples which are really up to date.

Examples

Abstracting and Indexing Services Directory, edited by J. Schmittroth. 3 vols. Detroit, Gale, 1982–83.
○ Entries, which are arranged alphabetically by title in each volume, include details of current publisher (with address and telephone number), year of first publication, and general notes on scope, subject coverage (mentioning any classification scheme or thesaurus associated with the service), sources scanned, contents and arrangement, frequency/cumulations, subscription information, former titles, document delivery, computer access, and microform availability of the individual abstracting and indexing journals. A single index, cumulated in each volume, combines entries for titles (both current and former), organizations connected with the services (including publishers), and subject keywords (which may or may not be present in the titles themselves). This is probably the most up-to-date and detailed guide of its kind. Coverage is not restricted to printed material, but includes some titles only available as online databases and as microform or card services. Neither is the scope of this guide limited to science, engineering and medicine; business, law, social sciences, education and humanities are also featured.

Abstracting and Indexing Periodicals in the Science Reference Library, by B. A. Alexander. 3rd ed. London, Science Reference Library, 1985.

○ This publication lists more than 1400 current abstracting and indexing journals held by the Science Reference Library (formerly the National Reference Library of Science and Invention, before that the Patent Office Library, but now part of the British Library). Brief entries, which are arranged alphabetically by title, include the starting date of the run and indicate when the item concerned is not available at the SRL's main (Holborn) site. A subject index, based on fairly broad headings, is also provided.

A Guide to the World's Abstracting and Indexing Services in Science and Technology. Washington DC, NFSAIS, 1963. Report no. 102 of the National Federation of Science Abstracting and Indexing Services (NFSAIS).
○ Entries, which include publication details and an indication of coverage, are arranged alphabetically by title. There is a list of services arranged by Universal Decimal Classification (UDC), together with country and subject indexes.

Abstracting Services . . ., compiled by the International Federation for Documentation (FID). 2nd ed. 2 vols. The Hague, FID, 1969.
○ Volume 1 covers science, technology, medicine and agriculture. Volume 2 covers social sciences and humanities. Entries, which include details of publication and coverage, are arranged alphabetically by title. There is a list of titles arranged by subject according to UDC, and an alphabetical list of subjects referring to UDC subject headings used. The International Federation for Documentation also produced *Index Bibliographicus*, Vol. 1 of which (4th ed., 1959) covered abstracting/indexing services in science and technology.

Keyword Index of Guides to the Serial Literature, by A. G. Myatt. Boston Spa, British Library Lending Division, 1974.
○ Previously published as *A KWIC Index to the English language Abstracting and Indexing Publications currently being received by the National Lending Library*. Entries consist only of title and BLLD (formerly NLL) shelfmark.

Inventory of Abstracting and Indexing Services Produced in the UK, by J. Stephens. London, British Library, 1983. Library and Information Research Report No. 21.
○ Entries, which are arranged alphabetically according to the name of the service, include the following details: name/address/telephone number of producer, name of a person to contact, subject scope, types of material, languages and geographical area covered, form and cost of output, starting date, frequency of appearance and number of items per issue, whether online access is possible, availability, and notes on other features. Indexes of subjects (both broad and narrow), organizations, and online databases (arranged by processor) are provided.

Further information about indexing and abstracting journals may be found in **General guides** (p. 40) and **Subject guides** (p. 42).

Chart 4. Subject coverage of some better known abstracting and indexing journals

Whilst it is true to say that material on virtually any topic can occur in almost every abstracting/indexing journal, most of these guides to the literature tend to concentrate on a particular range of subjects (which is generally much broader than their titles suggest).

The chart is designed to show the subject coverage of a few better-known abstracting/indexing journals: good coverage of a particular field is indicated by the larger circles; limited, or narrow specialized coverage, by the smaller.

If you wanted an abstracting journal which might guide you to information about 'the generation of electricity using solar energy', you might decide this topic would be covered under electrical engineering (there are other possibilities, though: for example physics and/or mechanical engineering). The chart then shows you should definitely try *Electrical and Electronics Abstracts*, and *Engineering Index*; also that you have a good chance of finding relevant material in one or two others, such as *Physics Abstracts* and *STAR*. In a case like this, the specialized coverage of electrical engineering offered by guides such as *Computer and Control Abstracts* can probably be rejected without great loss.

Bibliographies

A bibliography is a list of books, chapters in books, periodicals, journal articles, reports, conference papers . . . (i.e. any kind of literature), concerned with a given subject or wide range of subjects, and generally arranged in some kind of order (perhaps subject, author or chronological).

This definition embraces the 'list of references' found at the end of a journal article or book, and the output of a computer search, as well as works occupying many volumes on their own account.

The word 'bibliography' is also used to denote the study of books as physical objects—but it is *never* so used in this book.

Entries in individual bibliographies vary in detail from a simple author/title/publication data statement to a comprehensive description of each item considered, together with critical comment.

According to our definition of a guide (p. 39), all bibliographies are also guides (though not all guides are also bibliographies). For this reason many examples mentioned in this book have been dealt with as guides to particular types of literature.

Apart from this, it is convenient to divide bibliographies into two categories: general bibliographies and subject bibliographies—the basis of this division being that the former cover material from a wide range of subjects while the latter deal only with material on a particular topic.

General bibliographies

These are usually substantial works running into several (or many) volumes, including printed catalogues of large libraries, and may be concerned with special forms of literature (for instance, catalogues of books, or periodicals, or government publications).

Chart 4. Subject coverage of some better-known abstracting and indexing journals

Legend:
- ● Good coverage
- • Limited or specialized coverage

Subject coverage	Applied Mechanics Reviews	Applied Science & Technology Index	Bibliography & Index of Geology	Biological Abstracts	British Reports, Translations and Theses	Chemical Abstracts	Computer & Control Abstracts	Current Technology Index	Dissertation Abstracts International	Electrical & Electronics Abstracts	Engineering Index	Government Reports Announcements & Index	Index Medicus	INIS Atomindex	International Aerospace Abstracts	Mathematical Reviews	Metals Abstracts	Physics Abstracts	Science Citation Index	Scientific & Technical Aerospace Reports (STAR)
Aeronautics/astronautics	●	•		•	•	•	•	•	•	•	●	•	•	•	●	•	•		•	●
Agriculture			●		•	•	•	•	•		•	•		•	•				•	•
Astronomy/space science		•		•	•	•	•	•		•	•	•	•		●	•		●	•	●
Biochemistry/pharmacology			●		•	●	•		•		•	•	•	•	•				•	•
Bio-engineering	•	•		•	•		•	•	•	•	●	•	•	•	●			•	•	●
Botany			•	●	•	•	•		•		•	•	•	•					•	•
Chemical technology	•	•		•	•	●	•	•	•	•	•	•		•	•		•		•	•
Chemistry			•	•	•	●	•		•	•	•	•	•	•	•		•	•	•	•
Civil engineering	●	•	•		•	•	•	•	•	•	●	•		•					•	•
Computer science/technology	•	•		•	•	•	●	•	•	•	•	•	•	•	•	•			•	•
Control	●	•			•	•	●	•	•	•	•	•		•					•	•
Economics/management science				•	•	•	•	•	•	•	•	•		•	•				•	•
Electrical engineering		•		•	•	•	•	•	•	●	●	•		•	•		•	•	•	•
Electronics		•		•	•	•	•	•	•	●	●	•	•	•	●		•	•	•	●
Environmental sciences		•	•	●	•	•	•		•		•	●		•				•	•	•
Geology/earth sciences	•	•	●	•	•	•			•	•	•	•		•	•			●	•	•
Marine engineering	•	•			•	•	•	•	•		●	•		•			•		•	•
Materials science/technology	•	•			•	●	•	•	•	•	●	•		●	●		•	●	•	●
Mathematics/mechanics	●	•			•	•	•	•	•	•	●	•		•	•	●		●	•	•
Mechanical engineering	●	•		•	•	•	•	•	•	•	●	•		•			•	•	•	•
Medicine				•	•	•	•		•	•	•	•	●	•	•			•	•	•
Metallurgy	•	•			•	●	•	•	•	•	●	•		•	•		●	•	•	•
Microbiology				●	•	•	•		•		•	•	•	•					•	•
Nuclear science/technology		•			•	•	•	•	•	•	•	•		●	•		•	●	•	•
Oceanography	•		•	•	•	•	•	•	•	•	•	•		•	•			•	•	•
Physics	•	•	•		•	•	•	•	•	•	•	•		•	●	•	•	●	•	●
Physiology				●	•	•	•		•	•		•	●	•	•				•	•
Psychology/psychiatry				•	•		•		•	•		•	●	•		•	•		•	•
Social/behavioural sciences				•	•		•		•	•		•	•	•	•	•	•		•	•
Telecommunications		•		•	•	•	•	•	•	●	•	•		•	●	•			•	●
Transportation		•		•	•	•	•	•	•	•	●	•		•	•				•	•
Zoology			•	●	•	•	•		•		•	•	•	•					•	•

Uses

- To identify, or obtain details of, a particular piece of literature (e.g. book or periodical), especially older or out-of-print items.
- To search for material in a given subject area.
- To see whether a specific item is held by one of the major libraries.

Access

Via libraries.

Caution

Printed library Catalogues can be difficult to use, and may not provide for a subject approach; they are generally not very up to date.

Examples

Most of the important examples are dealt with elsewhere in this book under the heading of guides to the type of literature which they cover, e.g. **Guides to books** (p. 62). However, printed library catalogues are best considered here, since they cover more than one kind of literature.

British Museum General Catalogue of Printed Books . . . to 1955. 263 vols. London, Trustees of the British Museum, 1959–66.

○ This catalogue and its supplements list the holdings of Britain's largest library (formerly the British Museum Library, now part of the British Library). Entries for books are arranged alphabetically by author, including details such as title, edition, number of pages, publisher, place and date of publication, and the library's own shelfmark. Periodicals are listed under the general heading Periodical Publications, which is subdivided (alphabetically) by town of publication—within which titles are arranged alphabetically. The rules for entering certain kinds of material are far from obvious: for example, publications of the Institution of Civil Engineers are listed under LONDON–III [Miscellaneous Institutions, Societies, and other bodies] Institution of Civil Engineers. Fortunately, some cross-referencing is provided. A biography can be found by looking under the name of the person concerned, but, in general, subject approach is not provided for. The basic catalogue has been updated by a 50-volume supplement covering 1956–65, and a 26-volume supplement covering 1966–70. A 13-volume supplement covering 1971–75 was issued by British Museum Publications under a slightly different title, *The British Library General Catalogue of Printed Books*, in 1978–79, and K. G. Saur of Munich are in the process of publishing a cumulated version of the original catalogue plus its first three supplements as *The British Library General Catalogue of Printed Books to 1975* (1980–   ). The work has been further extended by a set of 402 microfiche produced in 1982 by the British Library under yet another title, *The British Library Reference Division General Catalogue of Printed Books, 1976–82*, in which material has been catalogued according to rules different from those used for its predecessors.

A Catalog of Books represented by Library of Congress printed Cards . . ., 167 vols. Ann Arbor, Michigan, Edwards Brothers, 1942–46.

○ This is the catalogue of the largest library in America. Entries are arranged in one alphabetical sequence of authors (for books, etc., including organizations where appropriate) and titles (for periodicals—though these may be listed under the name of an organization if it is responsible for them). The usual bibliographical details are given (author, title, edition, publication data, size) together with the Library of Congress subject classification and, in the case of periodicals, information about changes of title. Subject approach is not provided for, however. This publication is easier to use than the British Museum catalogue. It has been updated by a supplement covering 1942–47, and continued as *The Library of Congress Author Catalog* (1948–52), then being replaced by *The National Union Catalog*: a cumulative author list representing Library of Congress printed cards and titles reported by other American libraries (various hard-copy and microfiche cumulations plus annual volumes, issued by several different publishers, cover the period 1953–82). A subject approach is available firstly from *Library of Congress Catalog—Books: subjects* (various hard-copy and microfiche cumulations, issued by several different publishers, cover the period 1950–74) and then from *Subject Catalog* (1975–82, mostly as annual cumulations). From 1983–   the Library of Congress has published, as microfiche only, *National Union Catalog. Books* and *National Union Catalog. US Books*; both appear monthly with cumulating indexes—name (author), title, subject, and series. Bibliographic records for books held by the Library of Congress may also be sought in the *LC MARC* and *REMARC* databases, accessible online through the DIALOG service (p. 153).

Subject bibliographies

These give details of the literature covering a particular subject, often taking the form of a list of references at the end of a journal article, report, thesis, book (or chapter), but can be more substantial, occupying a volume or two on their own account. They may also be the end product of searches made in computer databases. A subject bibliography nearly always forms part of a review article: this type of literature is dealt with on pp. 125–127.

Uses

● To save time by providing a ready-made literature search or survey on your subject.
● To find other material about a specific piece of research, including earlier related work by the same author (sometimes useful when trying to identify an inaccurate reference).
● To help with the problem of going from secondary literature (such as books) to primary literature (such as journal articles) when working in an unfamiliar field.

Access

Via libraries, or material in a personal collection: for instance, lists of references in textbooks or journal articles on your own (or a colleague's) shelves.

Caution

• They are usually constructed from a limited number of sources, and over a limited period of time: therefore never regard them as 'complete'. Check their comprehensiveness wherever possible, and be prepared to bring them up to date by further searching.
• Regrettably, not all authors check their references thoroughly; so be aware of the possibility of errors and omissions in the details of documents cited (especially reports and conference papers).

Guides to bibliographies

See also **Guides to reference material**, p. 115; **Guides to reviews and review serials**, p. 126; **Subject guides**, p. 42

Some organizations, professional associations for example, and some special libraries sponsor bibliographies and can therefore advise on their existence. Bibliographies may also be retrieved by using certain online databases.

The printed guides are sometimes referred to as bibliographies of bibliographies; these are lists of bibliographies (which may include journal articles, or chapters in books) providing for a subject approach.

Uses

To find out if there are any 'ready-made' literature searches covering your subject.

Access

Via libraries.

Caution

The guides mentioned below, even between them, do not provide anything like complete coverage of the scientific/technical field.

Examples

Bibliographic Index: a cumulative bibliography of bibliographies. New York, H. W. Wilson, 1938–   .
○ Entries, which include author, title and publication details, are listed under fairly narrow subject headings (which are themselves arranged alphabetically). Author approach is not provided for. Coverage currently includes about 2600 periodicals (in all subject fields) as well as bibliographies published separately or appearing as parts of books. *Bibliographic Index* is published in April and August, with a bound cumulation each December; multi-annual cumulations are also available covering the period 1937–68. Online access should become possible via WILSONLINE during 1985 in the USA, but the date from which this service may be used by searchers in the UK is unknown at the time of writing. For the scientist/engineer, this example is probably of greater value than the others cited here.

A World Bibliography of Bibliographies, by T. Besterman. 4th ed. 5 vols. Lausanne, Societas Bibliographica, 1965–66.

○ Entries, which include author, title and publication details, are listed under fairly narrow subject headings (which are themselves arranged alphabetically). In addition, there is an index covering authors, titles of periodicals, names of libraries whose publications are included and, under the heading patents, subject entries for abstracts of some British patent specifications. Coverage is limited to separately published bibliographies. This work was reissued in 1971 by Rowman and Littlefield of Totowa, New Jersey.

A World Bibliography of Bibliographies, 1964–1974, compiled by A. F. Toomey. 2 vols. Totowa, NJ, Rowman and Littlefield, 1977.

○ Intended as a supplement to the previous example, this has entries based on Library of Congress printed catalogue cards arranged alphabetically under fairly broad subject headings. Author approach is not provided for, and again coverage is limited to separately published bibliographies.

Do not forget that, as subject bibliographies are often associated with books, journal articles and so on, guides to these forms of literature act indirectly as guides to subject bibliographies.

Further information about bibliographies and their guides may be found in **General guides** (see p. 40) and **Subject guides** (see p. 42).

Books
See also **Reference material**, p. 101

Books are familiar to practically everyone as a source of information. Little need be said about them here, but it may be helpful to remember that scientific and technical books fall into three main categories: textbooks, treatises and monographs.

Textbooks generally cover a fairly broad subject area: they are designed, primarily, for educational purposes (including self-instruction), but they are not necessarily easy to understand. Textbooks may be used as introductions to a subject and, if sufficiently advanced, as sources of reference.

Treatises give an extended, systematic exposition of the principles of a subject, usually with many references to both primary and secondary literature: their value lies in the large amount of information they contain. Generally written by more than one expert, treatises are used mainly for reference purposes: they are often issued in parts over a long period of time, so the earlier volumes may be out of date before publication is complete.

Monographs are 'mini-treatises' dealing with a single topic, or small group of topics, issued complete (as opposed to treatises), again giving a detailed account of their field, written by experts, with references to other literature.

Uses

For education and reference.

Access

Via libraries, or purchase through booksellers.

Caution

- Do not expect to find the latest information on a subject in books!
- Make sure (where appropriate) that the books you use are as up to date as possible, accurate, and written in a style, and at a level, you can understand.
- The author of a wide-ranging textbook may not be an expert over the entire field covered by that book.
- Obtaining information from a book is not only a matter of checking its index: it may be necessary to use the contents page and/or browse through appropriate sections.

Guides to books

See also **Bibliographies**, p. 56; **Guides to reference material**, p. 115; **Information services**, p. 194

The guides mentioned in this section are all examples of general bibliographies: they list books, giving details such as author, title, place of publication, publisher and date, with facilities for tracing individual items by subject, author and (in some cases) title. Books are also covered by certain online databases. For guides to book reviews, see p. 126.

Uses

- Not only for finding textbooks, etc., but for tracing almost anything published as a book (including reference material, guides, conference proceedings).
- To find a representative selection of books on a given subject (to supplement a library catalogue, perhaps, which naturally only lists the items available in that library).
- To see which is the latest edition of a book, or whether it is still in print.
- To check the existence of a book, or to confirm publication details, before ordering it or requesting it from a library.

Access

Via libraries (or, occasionally, some of these guides may be available for consultation at a bookshop).

Caution

- Do not rely on finding all relevant books under just one subject heading.
- Be prepared to consult more than one of the examples given below.

Examples

The British National Bibliography. London, British Library Bibliographic Services Division, 1950– .
○ *BNB* is compiled from books deposited under the copyright acts and gives virtually complete coverage of material published in Britain. Entries include details of author(s), title, edition, publisher, date of publication, size (including number of pages), whether there is an index and/or bibliography, ISBN (International Standard Book Number—see pp. 263, 264) and price.

They are arranged by subject according to Dewey Decimal Classification. Some entries, derived from advance information supplied by publishers, are designated 'CIP entry', where CIP stands for Cataloguing In Publication. These are amended after publication of the items concerned, then being annotated 'CIP rev'. *BNB* is published weekly, each issue containing an author/title index (a single alphabetical sequence with entries which include publisher, ISBN and price). The last issue of every month (green wrappers) has author/title and subject indexes covering the whole month. These subject indexes only provide the classification numbers used in the body of the work. There are two 4-monthly interim cumulations eventually superseded by annual volumes, the latter also being available as microfiche. A series of multi-annual indexes and subject catalogues covering the periods 1950–73 and 1951–70 respectively has been produced as well. Online access to *BNB* data is available through BLAISE-LINE (see p. 152).

The British Library Bibliographic Services Division further publishes *Books in English* on microfiche. This appears bi-monthly in the form of progressive cumulations, until each year is complete. The entries, based on British Library and Library of Congress cataloguing data, are arranged in a single alphabetical sequence of authors and titles. *Books in English* is available from 1970–80 as ultrafiche, requiring a special viewer, but as standard microfiche thereafter. (The British Library issued a standard microfiche cumulation covering 1971–80 late in 1984.)

British Books in Print. 2 vols. London, Whitaker, annually.
○ Entries, which include details of author, title, size, edition, price, publisher, date of publication and ISBN, are arranged in a single alphabetical sequence of authors and titles (with subject entries based on words appearing in titles). There is an alphabetical list of publishers giving addresses, telephone numbers and ISBN prefixes, whilst a numerical index of the latter enables a publisher to be identified when only an ISBN is known. *British Books in Print* is also available in a microfiche version, updated every month. The same publisher is responsible for: *Whitaker's Books of the Month & Books to Come* (monthly), *Whitaker's Classified Monthly Book List* (January 1983–) where material is arranged under broad subject headings such as 'Chemistry and Physics', *British Paperbacks in Print* (annually), and *The Bookseller* (weekly), a trade journal announcing British books as they are published. Online access to the *WHITAKER* database is available through BLAISE-LINE (see p. 152).

Cumulative Book Index: a world list of books in the English language. New York, H. W. Wilson, monthly (except August).
○ *CBI*, which has been published since 1898, provides relatively good coverage for material of American origin (though it has not the advantage of being compiled from copyright deposits, as in the case of *BNB*). Entries, which include details of author, title, edition, number of pages, price, date of publication, publisher and ISBN, are arranged in a single alphabetical sequence of authors, titles and subjects. A directory of publishers and distributors (with addresses) is included in each issue. There are various cumulations, covering from 3 months to 6 years. Online access is possible via WILSONLINE, which is available in the USA but not in the UK at the time of writing.

Books in Print. 6 vols. New York, Bowker, annually.
○ Volumes 1–3 are arranged alphabetically by author, Volumes 4–6 alphabetically by title (with a list of publishers and distributors in the United States, including addresses and telephone numbers, at the end of v. 6). Entries give details of author, title, edition, date of publication, price, ISBN and publisher. *Subject Guide to Books in Print* (3 vols., annually) is a companion publication listing material under relatively narrow subject headings based on those used by the Library of Congress with appropriate cross-references. Both are updated by *Books in Print Supplement* (2 vols., annually), which has author and subject sequences in the first volume, titles and a list of publishers in the second. Yet another Bowker guide is *Forthcoming Books; now including New Books in Print* (bi-monthly), with separate author and title sequences in each issue. *Books in Print* is also available as microfiche, updated monthly, again with separate author and title sequences. Furthermore this database may be searched online. Bowker are responsible for three more specialized compilations (whose titles are self-explanatory): *Scientific and Technical Books and Serials in Print* (annually), *Medical Books and Serials in Print* (annually), and *Scientific, Engineering and Medical Societies Publications in Print* (bi-annually).

Pure & Applied Science Books, 1876–1982. 6 vols. New York, Bowker, 1982.
○ Entries, which include full bibliographic details, are arranged by author within an alphabetical sequence of Library of Congress subject headings. Coverage is limited to material either published in the United States or distributed there (although originating elsewhere). Certain categories of book are excluded, for instance elementary textbooks and government publications. Bowker have also produced *American Book Publishing Record, Cumulative 1950–1977: an American national bibliography*. This 15-volume compilation issued in 1978 has entries listed (mainly) according to Dewey Decimal Classification; here, too, full bibliographical details are given. Author and title indexes are provided, and there is a subject guide employing narrow subject headings.

Bibliographic Guide to Technology. 2 vols. Boston, MA, G. K. Hall, annually.
○ Based on publications catalogued by the Research Libraries of the New York Public Library and the Library of Congress, entries are arranged in a single alphabetical sequence which includes authors, titles and subject headings. Full bibliographic details are given in the main entry for each item. This guide also covers some non-book material such as videorecordings.

International Books in Print: English-language titles published outside the United States and the United Kingdom. 2nd ed. 2 vols. Munich, K. G. Saur, 1981.
○ Entries are arranged in a single alphabetical sequence of authors and titles, generally with full details given in the author entry and only a cross-reference under title. A list of publishers, which includes addresses and telephone numbers, is provided too.

There are many other national bibliographies and guides to the book trade: these are listed by country in *Guide to Reference Material*, edited by A. J. Walford, 3rd ed. Vol. 3, pp. 17–51: or in *Guide to Reference Books*, compiled by E. P. Sheehy, 9th ed., pp. 39–81.
   Information about recent books (including book reviews) can be obtained from some periodicals, and those containing good bibliographies may be covered by *Bibliographic Index* (see **Guides to bibliographies**, p. 60).

Individual publishers' catalogues can also be useful, and may contain more information (e.g. chapter headings or a synopsis) than the other guides mentioned in this section.

Details of older books may be more easily found by use of one of the large, multi-volume, printed library catalogues; see under **General bibliographies** (p. 58) for further details.

This guide also has a section on **Guides to the book trade** (p. 263). Further information about books may be found in **General guides** (see p. 40) and **Subject guides** (see p. 42).

Conference proceedings
See also **Conferences**, etc., p. 288

Conference proceedings can be extremely difficult to identify or obtain. They may be published as a book or as part of a periodical, be mimeographed, or exist only in manuscript or on tape—if they exist at all. They may appear a few weeks or a few years after the conference, which might be called a colloquium, congress, convention, seminar, study group, summer school, symposium, workshop . . ., sometimes preceded by 'International' or another term.

Proceedings may consist of papers (or a selection of them) presented (or intended for presentation) at the conference, but can include the discussions in a condensed version or otherwise.

Many conference papers are similar in nature to journal articles, though they can be more speculative than the latter. Of particular interest are reviews, or state-of-the-art surveys, which feature at most conferences. Sometimes preprints of papers are circulated before the conference is held and these may wholly, or in part, take the place of published proceedings.

A British Standard (BS 4446: 1969) offers guidance on the presentation of conference proceedings, so anyone who incurs the responsibility for editing them should consult it.

Uses

● Recent conference proceedings can give a good indication of the present position of work in a subject field, especially with regard to the various kinds of approach made by different workers.
● Some conferences are (intentionally) multi-disciplinary in character: the proceedings of these bring together ideas from people in widely differing subject areas, and can stimulate new lines of enquiry.

Access

● Via libraries, or purchase.
● Often the inter-library loan system will have to be used.
● When conference proceedings are published in a periodical, they may appear in its regular issues, or in a special supplement, or (occasionally) they are scattered over a number of different journals.
● It will sometimes be necessary to make direct contact with the author(s) of a conference paper, or organization(s) concerned, in order to see whether that paper is available.

Caution

- They can be very difficult to identify and obtain.
- They may be hard to find in library catalogues, as the rules for entering them are complex and not self-evident to the catalogue user.
- Conference proceedings are often carelessly cited in the literature.
- They may be poorly processed, exclude any discussions and be without an index.
- Conference papers are not usually subject to refereeing (see **Journal articles**, p. 74): they may repeat information which has appeared in the literature before and, on occasion, display a sales-oriented bias.
- Be prepared for a considerable delay (perhaps a year or two) between the holding of a conference and the publication of its proceedings.
- Do not assume that just because an important conference has taken place, its proceedings are *certain* to be published.
- Conference proceedings may be biased towards the interests of a sponsoring body, e.g. practical rather than theoretical, or vice versa.

Guides to conference proceedings
See also **Information services**, p. 197

Individual conference papers are covered by some abstracting/indexing journals (see p. 46) and online databases, but, in general, this is a poorly guided field.

Some of the guides mentioned below cover only the collected proceedings of conferences, and cannot be used to trace contributions by particular authors.

Uses

- To find what meetings have been held in the past in your subject field.
- To see whether the proceedings of a particular conference have been published.
- To trace individual papers by author or subject.

Access

Via libraries.

Caution

- Coverage tends to be incomplete.
- Most guides are of comparatively recent origin, so older conference proceedings can present considerable difficulties as they may only be traceable through **Guides to books** (p. 62) or **Bibliographies** (p. 56).

Examples

InterDok Directory of Published Proceedings. New York, InterDok Corporation, 1966– .
○ This is currently available in three versions: *Series SEMT—Science/ Engineering/Medicine/Technology* (since 1967), *Series SSH—Social Sciences/ Humanities* (since 1968), and *Series PCE—Pollution Control/Ecology* (since 1974). Entries are arranged chronologically according to the date (month/year)

when the conference was held, starting from January 1964. Information given includes the place at which the meeting occurred, the conference title and sponsor(s), the name and address of the publisher of the proceedings, the editor(s), date of publication and price. Three indexes are provided: editor, location and subject/sponsor. There is a list of publishers, with addresses, too. *Series SEMT* is issued, monthly, ten times a year from September to June; these issues are then cumulated into one volume. *Series PCE* comes out semi-annually, with bi-annual cumulative indexes. The InterDok Corporation provides an acquisition service for subscribers and will supply all conference proceedings listed in the directory which are in print. Details of *MInd: the meetings index*, announced by InterDok for 1984– , will be found on p. 289.

Proceedings in Print. Arlington, MA, Proceedings in Print Inc, 1964– .
○ In each issue, entries are arranged alphabetically according to a 'subject title' by omitting words such as 'Conference on' where necessary. Until 1980 the entries were divided between two sections: current (published within the last 2 years) and retrospective (published previously). Information given includes place and date of conference, name of editor(s), and publication details. The index has entries for subjects, editors and sponsors. If it is known that the proceedings of a particular conference will *never* be published, an entry is inserted to this effect. This guide is issued bi-monthly, and the index is cumulated annually.

Index of Conference Proceedings Received. Boston Spa, British Library Lending Division, 1973– .
○ This publication, which appears monthly and cumulates annually, records the relevant holdings of the British Library Lending Division. Entries are arranged in a single alphabetical sequence of subject keywords generally chosen from either the titles of conference proceedings or the names of organizations associated with the conferences concerned. However, an entry will not always be found under the first word of the title itself, even when this might have been expected from the subject point of view. Details given include name, place and date of the conference, together with a BLLD shelf mark; for everything in this list can be obtained from the British Library Lending Division through the inter-library lending network. A 3-volume cumulation covering 1974–78 has been published, and the contents of an earlier guide were cumulated in 1974 as *BLL Conference Index, 1964–1973*. A set of microfiche entitled *Index of Conference Proceedings Received 1964–81* is also available. Computer searching of the equivalent database is possible via BLAISE-LINE (see p. 152).

Bibliographic Guide to Conference Publications. 2 vols. Boston, MA, G. K. Hall, annually.
○ Based on publications catalogued by the Research Libraries of the New York Public Library and the Library of Congress, entries are arranged in a single alphabetical sequence which includes editors, titles (of both conferences and their proceedings), subject headings, and sponsoring organizations. Full bibliographic details appear in the main entry for each item.

Index to Scientific & Technical Proceedings. Philadelphia, Institute for Scientific Information, 1978– .
○ This guide to individual conference papers as well as complete volumes of

conference proceedings is published monthly and cumulates annually. Main
entries, which are arranged numerically according to proceeding numbers
assigned by the Institute for Scientific Information, each consist of the
conference title, where and when it was held, names of sponsors, full
bibliographic details of the proceedings as a whole (whether issued in book or
journal form) including price if available, followed by a title/author/page
number listing of the individual papers. Various indexes are provided:
category (allowing for a fairly broad subject approach), author/editor,
sponsor, meeting location, permuterm subject (based on pairs of significant
words chosen from the conference paper titles), and corporate divided into two
sections—geographic (arranged by city within state for the USA and then by
place within country for the rest of the world) and organization (names or
organizations entered alphabetically, though this must be used in conjunction
with the preceding geographic section). The equivalent database may be
searched online as *ISI/ISTP&B*, Index to Scientific & Technical Proceedings
and Books, and there is a companion publication *Index to Social Sciences &
Humanities Proceedings* (1979–   ).

Ei Engineering Conference Index. New York, Engineering Information Inc,
1983/4–   .
○ This is another guide covering individual conference papers as well as
complete volumes of proceedings. The contents are arranged as follows.
Pt. 1 Civil, environmental and geological engineering.
Pt. 2 Mining, metals, petroleum, fuel and nuclear engineering.
Pt. 3 Mechanical, automotive and aerospace engineering.
Pt. 4 Electrical, power, optical, acoustical engineering and applied physics.
Pt. 5 Electronics, information science, communications, computer and
       control engineering, applied mathematics and instrumentation.
Pt. 6 Chemical, agricultural, food and bioengineering.
Pt. 7 Cumulative index for parts 1–6.
The first six parts are complete in themselves and contain the following
indexes: subject, author, sponsor, author affiliation, conference title, and
conference code/conference book number.
An entry for the proceedings of each conference records the publication's title,
the name, location, date and sponsor(s) of the conference, the editor(s) of the
proceedings, and relevant bibliographic details. This is followed by entries for
individual papers presented at that conference, which include titles, authors,
and first authors' organizational affiliations. Online searches are possible
using the associated *EI ENGINEERING MEETINGS* database.

The proceedings of periodic international congresses are given in *World List of
Scientific Periodicals*, Vol. 3, pp. 1789–1824 (see p. 98 for further details of this
guide to periodicals). Other international congress proceedings may be traced
through *Yearbook of International Congress Proceedings* 1960/67–   (Brussels,
Union of International Associations, 1969–   ) or its predecessor *Bibliography
of Proceedings of International Meetings*, covering 1957–9.
From 1969 to 1973, individual conference papers were covered by *Current Index
to Conference Papers*, produced by CCM Information Corporation (New York).
This began publication as three series: *Chemistry*, *Engineering* and *Life Sciences*,
which amalgamated in 1971.

From 1973 to 1977 a monthly current-awareness service entitled *Current Programs* was issued, which included details of papers to be presented at forthcoming conferences. This has since been replaced by *Conference Papers Index* (Bethesda, MY, Cambridge Scientific Abstracts, 1978–   ); see p. 288.

Further information about conference proceedings may be found in **General guides** (see p. 40) and **Subject guides** (see p. 42).

Government publications
See also **Patents,** p. 78, and **Reports,** p. 117, for accounts of government-sponsored research, etc.

Government publications, or official publications (in practice these terms are interchangeable), provide reliable, factual and up-to-date sources of information.

British official publications fall into three categories: Parliamentary, Non-Parliamentary and Non-HMSO. The first two categories are published by HMSO (Her Majesty's Stationery Office) and constitute about 20 per cent of the total. Libraries holding items in more than one of these categories may treat them quite differently.

Parliamentary publications include Sessional Papers (see below), Acts of Parliament (which are of two kinds: Public General Acts, Local and Private Acts), and Hansard (the official record of what is said during Parliamentary debates).

A session is the period of time between a State Opening of Parliament and its subsequent prorogation, usually almost 12 months later. Sessional Papers consist of three categories:

(1) Command Papers. These are presented to the House 'by Command of Her Majesty' from outside, by the Minister responsible.
(2) House of Commons Papers. These arise out of the deliberations of the House, or are needed for its work, and are 'Ordered by the House of Commons to be printed'.
(3) House of Commons Bills. These are draft Acts of Parliament.

The terms Blue Book, Green Paper, and White Paper are frequently mentioned in connection with British government publications. *Blue Books* are Sessional Papers, particularly Command Papers, many of which are issued in blue covers, but the term has no precise meaning. *Green Papers* set out proposals for changes in government policy and are intended to form a basis for discussion by interested parties: they are mostly Command Papers. *White Papers* generally contain statements of government policy (issued as a Command Paper), but, in a wider sense, the term is applied to Sessional Papers of insufficient length to need a blue cover.

Non-Parliamentary publications (published by HMSO) can include those of government ministries and departments, and bodies such as the Forestry Commission, the British Geological Survey, and the National Physical Laboratory. However, it is more convenient in the present context to consider technical reports produced by organizations like the Building Research Establishment or the Forestry Commission under the general heading of **Reports** (see p. 117), even if they are also government publications.

Non-HMSO publications are those published independently of HMSO by organizations financed or controlled wholly or in part by the British government. These organizations comprise government departments, nationalized industries, research institutes, quangos (quasi-autonomous non-governmental organizations) and other official bodies. Examples here are the British Railways Board, the Engineering Industry Training Board, and the Independent Broadcasting Authority.

It so happens that certain organizations such as the aforementioned Building Research Establishment and the Forestry Commission publish in both the Non-Parliamentary and Non-HMSO categories!

Many different committees and other types of enquiring body have been set up over the years to carry out specific investigations or advise the government. Their reports, which can be of considerable importance, may be issued as Parliamentary, Non-Parliamentary, or Non-HMSO publications. Often, such documents become popularly known by the name of the committee chairman, a method of citation which could prove inexact. For instance, 'The Merrison Report' might refer to the enquiry concerning the 'Basis of design and method of erection of steel box girder bridges' (1971–74), or the 'Regulation of the medical profession' investigation (1975), or even the working party on 'Support of university research' report (1982).

Though this section deals almost entirely with British government publications, other countries have an official publications literature of a similar kind. It may be worth noting that Pergamon Press publishes a series entitled 'Guides to Official Publications' which includes volumes covering the following: Australian, Canadian, French, Irish, Japanese, and US Federal official publications.

Uses

For official information of every type; including statistics, economics and finance, law, government policy, Parliamentary matters, all of which can affect most areas of work.

Access

● Via libraries, or purchase.
● There should be a library which holds a reasonable selection of British government publications near most major centres of population in the UK; they may also be available through the inter-library loan system (see p. 254).
● British Parliamentary and Non-Parliamentary publications can be purchased from the Government Bookshop, 49 High Holborn, London (callers only). Orders by post and general enquiries should be sent to HMSO Books, PO Box 276, London SW8 5DT (tel. 01-622 3316 when placing an order, otherwise 01-928 6977). Outside the London area orders may be given either to a bookseller acting as an agent for government publications (details are available from HMSO monthly or annual catalogues), or to a local branch of the Government Bookshop (Edinburgh, Manchester, Bristol, Birmingham and Belfast). Requests for free catalogues of material published by HMSO should be addressed to HMSO Publications Centre (P10A/1), 51 Nine Elms Lane, London SW8 5DR (tel. 01-211 5266). Enquiries concerning photocopies

of out-of-print government publications should be directed to HMSO Publications Centre (PC13B/2), 51 Nine Elms Lane, London SW8 5DR.

● Non-HMSO publications cannot be obtained in their entirety from a single source. The *Directory of British Official Publications: a guide to sources*, compiled by S. Richard (2nd ed., London, Mansell, 1984), lists by area and type a wide variety of official organizations, in each case outlining the range of publications available and saying how these may be acquired (entries include relevant addresses and telephone numbers); indexes of organizations and subjects are provided. Reference should also be made to the *Catalogue of British Official Publications not published by HMSO* produced by Chadwyck-Healey (see p. 73).

● Information about ordering certain American official publications will be found in current issues of the *Monthly Catalog of United States Government Publications* (see p. 73), where Federal Depository Libraries are also listed. For documents of a military nature, see *How to Get It: a guide to defense-related information resources* (p. 129).

Caution

● Some tend to be pitched at an administrative level, and can be dull!

● As their means of identification are so varied, they can be difficult to locate, especially when given nicknames such as 'the Cogwheel Reports'—so named from the design on the cover.

● For Parliamentary papers, the date of the session of Parliament is an essential part of the reference.

● When dealing with Command Papers, the reference must be noted and quoted *exactly* as it is officially written, e.g. Cmd 1 is *not* the same as Cmnd 1: the first relates to the 1919 session of Parliament, the second to 1956/7.

● For Non-Parliamentary papers, the name of any associated department or ministry, and the calendar year of publication, are essential parts of the reference.

● For Non-HMSO publications, the name of the issuing body and the date are essential parts of the reference.

● When dealing with committee reports, try to obtain more information than just the name of the chairman: certain chairmen have been associated with more than one committee, and sometimes the chairman who actually signs the report has taken over from the chairman originally appointed to the committee—in which case the report is likely to be known by the name of the first chairman!

Government publications present so many complications that most scientists/ engineers needing to make use of them will be well advised to consult a librarian or information officer with expertise in this field.

Guides to government publications
See also Information services, p. 200

These either take the form of official lists of government publications (and it should be noted that not every country produces a list of this kind), or books indicating the nature, scope, organization and availability of such literature. Government publications are also covered by some online databases, such as *POLIS* (Parliamentary On-Line Information System) developed for the House of Commons Library in the UK.

Uses

- To find out whether any government work has been carried out in your field.
- To see if government policy/regulations affect your work.
- To identify publications precisely (i.e. obtain the exact reference) before borrowing or purchasing.
- To determine the sources of supply of some official publications.

Access

Via libraries.

Caution

- Even officially published lists may not be comprehensive.
- Subject indexes are often general rather than specific, which means searching under broad headings.
- Technical reports may not be indexed individually.

Examples

The first set of examples is concerned with British government publications.

Government Publications. London, HMSO, annually.
○ This is basically an HMSO sales list. It began in 1922 as *Consolidated List of Parliamentary and Stationery Office Publications*, then became *Consolidated List of Government Publications* (1923–50), changed its title to *Government Publications: Consolidated List* (1951–53), becoming *Government Publications: Catalogue* (1954–55), and then *Catalogue of Government Publications* (1956–71).
The information cumulated here appears first in the *Daily List of Government Publications* from HMSO, and then in the monthly catalogue entitled *Government Publications* [month, year], popularly known as the *Monthly Catalogue*.
In the annual volumes Parliamentary publications are listed numerically within their various series; then comes a classified list with Non-Parliamentary and virtually all Parliamentary papers arranged by issuing ministry, government department, or other appropriate organization (including some material sold but not published by HMSO); a third section is devoted to periodicals, giving details such as subscription prices. The index has entries for subjects and personal names (chairmen of committees, for example). Cumulated 5-year indexes to these catalogues have been produced since 1936.
HMSO also produces a series of *Sectional Lists* covering the publications of a particular ministry, government department or associated organization, or in a specified subject area; most of these include only 'in print' publications.

Sessional Indexes to Parliamentary Papers. London, HMSO, 1828– .
○ These list Bills, House of Commons Papers and Command Papers both numerically and under broad subject headings (each volume dealing with the papers of one session of Parliament). Sessional Indexes are cumulated decenially, omitting the numerical lists, and finally appear as 50-year General Alphabetical Indexes.

Catalogue of British Official Publications not published by HMSO. Cambridge, Chadwyck-Healey, 1981–   .
○ This guide is issued bi-monthly, with annual cumulations. Certain classes of publication are specifically excluded: for example, Ordnance Survey maps, patents, ephemeral items, internal memoranda and reports. Material is listed alphabetically by title under the names of the appropriate publishing bodies, which may themselves appear within the entries for their parent organizations. Thus the Freshwater Biological Association will be found under the Natural Environment Research Council. An alphabetical index of contributing organizations links such subsidiaries with their parents, and supplies initial entry numbers referring to the main body of the catalogue. The information accompanying each title listed includes date of publication, price, and a code indicating the source from which it may be obtained (an index to these sources contains full postal addresses for the bodies concerned). A combined author/subject index is provided as well as a cumulating keyword index on microfiche (1983–   ). In many cases microfiche versions of documents covered can be purchased either by standing order or individually. Enquiries should be directed to: Chadwyck-Healey Ltd, 20 Newmarket Road, Cambridge CB5 8DT (tel. 0223 311479), or Chadwyck-Healey Inc, 623 Martense Avenue, Teaneck, NJ 07666, USA (tel. 201 692 1801).

An Introduction to British Government Publications, by J. G. Ollé. 2nd ed. London, Association of Assistant Librarians, 1973.
○ This guide provides a brief account of the entire field, with particular emphasis on the role played by HMSO.

British Official Publications, by J. E. Pemberton. 2nd ed. Oxford, Pergamon, 1973.
○ A more detailed account than that given in the previous example, with chapters devoted to the main types of publication.

Current British Government Publishing, by E. Johansson. London, Association of Assistant Librarians, 1978.
○ Deals with the growing problem created by changes in publishing practices of the British government, especially the declining role of HMSO.

Official Publications in Britain, by D. Butcher. London, Clive Bingley, 1983.
○ A current survey of the entire field, with a chapter on official publications in libraries.

The next few examples deal with United States government publications.

Monthly Catalog of United States Government Publications. Washington DC, US Government Printing Office, 1951–   .
○ This began in 1895 as *Catalogue of Publications issued by the Government of the United States.* It then became *Catalogue of the United States Public Documents* (1895–1907), changed its title to *Monthly Catalog, United States Public Documents* (1907–39), and was then known as *United States Government Publications: a monthly catalog* (1940–50). Entries are arranged under the US government department, or other organization, responsible for the documents. Indexes currently provided include: author, title, subject, series/report, and title keyword; these cumulate annually. The degree of indexing was less

comprehensive before 1976. Monthly issues contain instructions for ordering items listed therein. Although not giving complete coverage of material published by the US government and its agencies, the *Monthly Catalog* is a useful guide to report literature. The associated database may be searched online as *GPO MONTHLY CATALOG*. A 15-volume cumulative subject index spanning 1900–71 (originally published by the Carrollton Press) is available from Research Publications Inc. of Woodbridge, Connecticut, and cumulative personal author indexes (1941–75) have been published by Pierian Press of Ann Arbor, Michigan.

Introduction to United States Public Documents, by J. Morehead. 2nd ed. Littleton, Colorado, Libraries Unlimited, 1978.
○ A detailed survey of the field, which both explains the structure and organization of US official publications literature, and describes the related bibliographical guides.

Recent information on US departments and agencies will be found in *United States Government Manual*, published annually and obtainable from the Superintendent of Documents, US Government Printing Office, Washington DC 20402 (price about $10).

Government Publications: a guide to bibliographic tools, by V. M. Palic. 4th ed. Washington DC, Library of Congress, 1975 (reissued Oxford, Pergamon, 1977).
○ Earlier editions were written by J. B. Childs and published under different titles. Nearly half of this work is concerned with US official publications, the remainder dealing with those of some international organizations and foreign countries.

Details of the official lists and catalogues of government publications from other countries may be obtained from *Guide to Reference Material*, edited by A. J. Walford, 3rd ed., Vol. 3, pp. 181–92.

Journal articles

Information about much important research is first given to the scientific community in the form of a journal article. There are three main reasons why scientists and engineers (who naturally wish to establish a record of the priority of their original work) prefer to publish their latest results in this way: (1) acceptable speed of printing and distribution; (2) guaranteed circulation, probably world-wide; (3) material not suitable for book or monograph treatment.

Some delays in publication are, however, inevitable, especially if the article is submitted to a 'high prestige' journal with a long waiting list of would-be contributors.

Another major cause of delay is the system of refereeing employed by many journals: articles are sent to independent experts in the fields concerned, who are asked to pronounce on their fitness for publication. This process may take several weeks, further delay being incurred should the referees suggest substantial alterations before the papers are accepted for publication. Referees are not infallible: there are classic cases of important scientific papers being rejected including that describing T. H. Maiman's realization of the first laser! But, in general, refereeing is necessary to maintain high standards of quality and reliability of published information.

Most journals expect contributors to adopt a specified style, covering layout, method of citing references, presentation of diagrams, etc. These requirements are sometimes printed on the cover or preliminary page(s) of a current issue, but they may be available as a 'Notes for Contributors' leaflet from the editorial office of the journal concerned.

Unfortunately, some major journals operate a system of 'page charges'. In the case of *Physical Review*, for example, authors' institutions are requested to pay $45 per page (which entitles them to 100 free reprints). Non-payment of this charge may lead to delays in publication. An additional $30 is requested for advance publication of the abstract in *Physical Review Abstracts*. This practice has reduced the number of British papers appearing in such journals.

Frequently there is information about a research project which is not worth printing in an article intended for wide circulation, although it may be of considerable interest to a few people working along similar lines (minor experimental details, quantities of supporting numerical data, and computer programs might be relevant here). This problem can be solved by placing the extra material in a depository where it will be readily accessible, or perhaps issuing it as a microfiche supplement to the journal carrying the original article. Taking the case of certain journals published by the American Chemical Society, for instance, supplementary material may be ordered from the publisher either as photocopies or on microfiche, and appears anyway in microfiche editions of the journals concerned. In the UK, the BLLD (see p. 259) is involved with storing and supplying supplementary material of this kind. Future developments in electronic publishing and electronic document delivery may well affect the ways in which authors prepare material for (and research workers use) journal articles.

Finally, a word about preprints and offprints. There is a tendency for authors to distribute mimeographed copies of papers submitted, or intended for submission, to journals, either to obtain comments, or for the information of selected colleagues—these are generally known as preprints. It should be noted that the 'preprint' paper may never achieve publication in a journal, or may be substantially altered before it is so published. Offprints, or reprints, on the other hand, represent extra copies of an article as published in a journal: again these may be circulated among interested colleagues, to keep them informed of recent work.

Uses, etc.

See the relevant sections under the heading **Periodicals**, p. 96.

Guides to journal articles

The principal guides to journal articles (as opposed to the periodicals in which they appear) are abstracting and indexing journals, dealt with in detail on pp. 46–53, and numerous online databases. Other guides include subject bibliographies (p. 59), review articles (p. 125), annual indexes to individual periodicals, books, reports, conference proceedings, and theses.

Newspapers and newspaper articles

These are a well-known source of information. Newspapers are (by definition) periodicals, so much of what is said in a later section (pp. 94–101) applies to them. However, most scientists and engineers think of newspapers, and use them, in a different way from other periodicals.

Uses

- For very recent technical/scientific news.
- For company news.
- Advertisements are useful when job-hunting, or seeking information about a company and its products.

Access

- Via newsagents, personal subscription, or libraries.
- Publishers often keep and sell back numbers.
- Press-cuttings agencies can supply newspaper articles, advertisements, etc., relating to your specified requirements; see p. 290 for further details.
- The British Library Newspaper Library, Colindale Avenue, London NW9 5HE (opposite Colindale Underground Station), tel. 01-200 5515, houses the largest collection of newspapers in the UK. Admission to the reading room is by pass only, a leaflet setting out the relevant conditions being available from the library (and it may be worth noting that persons under the age of 21 are not normally admitted); tickets for the British Library reading room at Bloomsbury are also valid for use at Colindale. Readers should remember that requests for material are not processed after 16.15 each day. Non-ticket holders may order photocopies of newspaper articles (the *exact* reference must be given), subject to the provisions of the Copyright Act for material published within the last 50 years, on *written* application to the library.

Caution

- Details given in newspaper articles can be hopelessly inaccurate. 'Don't believe everything you read in the papers!'
- Different editions of a paper on the same day may carry different news items.

Examples

Most people have a favourite newspaper. The authors of this book have found *The Times* (with its various supplements) and the *Financial Times* valuable sources of information. Some newspapers (including *The Times*) are also available on microfilm.

Guides to newspapers and newspaper articles
See also Guides to periodicals, p. 97; Information services, p. 205

Some guides mentioned in this section extend their coverage to periodicals in general; however, they have been considered here as scientists and engineers use them mainly in connection with newspapers.

A few newspapers (*The Times* is one example) publish detailed indexes to their contents. Otherwise, the searching of a file of newspapers can be an extremely tedious business—some idea of the date at which the information is likely to have been published being essential.

Certain newspaper publishers run library/information services, which can answer enquiries about information published in their own papers; the *Financial Times* is an example.

Uses

- To obtain publication details relating to a particular newspaper.
- To find newspapers in a specific locality or subject area.
- As a way of finding out about information published in a few newspapers.

Access

Generally, via libraries.

Caution

- Make sure you have the most recent edition of the guide.
- Remember to search indexes under all possible subject headings.

Examples

Willing's Press Guide: a guide to the press of the United Kingdom and to the principal publications of Europe, Australasia, the Far East, Middle East, and the USA. East Grinstead, Thomas Skinner Directories, annually.
○ Entries, which include names and addresses of publishers together with subscription details, are arranged alphabetically by title within country, the UK section occupying about half of each volume. A classified index lists publications under their countries of origin within broad subject divisions, whilst a newspaper index enables material associated with particular places to be identified. Lists of 'publishers and their periodicals' and 'reporting, news and press-cutting agencies' are also provided. Coverage extends beyond newspapers to magazines and more scholarly periodicals.

Benn's Press Directory. 2 vols. (UK and International). Tunbridge Wells, Benn Business Information Services, annually.
○ The UK volume includes the following sections: publishing houses, national/London newspapers, provincial papers for the rest of the UK (according to place of publication), local free-distribution publications, periodicals (both general and scholarly, listed alphabetically by title with a classified index using broad subject headings), agencies and services for the communications industry (arranged alphabetically by name of organization); an index of publications is also provided.
The international volume basically lists publications according to country of origin (preceded by a short international section); newspapers are followed by other periodicals (recorded under various classified headings), but details of certain official organizations (such as embassies) and media agencies are provided too.

Ayer Directory of Publications. Philadelphia, Ayer Press, annually.
○ Primarily a guide to the North American press, whose entries are arranged by city or town within state for the USA, and by city or town within province for Canada. Details include the title of each publication, when first established, frequency of appearance, advertising rates, circulation figure, name of editor, name/address/telephone number of publisher. A classified section is provided as well, together with lists of 'daily newspapers' and 'weekly, semi-weekly and tri-weekly newspapers'. An alphabetical index of titles completes the volume.

The Times Index. Reading, Research Publications Ltd, monthly (with annual cumulations).
○ Entries are arranged in a single alphabetical sequence of names and subjects. *The Times* first appeared in 1785, and indexes covering its contents from 1790–   are available. Before 1977 there were few annual cumulations, so several sequences must generally be consulted for each year of interest.

Annual Index to the Financial Times. London, Financial Times Business Information, 1981–   .
○ Cumulative edition of *Monthly Index to the Financial Times*. The contents are arranged in three sections: corporate information (dealing with companies), general information (entered under subjects, countries, or subjects within countries), and personalities.

The British Library; Catalogue of the Newspaper Library, Colindale. 8 vols. London, British Museum Publications, 1975.
○ This catalogue includes new UK titles and amendments up to the end of 1970, overseas titles and amendments up to the end of 1971. In volumes 1–4 entries are arranged alphabetically by title according to the places where the newspapers were either published or circulated: v. 1 is devoted to London; v. 2 covers England and Wales, Scotland and Ireland; v. 3–4 deal with the rest of the world, the arrangement being by town within country. A single alphabetical sequence of titles occupies the remaining volumes. The entries themselves give the title of each publication, changes of title being cross-referenced, with an indication of the numbers and dates held by the Newspaper Library. Coverage extends as far as magazines and trade journals.

World List of National Newspapers: a union list of national newspapers in libraries in the British Isles, compiled by R. Webber. London, Butterworths, 1976.
○ Newspapers are listed alphabetically by title, with codes indicating the relevant holdings of major British libraries. An index of titles arranged by country is also provided.

Patents, trade marks, trade names and industrial designs

These are sometimes known collectively as 'industrial property', or as 'intellectual property' when copyright is also included. Trade marks, trade names and industrial designs are considered towards the end of this section, most of which is devoted to patents.

A patent is a sole or monopoly right to make, use or sell an invention for a given period of time, granted by a Crown, State or Government, and protected by law. The law may vary in detail from country to country. The word 'patent' is often used in a vague way to describe patent applications, patent specifications, or the legal documents conferring the patent (see later). Individuals or organizations (applicants) may apply for patents in order to control the making, use and selling of inventions which they rightfully own. This prevents anyone else from doing so without permission, during the life of the patent (a temporary monopoly right). In the UK, this protection may be obtained for up to twenty years, provided that the appropriate conditions are met, including the payment of fees. In return, the applicant must reveal or disclose sufficient information for someone in the same field to reproduce the invention. Thus apart from the legal aspect, patent specifications, which contain the detailed descriptions of the inventions, can serve as a major source of information, not only to spread the knowledge of recent advances, but also to encourage others to even greater advances. The patent applicant gets protection; the State, in return, gets information. Applicants who are not themselves the inventors must provide details of the inventor and of how they derived the right to apply.

To be patentable, an invention must be new, not obvious, capable of industrial application (in the broadest sense) and concerned with the composition, construction or manufacture of an article, substance or apparatus, or with any industrial process. The following are not patentable: aesthetic creations; mathematical models; discoveries; surgical or therapeutic methods; new animal or plant varieties; rules and methods for performing mental acts, playing games or doing business; and computer programs. Any anti-social implications would normally invalidate the application.

A patent does not offer automatic protection in the sense that the proprietor must initiate High Court action against any infringers. Proprietors of patents may sell their patents by means of an assignment, or allow another person or organization to exploit the invention on payment of royalties under a licensing agreement.

Patent procedures and documentation are very complicated, and much of this section has probably been oversimplified. On the other hand, because the information in patents is usually very detailed, up to date, often never published in any other form, and is becoming an increasingly important source of information, the list of guides provided here is more extensive than in some of the other sections. Anyone particularly interested in patents is urged to study these more authoritative guides.

In the UK we have had two sets of legislation which affect existing patents, the 1949 Patents Act and the 1977 Patents Act. This has resulted not only in two different systems that need to be understood, but also in the complexities of having two procedures running in parallel until all the patents issued under the 1949 legislation have passed through the system. Unless otherwise stated, this section covers UK documentation for the 1977 legislation and procedures. Some 400 specifications of patent applications and some 400 specifications of granted patents are published each week under the 1977 legislation. All newly published specifications are displayed on tables in the Science Reference Library (see p. 259) for one week, after which time they are shelved alongside earlier specifications.

Uses

- National sources of subject information.
- Prevention of duplicated effort.
- Checking the novelty of new ideas.
- Protection of inventors' rights.
- Keeping an eye on competitors' inventions.
- Ideas for licensing opportunities.

Access

- UK and foreign patent specifications and associated literature may be consulted, free of charge, at the British Library's Science Reference Library (SRL, Holborn Reading Room, 25 Southampton Buildings, Chancery Lane, London WC2A 1AW), and more limited collections of patent literature at other libraries in:

Aberdeen	Huddersfield	Nottingham
Aberystwyth	Hull	Plymouth
Belfast	Leeds	Pontypridd
Birmingham	Leicester	Portsmouth
Bradford	Liverpool	Preston
Bristol	Manchester	Sheffield
Coventry	Middlesbrough	Swindon
Edinburgh	Newcastle upon Tyne	Wolverhampton
Glasgow	Norwich	

These libraries, with collections of UK and, sometimes, foreign patent literature, form the UK 'Patents Information Network' (PIN Libraries). Each library provides public access to its own holdings, gives practical help and advice, can arrange for photocopies of patents, and can check other libraries in the network for material that it cannot supply itself (note that patents are not normally borrowable).
- Further details of these libraries and their holdings may be found at the SRL, in the *Official Journal (Patents)* (see p. 84), or in *Patents: a source of technical information* (see p. 87).
- UK patent specifications (applications and granted patents) may be purchased from the Patent Office at £1.95 each: by post from the Sale Branch, The Patent Office, St Mary Cray, Orpington, Kent BR5 3RD (tel. Orpington 32111, ext. 302) or by direct application, for very recent items, to the Sale Counter, The Patent Office, 25 Southampton Buildings, London WC2A 1AY (tel. 01-405 8721). Foreign specifications may be obtained direct from the countries concerned, or as photocopies from the SRL at prices from £1 upwards depending on country and number of pages. There is a while-you-wait photocopy service at the SRL for around 10p per sheet. The Patent Office Sale Branch will also post copies of all current UK specifications in chosen subject fields, each week, to subscribers who have a deposit account. Specifications are charged at the current price. Tabulated lists of specification numbers in chosen subject fields are also available. Derwent Publications Limited also provide a patents copy service; this includes photocopy/printed copy, fax, express and translation services (for address see p. 90).

Caution

● Titles can be vague and do not always give a clear indication of the subject of the invention.
● The text of patent specifications is couched in a specialized language (patentese) and can be difficult to follow especially if the emphasis has been to protect the invention rather than to describe it.
● Check that a patent is still in force before assuming that you cannot copy the invention. In the case of UK patents this may be done via the progress registers kept by the SRL, but it must be noted that the registers provided by the SRL are not official records. Official, statutory registers can be checked in the Patent Office (State House, 66–71 High Holborn, London WC1R 4TP, tel. 01-831 2525) on payment of a fee, and should be if you are thinking of copying an invention.
● Patent law may change from time to time, resulting in different types of procedure and publication.

UK patent specifications

These are the printed descriptions or disclosures of inventions which are made widely available for anyone to see or purchase. They are the documents which most people think of as patents, but they actually represent the information that the State receives from patent applicants in return for the protection it may grant (cf. British Letters Patent/Certificate of Grant, below). Each specification must give sufficient detail to enable anyone, skilled in the art, to reproduce the invention. Until the specification is published, only the title of the application, the date of application, the name of the applicant, the priority details if any, and the application number are available to the public (p. 85).

Two specifications are published for each successful application. The first 'early publication' specification of the patent application is published after a preliminary examination and has a seven-figure number suffixed by the letter A (for example, *UK Patent Application GB2468500A*). The second specification, amended if necessary, of the granted patent is published after the 'substantive' examination, acceptance and granting, and has the same serial number suffixed by the letter B (for example, *UK Patent GB2 468 500B*). There was only one specification published for each invention under the pre-1977 legislation.

It is useful to know certain details of layout and content (items a to f appear on the front page of the specification):

(a) The patent or 'document' number is a unique national number which enables the patent to be easily identified; it also gives a rough idea of the date of publication (the application number is only unique for a given year). Numbers for the 1977 Act patent specifications started from 2000001. Numbers for pre-1977 legislation are up to about 1605000.

(b) Each patent specification is given one or more national classifications (letters and numbers or codes which stand for subject headings), enabling patents on a particular subject to be identified by use of the *Classification Key* and the *Catchwords Index* (see **Guides to patents**, p. 86). An international classification (IPC) is also given.

(c) Every patent specification gives various dates, each indicating a stage in the process of obtaining the patent. The earliest date indicates when the first application was made, in order to establish the applicant's entitlement to his monopoly rights, and is called the 'priority date'. Any information available to the public before the priority date may invalidate the application. An invention must not normally be publicly used in the UK before this priority date.

'Prior art' is that art which is already known and leads up to the invention without actually including it (see also g below).

(d) The name of the applicant (individuals or organization). The patent applications published under the name of a particular applicant may be found by use of the *Applicants' Name Index* (see **Guides to patents**, p. 86). If the applicant is not the inventor, the inventor's name is also given.

(e) A list of documents referred to or cited.

(f) An abstract and, if appropriate, a small drawing in the case of the suffixed 'A' series.

(g) The rest of the text is divided into sections:

An amplification of the title.

If appropriate, an explanation of the prior art, i.e. of what is already known, as a 'lead up' to the invention.

The object of the invention and the method of achieving it.

The practical details of the invention, possibly illustrated by examples or drawings.

The claims of the patent (numbered 1, 2, 3, etc.), which define the scope of the monopoly of the invention and its variants which are protected.

British Letters Patent/Certificate of Grant

British Letters Patent are the legal documents in which, under the 1949 legislation, the state granted to the patentee a temporary monopoly right to make, exercise, use or sell his invention. The patentee has to encourage the exploitation of his invention, and cannot suppress the invention by completely refusing to allow its production and use, and there are provisions for compulsory licences to be issued if it can be shown that the patentee is not exploiting it to the fullest extent practicable. Under the current 1977 Patents Act, British Letters Patent have been replaced by grant certificates, and patentees are usually referred to as applicants or proprietors, depending on whether or not the patent has been granted. There have been a number of other changes, the most obvious being in the procedures for publication and examination.

The published patent may be opposed by interested parties at any time during its 'life', normally after it has been granted. Anyone can apply to the Court or to the Patent Office for revocation of a patent. If this opposition is successful, the patent is revoked and the proprietor loses his monopoly rights.

Non-UK patents

Many countries have their own patenting systems, which, although basically similar to that in the UK, may involve significant differences in practice, such as no early publication, publication of unexamined specifications, and different classification schemes. The USA, for example, has its patent office situated in Arlington, Virginia, where copies of issued patent specifications are kept, together with appropriate guides. There is also a system of regional patent copy depository libraries. The US patent procedure does not include early publication, and normally no information about pending applications is published. There is also a system of defensive publication where, effectively, an applicant abandons a patent application but allows official publication in order to prevent anyone else from patenting the invention. Some other countries have different types of patents, including 'petty' patents which give more limited protection than a full patent. There has been some interest in this system for the UK. Apart from UK and US specifications, important sources of patent information for scientists and engineers frequently include German, French and Canadian specifications.

International patents

There is no such thing as a truly international patent, but the examples below indicate a certain movement in that direction.

European Patents

The European Patent Convention (EPC) permits people in eleven of the West European countries to apply for a European Patent through their national patent office. The main advantage for this system is that it provides a single procedure for taking out a patent in a number of the contracting countries, which also may be cheaper than taking out separate patents in each different country. The countries in which protection is required are 'designated' by the applicant. The more countries designated, the greater the total cost, but the cheaper it becomes compared to taking out separate patents in each different country. The single application and procedure results in a European patent, which is effectively a bundle of national patents for the designated countries. A booklet, *How to get a European Patent*, may be consulted at the SRL or obtained from the European Patent Office, Erhardstrasse 27, D-8000 München 2, West Germany.

International patent applications

The Patent Cooperation Treaty (PCT) has resulted in the nearest thing to a truly international patent application system, covering over thirty countries including the USA and the USSR. The remarks made about the European Patents apply, more or less, to the PCT applications except that the process directly involves national applications procedures, and the end result is an actual set of national patents derived from one application. The booklet *PCT Applicants' Guide* may be consulted at the SRL or purchased from the World Intellectual Property Organization (WIPO), 34 chemin des Colombettes, 1211 Geneva 20, Switzerland.

Patent families

Sets of equivalent specifications/applications from a number of countries are known as patent families. Some guides and most patent databases include a facility for identifying equivalent patents. The earliest member of a patent family is the one with the earliest priority date, and is sometimes referred to as the priority application, but this can have a different meaning. Derwent use the term 'basic' to describe the first member of a patent family to enter the Derwent premises.

Guides to patents (printed and online)

These include guides to the nature, scope, availability and law of patents, special indexing and abstracting publications and databases (see also **Abstracting and indexing journals**, p. 46, and **Information services**, p. 139).

Uses

● To enter the patent literature of a particular country or countries by subject, patent specification or application number, or name of a person or organization.
● To find out about the procedure for taking out a patent.
● To find out about techniques for searching patent literature.

Access

● Apart from the SRL, various other public and industrial libraries subscribe to these guides, and make them available for reference (see locations of network libraries above, and the examples of guides below, and check with the library concerned, before you go).
● All the Patent Office publications may be purchased from the Sale Branch in Orpington or the Sale Counter in London (see p. 80; some of the descriptive pamphlets are free).

Caution

● Classification schemes can split subjects, so that you may have to look in more than one place.
● The British classification scheme changed during 1963, and the concordance or conversion charts may be used to convert from one scheme to another.
● Classification schemes are revised from time to time, and the classification schedules apply to specified time periods.
● Patent databases are not comprehensive and may be restricted to certain dates, countries and subject areas. They may also give a 'descriptive' title rather than the actual title of the specification.

Examples

UK PATENT OFFICE PUBLICATIONS

Where appropriate, references to SRL publications and guides are included here. See also Chart 5.

The Official Journal (Patents). Published weekly, this includes lists of

Chart 5. Searching for information in UK patents

		Subject searches	Name searches (applicant/inventor)
Patent application (titles only)	Current	Only indirect, by scanning the titles in weekly issues of the *Official Journal (Patents).*	Search in weekly issues of the *Official Journal (Patents)* for applicants.
	Retrospective (only back 5/6 years)	Only indirect by scanning the titles of applications in the annual card *Name Indexes of Applicants* (SRL), provided you know what names to look under, e.g. competitors.	Search in the annual card *Name Indexes of Applicants* (SRL).
Patent specifications (published)	Current	(1) In weekly issues of the *Official Journal (Patents)* using appropriate subject classification*. (2) In *Abridgments/Abstracts of Specifications* issued weekly in subject groups using appropriate subject classification*. (3) By scanning each week's patent specifications (approx. 800) as they become available, e.g. at the SRL. (4) In weekly lists of specifications available at the SRL.	(1) In weekly issues of the *Official Journal (Patents),* includes inventors. (2) Indirectly by checking whether known applications have been accepted using the card *Name Indexes of Applicants* and the *Applications Register.* (3) In weekly lists of specifications (includes inventors) available at the SRL.
	Retrospective	(1) In volumes of *Abridgments/Abstracts of Specifications* using appropriate subject classification* (each volume covers 1 year or a given number of specifications). (2) Using *File Lists* (purchased from the Patent Office), which give the numbers of the specifications on a particular subject. (3) In computer-searchable databases.	(1) Search *Applicants' Name Index* (includes inventors and each index covers one year or a given number of applications). (2) In computer searchable databases.

* The subject classification may be obtained by using first the *Catchwords Index* and then the *Classification Key.*
Note. The classification scheme changed from 1963 with Patent No. 940001, and the *Concordance* must be used for finding how the old classification corresponds to the new. The *Classification Key* is currently revised annually and different editions cover different time periods. The *Index, Key* and *Concordance* may be consulted at the SRL or at the PIN Libraries, or obtained from the Patent Office.

applications for patents, applications published under the 1977 Act, patents granted under the 1977 Act, specifications accepted under 1949 legislation, together with corresponding details such as numbers, titles, dates and names, and, where appropriate, concordances and name and subject indexes. Other information includes official notices, status of applications, designs, European patents granted which designate the UK, and some housekeeping information about the Science Reference Library.

Abridgments/Abstracts of Specifications. These form the abstracting journal of the UK patent literature. Currently the bulk of the records are abstracts of specifications published under the 1977 Act; the declining minority are abridgments of specifications accepted under the 1949 legislation. The abstracts are on single sheets, identical to the front page of the 1977 Act specifications/applications, and are published weekly at the same time as the specifications, but in pamphlet form. The abridgments are more like traditional abstracting journals, with as many entries as can be fitted in, printed on each page. The pamphlet series are published in 25 subject groups. Cumulated volumes of the two series, together with one combined name index and two separate subject classification indexes, are now published annually for each subject group. Subject searches, using classifications, for specifications published since the last cumulated volume indexes, may be performed using the subject indexes in the *Official Journal (Patents)* or via specially prepared lists in the SRL. These guides are mainly for retrospective subject searches via the subject (classification) indexes in each volume or via the *File Lists* (see below).

File Lists. These are lists of numbers for all the specifications on a particular subject (i.e. one classification or a combination of classifications). The span of years varies, depending on the subject, but may go back to 1911. Anyone wishing to use file lists should obtain further details from the Patent Office or the SRL.

The Catchwords Index to the Classification Key (formerly part of the *Reference Index to the Classification Key*, which is no longer published). This is used for the first stage in finding the correct classification for a particular subject, and is normally used in conjunction with the *Classification Key* itself. The index consists of an alphabetical list of 'catchwords' and shows the appropriate subject headings and corresponding broad subject classifications, which you then use to enter the *Classification Key* in order to identify the specific classifications for your subject.

The Classification Key. This is revised annually and care must be taken to use the correct one. It is available in separate parts, one for each of twenty-five subject groups, or as combined volumes. The *Key* displays the classification in the finest detail in alphanumeric order, showing corresponding fine subject headings. The *Catchwords Index* takes you to a particular broad subject area, and you can examine that area in detail in the *Classification Key*, and pinpoint the exact subject you want. The method is similar to using a street map and its index. A free pamphlet setting out the structure of the key is available.

Applicants' Name Index. This index only covers applicants whose complete specification has been published or accepted under either legislation, and is, in effect, a cumulated index of those published in the *Official Journal (Patents)*. Currently the *Index* is published annually, but weekly computer printouts cumulated quarterly are available in the SRL. Its main use is to identify the specifications published in the name of a known person or organization.

Details given include shortened form of title and cross-references to assignees (proprietors who are not the inventors). Details of applications before publication are available in the *Official Journal (Patents)* or in cumulated form in a card index in the SRL. The cards are kept for six years only. For each application year, cards are in alphabetical order of applicant and give brief title, priority data and application number. Unofficial SRL progress registers may then be used to obtain further information. For example, the register of applications may be used to find whether an application has resulted in a published specification, and if so, its number. The register of stages of progress may then be used to find whether a patent has been granted or the application has lapsed or expired.

There are various other publications and services available; some of these are described in the first of the pamphlets listed below, which are obtainable, free, from the Patent Office or the SRL, unless otherwise stated.

Patents: a source of technical information.
Instructions for the Preparation of Specification Drawings.
How to Prepare a UK Patent Application: and then apply for a patent.
Introducing Patents: a guide for inventors.
Basic Facts about Patents for Inventions in the UK.
The Patent Office: an introduction to the services of the Patent Office and the Trade Marks and Designs Registries.
Industrial Property: a selected list of publications (SRL).
Industrial Property Literature in the Science Reference Library: holdings for the UK, the European Patent Office and WIPO (PCT). SRL Aids to Readers No. 26.
Patent Licensing: a guide to the literature. SRL Guideline.

GUIDES TO NON-UK PATENTS

Other countries have publications similar to those in the UK, such as abstracts, indexes and official journals. for example in the USA there is a weekly journal called the *Official Gazette*, which lists newly issued 'utility' patent specifications in three broad subject groups: general and mechanical, chemical, and electrical, and then, within these groups, in numerical order by class and subclass. Each entry includes the principal claim and a diagram if appropriate. Reissues, defensive publications, plant patents and designs are also listed, together with various items of information similar to those published in the *UK Official Journal (Patents)*. There are other guides and indexes to help in subject or name searching. Anyone interested in foreign specifications is advised to consult the appropriate guides mentioned in this section, and also to pay a visit to the Foreign Patents Reading Room of the SRL (Chancery House Annexe; see p. 259).

GENERAL GUIDES TO PATENTS

The New European Patent System, by R. Singer. Translated and adapted by D. J. Devons. Lausanne, Seminar Services, 1981.
○ A good introduction to the European Patent Convention, the Community Patent Convention and the Patent Cooperation Treaty.

The United Kingdom Patent System, by N. Davenport. Havant, Kenneth Mason, 1979.

○ A helpful introduction to the development of the present patent system, with sufficient detail to answer many of those awkward questions. An extensive bibliography includes statutes, rules, law reports and official publications.

Patent Documentation, by M. Hill (derived from a German edition by A. Wittman and R. Schiffels). London, Sweet & Maxwell, 1979.
○ A good general work covering various national patent systems and publications and search procedures, including computerized services.

Guide to Official Industrial Property Publications, by B. M. Rimmer. London, British Library, Science Reference Library, 1985.
○ A looseleaf guide covering patent, trademark and design publications and about forty different patenting authorities. The author has written separate guides for individual countries including the United States, Germany and France.

Guide to Foreign-Language Printed Patents and Applications, by I. F. Finlay. London, Aslib, 1969.
○ This somewhat dated guide is still useful because it shows facsimile copies of foreign patents and applications, and explains the various formats so that you know, for example, which date is the priority date.

Foreign Patents: a guide to official patent literature, by F. J. Kase. New York, Oceana Publications, 1972.
○ Another older guide, but useful for its reproductions and explanations of patent literature.

How to Find Out about Patents, by F. Newby. London, Pergamon, 1967.
○ Although out of date, this book is still useful because it includes copies from actual pages of specifications, guides and indexes, and gives detailed explanations.

The Inventor's Information Guide, by T. S. Eisenschitz and J. Phillips. London, Fernsway Publications, 1983.
○ A useful guide for private inventors or employees of firms with no patenting facilities.

Patents, Trade Marks, Copyright and Industrial Designs, by T. A. Blanco White and R. Jacob. 2nd ed. revised by J. D. Davies. London, Sweet & Maxwell, 1978.
○ This book provides the reader with the basic information he needs before seeking specialist advice.

Manual for the Handling of Applications for Patents, Designs and Trade Marks throughout the World. 2nd ed. Amsterdam, Octrooibureau Los en Stigter, 1936– .
○ A looseleaf manual in four volumes, also known as the Dutch manual, updated about once a year, but with worldwide coverage.

Terrell on the Law of Patents, by W. Aldous and others. 13th ed. London, Sweet & Maxwell, 1982.
○ This is the classic textbook for people involved with any aspect of patents in the UK.

Patents for Inventions: and the protection of industrial design, by T. A. Blanco White. 5th ed. London, Stevens & Sons, 1983.
○ A standard work on the UK Patents Act of 1949.

CIPA Guide to the Patent Act 1977: texts, commentary and notes on practice, by the Chartered Institute of Patent Agents. London, Sweet & Maxwell, 1984–   .
○ A basic work on the current UK legislation, presently issued with cumulative supplements.

Encyclopedia of United Kingdom and European Patent Law, by T. A. Blanco White and others. London, Sweet & Maxwell, 1977–   .
○ An unusual reference work published 'bit by bit' in a looseleaf binder. A classic text for people involved in the legal area.

Patent Law of Europe and the United Kingdom, by A. M. Walton and others. London, Butterworths, 1978–   .
○ A similar work to the previous one and issued in the same way using looseleaf format.

Legal Protection of Computer Programs, by B. Niblett. London, Oyez, 1980.
○ Although somewhat out of place in this section, the subject is of sufficient importance to warrant inclusion, even though the law may change in this area in the near future. Methods of protection discussed include patents, copyright and contracts.

Intellectual Property: patents, copyright, trade marks and allied rights, by W. R. Cornish. London, Sweet & Maxwell, 1981.
○ This book is included for its general legal approach to 'intellectual property' and its specific coverage of trade marks and registered designs.

Intellectual Property Law and Taxation, by R. J. Gallafent, N. A. Eastaway and V. A. F. Dauppe. London, Oyez, 1981.
○ A practical introduction to the legal aspects of patents, designs, trade marks and copyright.

UK Patents Acts may be consulted in most large public libraries or law libraries. Frequently they are to be found as part of a series or are published in some annotated form. It is best to ask the advice of the library staff.

Derwent Publications Limited provide various documentation services, which include the following English-language publications and other services.

● *World Patents Index (WPI)*: four gazettes published weekly, indexing general, mechanical, electrical and chemical patent specifications in the more important countries including European and PCT specifications. There is a separate *Priority Index*, also published weekly, to help in finding patent family relationships. No abstracts are included in this service, but the indexes are cumulated quarterly and annually on microfiche.

● *World Patents Abstracts (WPA)*: separate weekly reports giving complete coverage for each of Belgian, British, European, West German, PCT, Soviet Union and United States patent specifications, and all with abstracts. Only chemical patents are covered in the case of the French, Japanese and Netherlands reports. In addition, seven different subject journals are published covering the non-chemical patents of most of the more important countries excluding Japan. There are indexes for each weekly edition.

● *Central Patents Index (CPI)*: a whole set of services for the chemically related patents (e.g. polymers, drugs, food, chemicals, textiles, petroleum, ceramics and metals). Twelve 'major' countries and thirteen 'minor' countries are covered. There are alerting bulletins, basic abstracts journals, profile booklets, coded cards, microfilm, microfiche and magnetic tapes available. Technical assistance is also available.

● *Electrical Patents Index (EPI)*: a similar service to CPI, provided for the special needs of the electrical- and electronics-related industries and interests.

● Online Search Service, the *WPI* database (see below), is available online direct to users through certain hosts/service providers or as a bureaux service direct from Derwent, who will perform searches on their clients' behalf.

For full details contact Derwent Publications Limited, Rochdale House, 128 Theobalds Road, London WC1X 8RP (tel. 01-242 5823).

PATENT DATABASES

There are many databases covering patents which are available for online searching. Some cover both patents and other kinds of documents or literature, such as *CA SEARCH* which corresponds to *Chemical Abstracts*, and this type of database is dealt with elsewhere (see p. 142). Some databases only cover patents, such as *WPI* and *INPADOC* which cover patents from a fair number of countries, while others such as *CLAIMS* and *PATSEARCH* concentrate on one country (the USA in this case). Details of all types of databases can be found in the general guides to databases (p. 186) or, in the case of databases specifically concerned with patent information, in *Patents Databases 1983*, compiled by the Online Information Centre, London, 1983. The following are a selection of patent databases, most of which give abstracts.

● *CLAIMS*: a set of databases covering US patents, including cited patents, some equivalent patents, US classification codes and chemical compound codes where appropriate (both *Chemical Abstracts* Registry Numbers and IFI/Plenum terms). Produced by IFI/Plenum Data Company, and available on DIALOG.

● *INPADOC*: covers nearly fifty countries, European and PCT patents including equivalents and IPC classifications. Produced by International Patent Documentation Centre and available on PERGAMON INFOLINE (the Centre also offers some services direct).

● *PATSEARCH*: covers US patents including defensive publications, PCT applications, with cited patents and US and IPC classifications where appropriate. Produced by Pergamon International Information Corporation and available via PERGAMON INFOLINE.

● *USPA* etc.: a set of databases covering US patents including cited patents, US and IPC classifications. Produced by Derwent Inc and available on SDC.

● *WPI/WPIL*: covers twenty-six countries, European and PCT patents, including IPC and special Derwent classifications. Items from the periodicals *Research Disclosures* and *International Technology Disclosures* are also included. Produced by Derwent Publications Ltd, and available on SDC, DIALOG and TELESYSTEMES QUESTEL. (A bureaux service is available direct from Derwent.)

• Two further patent databases are expected to become available. PERGAMON INFOLINE plan to mount a British patents database (GBPAT) going back to 1910 with abstracts from 1978, and MEAD DATA CENTRAL are planning to mount *LEXPAT*, covering US utilities.

Why and how to apply for a UK patent

If you think that you might have a patentable invention, remember that you may not be able to obtain a patent if you have already published or disclosed details of the invention.

A patent is only of value if it is necessary to prevent others from copying an invention, or you think that someone else is likely to patent the invention before you can show prior publication (before you can get details published in a journal, for instance) and thereby prevent you from exploiting the invention. A patent is not always a good idea. For example, if the techniques you use would be difficult to discover but capable of eventual improvement, you might wish to keep the techniques secret, so that your competitor could not use the knowledge from the patent specification as a stepping stone to an even better process, which he could patent to your detriment (say a chemical material that you had not had time to fully develop and that might be greatly improved by small changes in structure or substituents). Similarly, if you invented something that could supersede your existing product, but which would not be so financially rewarding in the long run, it could be unwise to rush into patent procedure (with an everlasting light bulb or razor blade, for example, you could lose income from the traditionally frequent purchase of replacements). Obviously you would have to think twice before spending large sums of money on the protection of an invention that was very difficult to police.

Alternatively you might delay taking out a patent if you wished to have the period of the monopoly commence at a later date, because either the product was not yet ready for the market, or the market could not be fully exploited within the immediate twenty-year period. Should you wish only to register a priority claim over an invention, without applying for a patent (because you did not care if anyone copied it or because you did not consider it possible that anyone else would have the time, inclination or means to copy it), then there are alternatives. You could obtain a fast publication in any traditional form, or alternatively by a special disclosure of the kind published in *Research Disclosures* (a periodical published by Industrial Opportunities, Havant, UK), comparable to defensive publication in the US.

If you wish to apply for a patent, you can ask the UK Patent Office to send you their various free pamphlets, forms and fee sheets, which are very helpful. If you are in employment, your employer may have an established procedure for dealing with patents, such as a patents section or a regular patent agent. Employees should, in any case, check their patent rights in relation to their employment contracts. The present legislation offers some compensation to employees in certain circumstances. If you are a private inventor it is usually best to employ a patent agent. A list of registered patent agents may be inspected at the SRL or any PIN Library, or purchased from the Registrar, Chartered Institute of Patent Agents, Staple Inn Buildings, London WC1V 7PZ. If you wish to sell your invention, prospective purchasers may insist that you should have made an appropriate patent application prior to any

Chart 6. A simplified guide to the UK patenting process. (Patent agents are usually used, and their fees may run to several hundred pounds in addition to the 1984 fees below.)

Action of applicant	Files specification/application	Files claims and abstract if not previously filed. Requests preliminary examination		Makes required amendments and requests substantive examination		Renews patent annually after 4th year if desired
Fees (without agent)	£10	£75		£90		£70 rising to..... ... £268
Time		max. 12 months	min. 18 months / max. 4½ years	max. 6 months		
Age of patent	0					4 years ... 20 years
Action of patent office	Title of application and name of applicant published. Application number and filing date given	Preliminary examination and search begin	Specification/application published with search report (suffix A)	Full examination begins	Patent granted if successful. Specification published (suffix B)	
Remarks	Priority established		Protection may be backdated to here		Infringement action may begin	Protection ceases

discussion of the invention. There are various organizations which specialize in helping inventors, and some advice is given in *Introducing Patents: a guide for inventors* and *The Inventor's Information Guide* (pp. 87, 88).

Trade marks and guides
See also Guides to patents, pp. 88–89

Trade marks are distinctive marks (devices, symbols, words, etc.) used by a trader to assist the purchasing public to distinguish his goods from rival products. Just as an inventor may apply for a patent, a trader may apply to register a trade mark.

Applying for a Trade Mark.
○ A free pamphlet available from the UK Patent Office, this gives all the necessary information concerning procedure, and should be read by anyone interested in British Trade Marks.

Trade Marks Journal.
○ This is published weekly by the Patent Office, and advertises marks which have been accepted by the Registrar. After the period of time allowed for opposition, these marks are entered in the Register.

A Practical Guide to Trade Marks, by A. Michaels. Oxford, ESC, 1982.
○ This is a guide to trade-mark law and practice for the non-specialist user, businessman or student, but it gives a comprehensive coverage of the topic.

Kerley's Law of Trade Marks and Trade Names, by T. A. Blanco White and R. Jacob. 11th ed. London, Sweet & Maxwell, 1983.
○ This is a classic textbook which will help people involved with any legal aspects of trade marks.

Note that anyone who wishes to make a search for conflicting trade marks may, for a small fee, search the classified representations and indexes of trade marks at the Trade Marks Registry of the Patent Office.

Trade names and guides

Trade names are those names by which products are known in the trade. Some are proprietary or owned, such as 'ASPRO', in which case the trade name is also a trade mark. Some are not proprietary, for example Aspirin, and these may be used by anyone.

UK Trade Names. 7th ed. East Grinstead, Kompass Publishers Ltd, 1982.
○ This index contains approximately 60000 trade names. There are two main sections: an alphabetical index of trade names giving a description of the product and the name of its user, and an alphabetical listing of companies giving addresses, telephone and telex numbers.

Chemical Synonyms and Trade Names, by W. Gardner. 8th ed., revised by E. I. Cooke and R. W. I. Cooke. Oxford, Technical Press, 1978.
○ A dictionary and commercial handbook containing over 35000 entries.

Trade Names Dictionary, edited by D. Wood. 4th ed. 2 vols. Detroit, Gale Research, 1984.

○ This American guide provides nearly 200 000 entries for trade names, trade marks and brand names, and their manufacturers, importers, marketers or distributors. Trade name entries give a brief description of the product, and the name of the corresponding company. The company entries give addresses. Gale Research publish companion volumes and supplements, including *Trade Names Directory: company index*, which provides a list of some 40 000 companies giving addresses and the corresponding trade names and product descriptions.

Note that the SRL, the commercial sections of major reference libraries (such as the City Business Library) and the various trade associations may be able to answer queries about trade names.

Industrial designs and guides
See also **Guides to patents**, pp. 88–89

A manufacturer may seek to protect his product, as far as its outward appearance is concerned, by applying for registration of its design.

Protection of Industrial Design.
○ A free pamphlet available from the UK Patent Office, this gives all the necessary information on procedure, as well as general information.

Copyright in Industrial Designs, by A. D. Russel-Clarke. 5th ed. revised by M. Fysh. London, Sweet & Maxwell, 1974.
○ A classic work with cross-references to cases and statutes.

The Protection of Industrial Designs: a practical guide for businessmen and industrialists, by G. Myrants. London, McGraw-Hill, 1977.
○ A practical guide for the UK.

Periodicals
See also **Journal articles**, p. 74; **Newspapers**, p. 76; **Reviews and review serials**, p. 125

For many people these are the most important source of information in the primary literature.

A periodical is a publication having a distinctive title, which is issued at regular intervals (weekly, monthly, quarterly, annually, etc.); alternative terms include journal, magazine, serial; or even bulletin, proceedings and transactions.

This section is concerned mainly with those periodicals which form part of the primary literature; newspapers have been dealt with separately (as a special case). However, periodicals are also to be found in the secondary literature (review serials) and in guides to the literature (abstracting/indexing journals, and some bibliographies). In this book the terms journal and periodical are employed synonymously (following common practice among users of the literature), so a reference to periodicals normally means those in the primary literature, unless the context indicates otherwise.

Scientific periodicals began by covering wide subject areas (indeed, some still do), but an evolutionary process has led to more specialized journals in narrow subject fields, and others which concentrate on printing short contributions quickly. Sometimes this may be observed in the fission of a single title. For example, *Physical Review*, which began in 1893, split (in 1970) into four sections (*General Physics*, *Solid State*, *Nuclear Physics*, *Particles and Fields*), an associated publication, *Physical Review Letters* (of the 'short contribution' variety) having started in 1958.

Periodicals with broad subject coverage (such as *Nature*) are valuable for their multi-disciplinary character, which helps break down barriers between different branches of science and technology. Some of these journals adopt a popular approach (*New Scientist*, for example) and are useful as the scientific equivalent of a weekly newspaper.

There are so many varied types of periodical that it is worth listing the kinds of information they may contain.

● *Abstracts.* Do not confuse with abstracting journals (p. 46). Some ordinary periodicals contain abstracts in their special subject field, sometimes including patents or trade literature.
● *Advertisements.* Usually for jobs; or apparatus, equipment and services.
● *Announcements.* Especially of meetings, conferences, symposia, courses, etc.
● *Bibliographies.* Lists of references to articles in a particular field, perhaps including patents or trade literature.
● *Book reviews.* It is often useful to consult a review before buying, or borrowing, the book. For guides to these, see p. 126.
● *Buyer's guides.* Certain periodicals devote a whole issue (generally once a year) to a survey of apparatus, equipment and services in their fields.
● *Correspondence.* Provides a quick way of announcing a new discovery, advertising for information or conducting a controversy.
● *Editorials.* A guide to the policies favoured by the periodical in question.
● *Journal articles.* These are dealt with in detail on pp. 74–75.
● *News items.* May be industrial, company and market news, or items from professional associations and societies.
● *Personalia.* Ranging from career details of scientists and engineers in the news, through elections to membership of learned societies, appointment of senior staff in government establishments and industry, to obituaries.

Periodicals contain some, or all, of these elements in varying amounts, ranging from pure 'research' journals (consisting almost entirely of journal articles) to trade journals, devoted mainly to advertisements.

It is difficult to devise a satisfactory structure into which all periodicals fit. One possibility is to classify them by origin. In this case they fall into two groups:

(1) Periodicals published by commercial publishers, for profit.
(2) Periodicals published by, or on behalf of, organizations. Following the classification of organizations used earlier in this book, periodicals may be associated with commercial organizations (e.g. company 'house journals', and 'controlled circulation' journals—mainly an advertising medium), with educational and official organizations (e.g. journals produced by a university or government department), or with professional and research organizations (a most important group, including the journals of learned societies, professional institutions, and so on).

It can be useful to bear these facts in mind when choosing a source of information. For example, an article which is acceptable to the editor of a learned society journal might be too esoteric for the publisher of a periodical in the first group, who must achieve a large circulation to ensure a profit and cannot, therefore, afford to publish articles with very limited appeal.

There is, however, no substitute for experience in recognizing and handling the different types of periodical: this may be gained in a large library, perhaps with the aid of a guide to the literature of the subject concerned.

Developments in electronic publishing will doubtless have a substantial effect on periodicals in the future. There are already instances of printed and online versions of specific journals existing side by side, and we may expect to see an increasing number of journals produced only in an electronic form.

Finally, something must be said about the very large number of scientific and technical periodicals in the world today, probably well in excess of 50 000 titles. Studies on the use of the literature seem to indicate that a high proportion of 'used' information is published in a comparatively small number of journals—sometimes referred to as core journals. The size of this core lies, perhaps, in the region of 3000–4000 titles. This phenomenon is presumably due to the fact that every subject has its high-prestige journals in which those working in the field would prefer to have their results published, and which they must regularly consult anyway.

Uses

- To keep up to date in the subject(s) covered by a periodical, consulting each issue as it appears (current-awareness browsing).
- As the source of information indicated by references in periodicals, books, abstracting/indexing journals and other kinds of literature.
- To see what work has been covered over a number of years by a particular periodical, using its own annual indexes (often supplied with the December, or following January, issue and usually found at the back, but occasionally at the front, of bound annual volumes).
- Publishers sometimes have their own library/information services, and can answer detailed enquiries about information published in their own periodicals, e.g. *New Scientist*, *Financial Times*.

Access

Via libraries, or personal/organizational subscription. Many organizations have a circulation list for current issues of significant journals. Special arrangements may have to be made for material which supplements published journal articles (see p. 75).

Caution

- No single periodical can cover everything, even in a very narrow subject field.
- Periodicals can change their names (or system of volume numbering) and also split into several new (separate) journals; a good library catalogue will cross-reference such changes.
- The title of a periodical may not appear where you expected to find it either

in an alphabetical index or on the library shelves; check the indexing rules for the arrangement in question and/or try under all possible headings.

• Older volumes of periodicals in some branches of science and engineering are subject to relatively little use. Certain libraries, therefore, divide the runs of these titles according to age, often storing the older material at a considerable distance from the remainder.

• Periodicals with foreign-language titles may nevertheless publish quite a high proportion of their contents in English, or provide English summaries of articles that have not been translated.

Examples

A few examples have been mentioned above. It is impossible, in a book of this size, to list even the major journals covering the principal areas of science and engineering. Readers requiring this information should consult the appropriate **Subject guides** (pp. 42–46).

Guides to periodicals
See also Guides to newspapers, p. 76; Guides to review serials, p. 126; Information services, p. 206

The guides which cover the *contents* of periodicals are **Abstracting and indexing journals** (pp. 46–53) and certain computerized databases.

Guides appearing in this section list the periodicals themselves, often alphabetically by title, together with useful information such as publishers, dates, locations, subjects covered, International Standard Serial Numbers (ISSN, similar to ISBN for books; see p. 62), depending on the guide involved (see examples below).

Uses

• To find periodicals covering a given subject.
• To find details of publisher, frequency or cost.
• To find which libraries hold a particular journal.
• To find out which abstracting/indexing journals cover a particular periodical.
• To help expand abbreviated periodical titles found in some references.

Access

Via libraries.

Caution

• Choose a guide that is as up to date as possible; the number of periodicals in the world is increasing all the time, and title changes of existing publications are by no means infrequent.

• Subject approach is not well provided for, perhaps because it is very difficult (often quite impossible) to classify a periodical by subject. The title of a periodical may not appear where you expect to find it in the list: check the indexing rules and/or look under all possible headings.

• Some of the information in these guides is condensed by abbreviation: read the 'how-to-use' section in order to make the most of them.

● Periodical title abbreviations may not be unique; for instance, *JASA* is commonly used for both *Journal of the Acoustical Society of America* and *Journal of the American Statistical Association*.

Examples

The first four examples give information about the periodical holdings of some British and American libraries.

British Union-Catalogue of Periodicals: a record of the periodicals of the world, from the seventeenth century to the present day, in British libraries. 4 vols. London, Butterworths, 1955–8.

○ Entries, with certain important exceptions, are arranged alphabetically by periodical title, changes of title being cross-referenced; but the indexing rules are complicated and the explanatory section should be read. The holding of a particular title by major British libraries is indicated by means of an alphabetic code, interpreted through the list of library symbols provided. There is no subject approach to the material covered.

BUCOP, as this publication is affectionately known, was updated by supplements until 1980; three cumulations cover the period up to 1973, with annual volumes thereafter. For changes after 1980, see *Serials in the British Library* below.

World List of Scientific Periodicals, published in the years 1900–1960. 4th ed. 3 vols. London, Butterworths, 1963–5.

○ Entries are arranged alphabetically by periodical title, considering only 'important' words, and cross-referencing title changes. Holdings of major British libraries are indicated by alphabetic codes. There is no subject approach to the material covered. This publication is also updated by the *BUCOP* supplements (see previous item).

Serials in the British Library. London, British Library Bibliographic Services Division, 1981–   .

○ This quarterly guide lists, alphabetically by title, periodicals published for the first time since 1976 newly acquired by the British Library, including changes of title. The relevant holdings of certain other UK libraries are given too. Microfiche cumulations are available which also contain information about earlier titles, and locations reported since the appearance of an entry in a quarterly issue. Periodicals which have ceased publication are noted as well. Subject approach is not provided for.

New Serial Titles: a union list of serials commencing publication after December 31, 1949. Washington DC, Library of Congress, 1950–   .

○ This guide effectively continues the *Union List of Serials in Libraries of the United States and Canada* edited by E. B. Titus, 3rd ed., 5 vols. (New York, H. W. Wilson, 1965). *New Serial Titles* appears monthly, with quarterly and annual cumulations. The years 1950–70 have been covered in a single cumulation, available as either hard copy or microfiche from Bowker of New York. Cumulations covering 5-year periods from 1971–   are published by the Library of Congress, Washington DC. Entries are arranged alphabetically by title, indicating the relevant holdings of major US and Canadian libraries. Subject approach is not provided for.

The next two examples provide a subject approach to periodicals.

Ulrich's International Periodicals Directory: a classified guide to current periodicals, foreign and domestic. 24th ed. 2 vols. New York, Bowker, 1985.
○ Coverage is limited to serials which are issued more frequently than once a year (for others, see the next example). Entries are arranged (mainly) alphabetically by title within fairly broad subject divisions. Details normally given include: ISSN, first year of publication, frequency of appearance, annual subscription, name and address of publisher, editor's name, main features of each periodical (whether it carries advertising, book reviews, or illustrations, for instance), circulation data, microform version availability, and a note of any coverage by abstracting and indexing journals. A list of titles which have recently ceased publication is also provided. An index of subjects will be found near the front of the first volume, and a title index at the end of the second. Both this and the next example are updated by *Ulrich's Quarterly* (same publisher), and online access to the *ULRICH'S INTERNATIONAL PERIODICALS DIRECTORY* database is available via the DIALOG service (see p. 153).

Irregular Serials & Annuals: an international directory. 10th ed. New York, Bowker, 1985.
○ The arrangement is similar to that of the previous item. Subject and title indexes are provided, as is a numerical index of ISSNs (International Standard Serial Numbers). For details of an updating service and online access, see the previous example.
Periodicals published in particular countries can be identified using another Bowker guide, *Sources of Serials: international serials publishers and their titles, with copyright and copy availability information* (2nd ed., 1981), where the arrangement is by publisher and/or corporate author within country, an index of the former being provided.

Some guides cover the periodical publications of a particular country, as in the following two examples.

Current British Journals, edited by D. P. Woodworth. Boston Spa, British Library Lending Division, 1982.
○ This is the 3rd ed. of a work previously published as *Guide to Current British Journals.* Entries, which are arranged alphabetically by title in UDC classified order (see p. 249), include date of first issue, name/address/telephone number of publisher, whether an index is produced, frequency of appearance, circulation data, coverage by abstracting and indexing journals, microform version availability, and ISSN. Title and subject indexes are provided.

The Standard Periodical Directory, 1983–84. 8th ed. New York, Oxbridge Communications, 1982.
○ Contains information about more than 60000 US and Canadian publications. Entries are arranged alphabetically by title in fairly broad subject divisions. The data given may include: name, address and telephone number of publisher, editor's name, brief description of contents, where indexed or abstracted; subscription, circulation, advertising and printing details, availability in other media, and ISSN. The main section is preceded by a list of cross-references from subjects not used as main headings, and the

volume concludes with an index of titles. Oxbridge Communications Inc (183 Madison Avenue, New York, NY 10016, USA) also offer a computerized information retrieval service via the associated *SPD* database.

Other guides cover the holdings of a particular library, or group of libraries.

Current Serials Received by the British Library Lending Division. Boston Spa, BLLD, annually.
○ The 1985 edition listed about 54 000 titles then currently taken by the BLLD at Boston Spa (see p. 259); a further 123 000 non-current serials were not included. Entries are arranged (with certain exceptions) in alphabetical order; the only information given, other than title, is the BLLD shelfmark. There are separate sections for cyrillic titles (e.g. Russian-language) and cover-to-cover translations of cyrillic serials. This guide is often helpful when trying to expand the abbreviated periodical titles found in some references.

Keyword Index to Serial Titles (KIST). Boston Spa, British Library Lending Division, quarterly (each issue supersedes its predecessors).
○ Only available as microfiche, this guide provides a keyword-out-of-context listing of over 200 000 serials held by two major branches of the British Library: the BLLD at Boston Spa, and the Science Reference Library in London (see p. 259). Titles are listed alphabetically (with certain exceptions) against keywords which appear to the left of the entries. The libraries' holdings are indicated by symbols, though the information is not always complete in the case of BLLD entries. Apart from its use in determining whether a particular journal is held by either library concerned, this guide is especially valuable for identifying a serial when the full or exact title is not known (since words within titles can serve as entry points into the system).

Many other libraries (university, public and company) produce lists of their periodical holdings. Finally, some more general items.

Directory of Title Pages, Indexes and Contents Pages. Boston Spa, British Library Lending Division, 1981.
○ Entries, which are arranged alphabetically by title, include a country of origin code, ISSN (where known), and notes on the availability of indexes, title and contents pages. Although comprehensiveness is disclaimed, the main value of this guide in the present context is that it may reveal whether individual indexes exist for particular periodicals.

There is, as yet, no universally accepted standard list of abbreviations for periodical titles. The British Standards Institution has published BS4148: 1985, which deals with general principles (an earlier word-abbreviation list, BS4148: Part 2: 1985, was withdrawn as obsolete in 1980). Other relevant standards have been produced by the American National Standards Institute (ANSI) and the International Organization for Standardization (ISO). These are ANSI Z39.5–1984: *Abbreviations of Titles of Publications*, and ISO4–1984: *Documentation—Rules for the Abbreviation of Title Words and Titles of Publications* (equivalent to BS4148: 1985).
Some periodicals, including abstracting/indexing journals, have introduced their own systems for abbreviating titles, one of which is featured in the next example.

Bibliographic Guide for Editors & Authors. Washington DC, American Chemical Society, 1974.
○ The publishers of *Biological Abstracts*, *Chemical Abstracts* and *Engineering Index* have jointly sponsored this guide, which lists nearly 28000 scientific and technical periodicals monitored by these abstracting services. Details include title abbreviations, CODEN (6-character codes, of which the first five are always letters, representing periodical titles in a convenient form for computer manipulation), indication of coverage by the three named abstracting journals, plus an annotated bibliography of US and international documentation standards.

Lastly, a wider-ranging example of similar character.

Periodical Title Abbreviations, compiled and edited by L. G. Alkire. 4th ed. 2 vols. Detroit, Gale, 1983.
○ Covers periodicals in many fields other than science and engineering. The intention is to record the way in which commonly used indexing and abstracting services (excluding, for the most part, *Chemical Abstracts*) abbreviate periodical titles. The first volume contains some 55000 abbreviations, arranged alphabetically, giving the corresponding titles, whereas v. 2 is an alphabetical listing of journals with their abbreviations. This work is supplemented between editions by *New Periodical Title Abbreviations* from the same publisher.

When expanding periodical abbreviations, the lists of journals covered produced by many abstracting and indexing services can be helpful. Further information about periodicals may be found in **General guides** (see p. 40) and **Subject guides** (see p. 42).

Reference material

The term reference material can embrace a wide range of literature, from favourite textbooks, through key journal articles, to personal manuscript notes. However, the term is used here in a narrower sense, to cover certain types of publication (such as dictionaries and handbooks) which are specially designed for reference purposes, having the following features in common: (a) they are arranged so that specific information can be quickly extracted; (b) they are not meant to be read from cover to cover, but consulted as the need arises.
   Each type of publication has its own peculiarities (illustrated in the examples below) but there are a few general points to bear in mind with all reference material.

Uses

● For 'factual' information of almost any kind; reference works exist for most fields of knowledge.
● For definitions, explanations, properties, numerical data, names and addresses, products, sources of information, brief introductions to subjects, etc.

Caution

● Reference books do not always contain the most recent information on a subject. Be prepared to search further (in primary journals, for instance) if you *must* have the latest information or if you are unable to find the particular data you require.

● Always be prepared to use more than one reference book, as coverage is often incomplete and errors can occur.

● Sometimes the method of presenting the information is complicated or unusual: remember to read the 'how-to-use' section before proceeding to the indexes.

● Be sure that you have the latest available edition.

Examples

See under the different types of publication below. These have been divided into:

Data compilations	Directories
Dictionaries	Encyclopaedias
Abbreviations	Handbooks
Biography	Mathematical tables
Language	Thesauri
Subject	

Only examples of a fairly general nature have been chosen, since this book is not concerned with the literature of individual subjects. If you are interested in the reference works for a particular subject, use the **Guides to reference material** (see p. 115), or **Subject guides** (see p. 42).

Data compilations

These contain information (generally numerical data) in a concise, often tabular, form arranged for immediate access. See also under **Mathematical tables** (p. 112).

Uses

For numerical data, constants, and properties of materials.

Access

Via libraries, or (occasionally) by purchase.

Caution

● Bear in mind the possibility of misprints, and obtain an independent check where you can.

● Most data appear first in the primary literature (journal articles, etc.); remember to search there if the particular data required cannot be found in any of the likely data compilations.

Examples

CRC Handbook of Chemistry and Physics: a ready-reference book of chemical and physical data. Boca Raton, Florida, CRC Press, annually.
○ Published under the auspices of the Chemical Rubber Company (CRC), and affectionately known to some as the 'Rubber Bible', this is probably the best single-volume book of constants, summarizing a wide range of physical/ chemical properties of elements and compounds, including a few mathematical tables. A detailed subject index is provided.

International Critical Tables of Numerical Data: physics, chemistry and technology, prepared by the National Research Council of the USA. 8 vols. New York, McGraw-Hill, 1926–33.
○ Although rather out of date, this compilation was valuable in its day and does give references to the primary literature for the data recorded (these might be useful as a starting point in a citation index search for more recent data). The last volume consists of a detailed subject index.

Landolt-Börnstein: Zahlenwerte und Funktionen aus Physik, Chemie, Astronomie, Geophysik und Technik. 6th ed. 4 vols. (in 28). Berlin, Springer-Verlag, 1950–80.
○ Although the title may be translated as *Numerical data and functional relationships in physics, chemistry, astronomy, geophysics and technology,* the text is almost entirely in German. There is, as yet, no detailed subject index to the contents, an outline of which follows.
   v. 1 Atomic and molecular physics: atoms and ions; molecules (nuclear structure); molecules (electron shells); crystals; atomic nuclei and elementary particles.
   v. 2 Properties of matter in its aggregated states: mechanical-thermal properties of states; equilibria except fusion equilibria; fusion equilibria and interfacial phenomena; caloric quantities of state; transport phenomena, kinetics, homogeneous gas equilibria; electrical properties; optical constants; magnetic properties.
   v. 3 Astronomy and geophysics.
   v. 4 Technology: material values and mechanical behaviour of non-metals; material values and behaviour of metallic industrial materials; electrical engineering, light technology, X-ray technology; heat technology.
This work, which includes extensive references to the primary literature, is updated by the next example.

Landolt-Börnstein: Zahlenwerte und Funktionen aus Naturwissenschaften und Technik/ Numerical data and functional relationships in science and technology. Neue Serie/New series. Berlin, Springer-Verlag, 1961–   .
○ A continuation of the preceding work, but with the text partially in English. Over 90 parts (from 6 groups) had appeared by February 1985. The contents are organized as follows:
Group 1 Nuclear and particle physics.
Group 2 Atomic and molecular physics.
Group 3 Crystal and solid-state physics.
Group 4 Macroscopic and technical properties of matter.
Group 5 Geophysics and space research.
Group 6 Astronomy, astrophysics and space research.
A comprehensive index covering both the 6th ed. and the New series of

Landolt-Börnstein has been announced as 'in preparation'. For further details of these works see the catalogues of their publisher, Springer-Verlag: Heidelberger Platz 3, D-1000 Berlin 33; or 175 Fifth Avenue, New York, NY 10010, USA.

Tables of Physical and Chemical Constants, and some Mathematical Functions, originally compiled by G. W. C. Kaye and T. H. Laby. 14th ed. London, Longman, 1973.
○ 'Kaye & Laby' is possibly the best known of all the compilations listed here, but it should not really be compared with the others, as its field is much narrower and there are hardly any references to the primary literature. It is useful as a relatively inexpensive publication for the laboratory, or personal, reference shelf; however, the research worker should be prepared to look futher.

Thermophysical Properties of Matter: a comprehensive compilation of data by the Thermophysical Properties Research Center (TPRC), Purdue University, edited by Y. S. Touloukian and C. Y. Ho. 14 vols (in 15). New York, IFI/Plenum, 1970–   .
○ The contents are arranged as follows.
v. 1–3   Thermal conductivity: metallic elements and alloys; nonmetallic solids; nonmetallic liquids and gases.
v. 4–6   Specific heat: metallic elements and alloys; nonmetallic solids; nonmetallic liquids and gases.
v. 7–9   Thermal radiative properties: metallic elements and alloys; nonmetallic solids; coatings.
v. 10   Thermal diffusivity.
v. 11   Viscosity.
v. 12–13 Thermal expansion: metallic elements and alloys; nonmetallic solids.
[v. 14]   Master index to materials and properties.
Apart from the last, each volume consists of three sections: some text (with references) on the theory, estimation and measurement of the property under consideration; numerical data in the form of graphs and tables, with references to data sources in the primary literature; and an index to the materials covered. Lists of corrigenda have appeared since these volumes were originally published.

Thermophysical Properties of High Temperature Solid Materials, compiled by the Thermophysical Properties Research Center, Purdue University, and edited by Y. S. Touloukian. 6 vols (in 9). New York, Macmillan, 1967.
○ 'High temperature solid materials' means those with melting points above 800 K (approximately 527°C or 1000°F) except for materials in the categories of polymers, plastics and composites. This compilation is of particular interest to engineers. Much of the information is presented in graph form, and there are references to the primary literature. An index of materials is provided at the end of each volume.

Thermophysical Properties Research Literature Retrieval Guide, 1900–1980, edited by J. F. Chaney and others. 7 vols. New York, IFI/Plenum, 1982.
○ This is not a data book, but a guide to specific references, mainly in the

primary literature, where data may be found. The substances concerned are distributed among the volumes (complete in themselves) as follows.

v. 1 Elements.

v. 2 Inorganic compounds.

v. 3 Organic compounds and polymeric materials.

v. 4 Alloys, intermetallic compounds, and cermets.

v. 5 Oxide mixtures and minerals.

v. 6 Mixtures and solutions.

v. 7 Coatings, systems, composites, foods, animal and vegetable products.

Within each volume the contents are divided into four sections.

A Materials directory. Substances are listed alphabetically under various categories of material and code numbers for use in the next section provided, the entries indicating which of fifteen physical properties have been covered (these include thermal conductivity and diffusivity, specific heat, viscosity, emittance, reflectance, absorptance, transmittance, Prandtl number, thermal linear and volumetric expansion coefficients, and thermal radiative properties).

B Search parameters. Here the substances are arranged numerically and corresponding reference numbers may be selected, taking into account the following additional parameters: physical state (including superconductive, doped, expanded, gas, liquid, powder, or solid), subject (data, experiment, general, survey, or theory), language, temperature range, and year.

C Bibliography. Full citations are recorded for the reference numbers chosen via the previous section.

D Author index. This may lead to other work by the authors identified using the subject approach outlined above.

Of similar construction is another guide, *Electronic Properties Research Literature Retrieval Guide, 1972–1976*, edited by J. F. Chaney and T. M. Putnam. 4 vols. New York, IFI/Plenum, 1979. Here the materials are divided thus:

v. 1 Elements.

v. 2 Inorganic and intermetallic compounds.

v. 3 Alloys and cermets.

v. 4 Mixtures, rocks and minerals, composites and systems, polymers.

The following are included (among others): dielectric, magnetoelectric, luminescence, magnetomechanical, photoelectronic, thermoelectric and piezoelectric properties, energy band structure, electric and magnetic hysteresis, electrical resistivity, Hall co-efficient, magnetic susceptibility, refractive index and work function.

Brief mention may also be made of the *Engineering Sciences Data* series issued by the Engineering Sciences Data Unit of 251–259 Regent Street, London W1R 7AD (tel. 01-437 4894). Over 750 items have appeared in 19 sub-series, mainly covering the fields of aeronautical and mechanical engineering, dealing with topics such as aerodynamics, tribology, heat transfer, fluid mechanics, noise, and structures. Updated and new data sheets are published from time to time (subscribers need take only those sub-series in which they are particularly interested), and a detailed subject approach is provided for in an annual index volume.

Dictionaries

These give, in alphabetical order, the spelling, pronunciation and definition of words in one or more languages, or of terms in one or more special fields.

Uses

For brief definitions, or for translating from one language to another.

Access

Via libraries, or purchase.

Caution

● A given set of initials often has more than one expansion, depending on the country of origin: always be prepared to consult more than one dictionary of abbreviations.
● Many 'translating dictionaries' only cover the technical terms of a particular subject: these should always be used in conjunction with a 'general' dictionary for the languages concerned.
● Subject dictionaries are unlikely to contain the most recent information.

Examples

DICTIONARIES OF ABBREVIATIONS

Abbreviations Dictionary, by R. De Sola. 6th ed. New York, Elsevier, 1981.
○ Contains more than 178 000 entries arranged alphabetically by abbreviation, drawn from all fields of knowledge and international in scope.

British Initials and Abbreviations, by I. Wilkes. 3rd ed. London, Leonard Hill, 1971.
○ Lists organizations in Britain known by their initials, and international organizations to which Britain belongs. Most entries include addresses and telephone numbers.

World Guide to Abbreviations of Organizations, by F. A. Buttress. 7th ed. Glasgow, Leonard Hill, 1984.
○ About 43 000 entries. Only the name of the organization is given, with, in some cases, the country.

Ocran's Acronyms: a dictionary of abbreviations and acronyms used in scientific and technical writing, by E. B. Ocran. London, Routledge & Kegan Paul, 1978.
○ This work is divided into two sections. In the first, arranged alphabetically by abbreviation, each entry contains an expansion of its abbreviation and notes the associated subject field(s). In the second section, abbreviations (with their expansions) are listed alphabetically under a series of broad subject headings such as aeronautics or computer sciences. Coverage is biased towards engineering rather than pure sciences.

Acronyms, Initialisms, and Abbreviations Dictionary, edited by E. T. Crowley. 8th ed. 2 vols. Detroit, Gale, 1982.
○ Over 250 000 entries covering terms drawn from many fields, technical and

otherwise. Although the emphasis is on American material, that from other countries has not been excluded. Entries are listed alphabetically by acronym, with the appropriate expansion. Between editions this work is updated by a supplement, *New Acronyms, Initialisms, and Abbreviations*. A third compilation from the same publisher, *Reverse Acronyms, Initialisms, and Abbreviations Dictionary*, whose 1982 ed. also occupies 2 vols., has entries arranged alphabetically by name, organization, or technical term, and facilitates identification of acronyms corresponding to the former.

DICTIONARIES OF BIOGRAPHY

Examples which cover 'live' scientists/engineers are dealt with as **Guides to people** (pp. 15–18). Those listed in this section are mainly concerned with figures of historic interest.

A Biographical Dictionary of Scientists, edited by T. I. Williams. 3rd ed. London, A. & C. Black, 1982.
○ The entries, which contain references to some sources of further information, are arranged alphabetically by biographee in a main sequence followed by a short sequence of additional biographies for recently deceased scientists, living persons being specifically excluded. A list of anniversaries (births and deaths) and a brief subject index are also provided.

Dictionary of Scientific Biography, edited by C. C. Gillispie (under the auspices of the American Council of Learned Societies). Vol. 1–16. New York, Scribner, 1970–80.
○ The entries are arranged alphabetically by biographee, each concluding with a bibliography usually containing both primary and secondary sources. Living scientists are specifically excluded. The fifteenth volume constitutes a first supplement, with additional material on some of the persons in volumes 1–14, new biographies, and several historical essays. The sixteenth volume is an index. A compact edition in 8 volumes has also been published.

Biographisch–Literarisches Handwörterbuch zur Geschichte der exacten Wissenschaften, gesammelt von J. C. Poggendorff. Berlin, Akademie-Verlag, occasionally (publication originally commenced at Leipzig in 1863).
○ 'Poggendorff' is the standard reference for scientific biography. The entries are arranged alphabetically by biographee, with brief biographical details and, usually, a comprehensive bibliography of works written by the biographee. Several separate sequences cover different periods in time, and some living scientists are included in the latest volumes.

LANGUAGE DICTIONARIES

Many dictionaries are available to help with the problem of translating the technical terms of a subject, or group of subjects, from one language to another. Some of these dictionaries are 'polyglot'—which means they cover several different languages (often as many as six, sometimes even more).

It is difficult to select general examples from the wide variety obtainable: ranging from the straightforward *A French-English Dictionary for Chemists* to the less obvious *Technical Terms, Symbols and Definitions in English, French, German, Swedish, Portuguese, Spanish, Italian and Russian, used in Soil Mechanics and*

Foundation Engineering. It may be more helpful to note that some of the major scientific publishers are active in this field. Elsevier, for example, publish over 70 language dictionaries (many of them polyglot) covering scientific/technical subjects—further details may be obtained from their catalogue. Other publishers which may be mentioned in this context include McGraw-Hill, Wiley–Interscience, Pergamon Press and Oscar Brandstetter Verlag.

SUBJECT DICTIONARIES

In addition to dictionaries covering specific subject areas, of the *Modern Dictionary of Electronics* or *Mathematics Dictionary* variety, there are some more general examples worth mentioning here.

Dictionary of Science and Technology, edited by T. C. Collocott. Edinburgh, Chambers, 1971.
○ Successor to *Chambers Technical Dictionary*, first published in 1940. Entries, which are arranged alphabetically, give a brief definition of the term (there is some cross-referencing) and occupy about 1300 closely printed pages.

McGraw-Hill Dictionary of Scientific and Technical Terms. 3rd ed. New York, McGraw-Hill, 1983.
○ A substantial work with over 115000 definitions (and some cross-referencing) plus more than 1800 illustrations.

Dictionary of Inventions and Discoveries, by E. F. Carter. 2nd ed. London, Muller, 1974.
○ Entries are arranged in one alphabetical sequence of inventions and discoverers, giving dates but no references to the literature.

A Dictionary of Named Effects and Laws, in Chemistry, Physics and Mathematics, by D. W. G. Ballentyne and D. R. Lovett. 4th ed. London, Chapman and Hall, 1980.
○ Entries arranged alphabetically by the name of the effect or law; however, there are no references to the literature.

It is convenient to deal with dictionaries of scientific units in this section; for conversion tables see under **Mathematical tables** (p. 112).

Quantities and Units of Measurement: a dictionary and handbook, by J. V. Drazil. 2nd ed. London, Mansell, 1983.
○ First published as *Dictionary of Quantities and Units* in 1971. This work is divided into three main parts, each arranged alphabetically. The first is a dictionary of units of measurement, including their symbols. The second presents a dictionary of quantities and selected constants, whereas the third provides a list of symbols denoting quantities and constants. Indexes to the names of quantities and constants (but not units) in French and German are included, as is a bibliography of relevant national and international standards. The introductory section contains a brief outline of systems and units.

A Dictionary of Scientific Units, including dimensionless numbers and scales, by H. G. Jerrard and D. B. McNeill. 4th ed. London, Chapman and Hall, 1980.
○ Approximately 800 entries, alphabetically arranged, are preceded by a short section dealing with systems of units in general. The appendices include

some conversion tables; a list of primary literature references and an index are also provided. A 5th edition is expected during 1986.

Units of Measurement: an encyclopaedic dictionary of units, both scientific and popular, and the quantities they measure, by S. Dresner. Aylesbury, Harvey Miller and Medcalf, 1971.
○ This work is basically in two parts: part 1, units of measurement; part 2, physical quantities; the contents of each are arranged alphabetically. A series of appendices cover systems of units, dimensions, obsolete units, and symbols, among other things.

Elsevier's Lexicon of International and National Units: English/American, German, Spanish, French, Italian, Japanese, Dutch, Portuguese, Polish, Swedish, Russian, compiled by W. E. Clason. Amsterdam, Elsevier, 1964.
○ International units are listed alphabetically with a brief definition (in English) and the name of the unit in the other languages. National units are listed under the name of the country concerned, the US, UK and metric equivalents being recorded. Indexes to the contents of both sections are provided.

Directories

These list people, organizations, products, research, etc., and are arranged in either alphabetical or subject order, generally giving useful information such as addresses, telephone numbers, names of personnel and/or organizations (as appropriate). In this book they are dealt with under **Guides to people** (p. 15) and **Guides to organizations** (pp. 20, 25, 29 and 33), but guides to directories are included among the **Guides to reference material** (p. 115).

Encyclopaedias

A general encyclopaedia summarizes, usually alphabetically, all fields of knowledge, well-known examples being *Chamber's Encyclopedia* and *Encyclopaedia Britannica*. It is worth bearing in mind that scientific articles in these works are often contributed by eminent authorities in the field, so they should not automatically be regarded as of value only to the non-scientist. Some encyclopaedias cover only one subject, or a group of related subjects: some examples of these are described below.

A good encyclopaedia article gives an extended summary of its topic including a few references to follow-up in the literature (secondary and/or primary). Larger encyclopaedias often have an index volume: this should *always* be consulted, as it may contain cross-referencing not repeated in the body of the work. The structure of most encyclopaedias is to some extent 'systematic': that is, related topics are grouped together under a particular subject heading rather than appearing in a strictly alphabetical sequence (as in a dictionary); the use of the index volume is essential to get round this problem. Again, with larger encyclopaedias, annual supplements or yearbooks may be published to keep the work up to date.

Uses

- For a brief account of an unfamiliar topic, or for background information.
- As a starting point when going into a new field.

Access

Via libraries.

Caution

- If an index volume is provided—use it!
- Remember to consult any supplementary volumes or yearbooks.
- Do not expect to find the very latest information on your subject.

Examples

Van Nostrand's Scientific Encyclopedia. 6th ed. 2 vols. New York, Van Nostrand Reinhold, 1983.
○ A well-known illustrated work with internal cross-referencing (no index), but without references to the literature.

McGraw-Hill Encyclopedia of Science & Technology. 5th ed. 15 vols. New York, McGraw-Hill, 1982.
○ An extensive, well-illustrated work in which the articles are signed by their contributors and references to the literature are included. An index volume lists the contributors and scientific notation used in the encyclopaedia, as well as providing a detailed index to the topics covered. This publication is updated by the *McGraw-Hill Yearbook of Science & Technology*.

Kirk-Othmer Encyclopedia of Chemical Technology. 3rd ed. 25 vols. New York, Wiley–Interscience, 1978–84.
○ A major work, whose field is perhaps wider than the title might suggest. The substantial articles, which are signed by their contributors, contain references to both primary and secondary literature. An index volume is also provided. This publication is associated with the first full-text scientific/technical encyclopaedia database to become available online (through DATA-STAR, see p. 153, and BRS, see p. 153).

There are many other examples too subject-oriented for consideration here; for further details see either the appropriate **Subject guides** (p. 42) or the **Guides to reference material** (p. 115).

Handbooks

These generally cover a particular subject area and contain the most frequently required data and subject information. Handbooks occupy a position somewhere between encyclopaedias and data compilations. They are useful for checking formulae, obtaining numerical data and definitions, and finding descriptions of a process, material, piece of apparatus, etc. Good handbooks include references to the literature.

Uses

- For facts, figures and formulae.
- To remind yourself of 'standard' information which you may have forgotten, or may not have used for some time.
- As a starting point when entering an unfamiliar field.

Access

Via libraries, or purchase.

Caution

- They may assume prior knowledge of the subject at quite a high level.
- The range of data given is unlikely to be wide: handbooks are not a substitute for data compilations.
- Some of them remain in print for a long time without revision.

Examples

Kempe's Engineers Year-Book. London, Morgan-Grampian, annually.
○ An authoritative, continuously updated work covering a very wide field, including some references to the literature. A detailed subject index is provided.

Handbook of Engineering Fundamentals, edited by O. W. Eshbach and M. Souders. 3rd ed. New York, Wiley, 1975.
○ A well-established compilation which first appeared in 1936. Some references to both primary and secondary literature are provided, and there is a detailed subject index.

Handbook of the Engineering Sciences, edited by J. H. Potter. 2 vols. Princeton, NJ, Van Nostrand, 1967.
○ Volume 1 covers the basic sciences (mathematics, physics, chemistry, engineering graphics, statistics, theory of experiments, and mechanics). Volume 2 covers the applied sciences (including heat and thermodynamics, electromagnetism and electronics, materials science, machine elements, control systems, information retrieval, report writing and computers). There are references to the literature (both primary and secondary) and each volume has its own index.

There are many other examples, covering narrower subject areas in science and technology. Rather than quote a few of these, it may be more helpful to note that some of the major scientific publishers are active in this field. McGraw-Hill, for example, publish well over 100 handbooks of this type—further details may be obtained from their lists and catalogues. Other publishers which may be mentioned in this context include Wiley–Interscience, Van Nostrand and Prentice-Hall.

Manuals (generally associated with a particular piece of apparatus or equipment) are dealt with under **Trade literature** (p. 136).

Mathematical tables

Individual volumes devoted to a collection of mathematical tables are well known. Less well known, perhaps, is the fact that many tables are not available in this form, but must be sought in primary journals, textbooks or other forms of literature, using the appropriate guides mentioned in the **Guides to reference material** (p. 116).

Uses

To assist with a particular calculation or problem involving the numerical value of special mathematical functions.

Access

Via libraries, or purchase.

Caution

● Be sure the tables have been calculated with sufficient accuracy for your purpose.
● There are slight variations in the way different authors define some special mathematical functions (particularly with regard to notation); make certain you understand, precisely, how each tabulated function you use has been defined.
● Misprints and errors in published tables are not unknown.
● Be prepared, in some cases, to look further than the handy one-volume collection of miscellaneous tables.

Examples

Chambers's Six-Figure Mathematical Tables, by L. J. Comrie. 2 vols. Edinburgh, Chambers, 1948.
○ Largely confined to elementary mathematical functions (e.g. logarithmic, trigonometric, exponential and hyperbolic).

CRC Handbook of Mathematical Sciences, edited by W. H. Beyer. 5th ed. Boca Raton, Florida, CRC Press, 1978.
○ Previously entitled *CRC Handbook of Tables for Mathematics*, and issued by the publishers of the well-known *CRC Handbook of Chemistry and Physics*, this is a balanced compilation covering elementary mathematical functions, a few higher functions and some statistical material. The same publishers also produce *CRC Handbook of Tables for Probability and Statistics* (2nd ed., 1968) and *CRC Standard Mathematical Tables* (27th ed., 1984), both edited by W. H. Beyer.

Jahnke–Emde–Lösch Tables of Higher Functions, by F. Lösch. 7th ed. Stuttgart, Teubner, 1966.
○ This volume, with its dual German/English text, does not cover elementary mathematical functions at all. References to the literature are included.

Handbook of Mathematical Functions; with Formulas, Graphs, and Mathematical Tables, edited by M. Abramowitz and I. A. Stegun. New York, Dover, 1965.
○ Initiated partly on account of the need for a modernized version of the classic tables of functions by Jahnke and Emde (original compilers of the

previous item). This work covers a wide range of the most frequently required functions, giving explanations, definitions and references to the literature. Some of the material has been brought up to date in *Mathematical Functions and their Approximations*, by Y. L. Luke (New York, Academic Press, 1975).

Tables of Integrals and other Mathematical Data, by H. B. Dwight. 4th ed. New York, Macmillan, 1961.
○ A useful collection of mathematical formulae, including many indefinite integrals. There are a few references, mainly to the secondary literature.

It is worth noting that some organizations specialize in the production of series of mathematical tables: for example the Royal Society and the National Physical Laboratory in Great Britain; the National Bureau of Standards and the Smithsonian Institution in the USA.

It is convenient to deal with conversion tables between different systems of units in this section. Examples of these are:

SI and Metrication Conversion Tables, compiled by G. Socrates and L. J. Sapper. London, Newnes–Butterworths, 1969.
○ Existing units are arranged alphabetically in a table which gives their SI equivalents. SI is the abbreviation for Système International d'Unités (or International System of Units).

Direct Reading Two-Way Metric Conversion Tables, including conversions to the International System (SI) Units, compiled by A. J. Biggs. London, Pitman, 1969.
○ The majority of tables in this book relate imperial and metric units.

SI Conversion Charts for Imperial and Metric Quantities, compiled by A. Parrish. London, Iliffe, 1969.
○ A series of charts and tables, with instructions for carrying out the conversion.

Thesauri

These are basically lists of words (often arranged like a dictionary), but instead of giving definitions they display relationships between words denoting similar concepts. A thesaurus may be used to control indexing-term vocabulary, especially in a computerized information storage and retrieval system: on looking up a given term, you may be told that it is 'used for' alternative words with the same meaning (synonyms), and you should also find lists of broader terms, narrower terms and related terms.

Uses

● To overcome the language barrier when looking for information (for instance, in conjunction with subject indexes).
● To assist with planning an online computerized search.
● As an aid to choosing indexing terms for a personal record system.

Access

Via libraries, or purchase.

Caution

● Many scientific/technical thesauri are of American origin and this is reflected in their vocabulary.
● Different computerized information systems will, of course, use different thesauri.
● Indexing language can change over the years, so subject headings shown in a current thesaurus will not necessarily be identical with those used in the indexes of a related abstracting journal published some time ago (for these, earlier editions of the thesaurus may be consulted).

Examples

Thesaurus of Engineering and Scientific Terms: a list of engineering and related scientific terms and their relationships for use as a vocabulary reference in indexing and retrieving technical information. New York, Engineers Joint Council, 1967.
○ In addition to the basic thesaurus section, there is a permuted index which covers significant words appearing in phrases (e.g. 'logic' in 'fluidic logic devices') and an alphabetical arrangement of terms under broader subject headings (similar to those used in *Government Reports Announcements & Index*—see p. 121).

NASA Thesaurus. 2 vols. Washington DC, National Aeronautics and Space Administration, 1982. NASA SP-7051.
○ This thesaurus governs the indexing terms used in *International Aerospace Abstracts* (see p. 51), *Scientific and Technical Aerospace Reports* (*STAR*—see p. 122) and the associated machine-readable *NASA* database (see p. 147). The first volume contains a hierarchical listing of subject terms, whereas volume 2 (used in conjunction with volume 1) consists of an access vocabulary which, for example, includes some cross-referencing between terms not present in the former.

INSPEC Thesaurus 1985. London, INSPEC–Institution of Electrical Engineers, 1984.
○ This thesaurus governs the indexing terms used in *Electrical and Electronics Abstracts* and *Computer and Control Abstracts* (see p. 50), *Physics Abstracts* (see p. 52), and the associated *INSPEC* database (see p. 146). An alphabetic display of terms is followed by a hierarchical listing. In the first section INSPEC classification codes (useful when searching online) are provided as part of the entry for each main subject heading.

SHE Subject Headings for Engineering. New York, Engineering Information Inc, 1983.
○ Indexing terms used in *Engineering Index* (see p. 50) and the associated machine-readable database *COMPENDEX* (see p. 146) are covered by this thesaurus. Because of the way in which *Engineering Index* is organized, *SHE* can be particularly valuable in planning a manual search when the appropriate subject headings are unknown.

The indexing journal *Index Medicus* publishes a thesaurus entitled *MESH (MEdical Subject Headings)* each year, which acts as a guide to the headings used in that journal as well as to the index terms in *MEDLINE* (the computerized information retrieval service based on the same material—see p. 147).

The American Society for Metals of Ohio and The Metals Society of London jointly publish a *Thesaurus of Metallurgical Terms* (5th ed., 1981) which covers indexing terms used in *Metals Abstracts* (see p. 51) and the *METADEX* database.

The Thesaurus for Fluid Engineering, edited by N. G. Guy and published at Cranfield, Bedford by BHRA Fluid Engineering in 1981, has been prepared for use with the *FLUIDEX* online database, but also offers a general guide to terminology in its field.

It is worth remembering the existence of *Roget's Thesaurus*, a general thesaurus covering the English language (many editions, both British and American), when writing reports, journal articles, conference papers, and even textbooks.

Guides to reference material
See also **Information services**, p. 207

These are lists of reference books, etc., often with annotation, usually arranged in subject order and including appropriate indexes. Some of them also cover guides to the literature and indexing/abstracting journals.

Uses

To find the reference material relating to a given subject.

Access

Usually via libraries.

Caution

- Be sure you have the latest edition of the guide.
- Do not skip the 'how-to-use' section.

Examples

Walford's Guide to Reference Material, edited by A. J. Walford and others. 4th ed. 3 vols. London, Library Association Publishing, 1980–   .
○ 'Walford' is one of the standard annotated guides in this field. Volume 1 deals with science and technology; volume 2 covers social and historical sciences, philosophy and religion; while volume 3 (3rd ed., 1977) deals with generalities, languages, the arts, and literature. Each volume has its own index. The same publisher issued a condensed version, *Walford's Concise Guide to Reference Material*, in 1981, where the English-language/British emphasis is more pronounced. The Library Association has also published *Printed Reference Material* edited by G. L. Higgens (2nd ed., 1984), a useful handbook for students of librarianship and library organizers.

Guide to Reference Books, compiled by E. P. Sheehy. 9th ed. Chicago, American Library Association, 1976.
○ Formerly associated with C. M. Winchell (among others), this is another standard annotated guide, perhaps with something of an American bias just as the previous example may lean towards British sources, but both try to be international in character. This work is updated by supplements between editions.

Guide to American Directories, edited by B. Klein. 11th ed. Coral Springs, Florida, B. Klein Publications, 1982.
○ Covers directories published in the US, and a selection of major foreign examples, with nearly 7000 entries arranged under some 300 industrial, technical, mercantile, scientific and professional headings. For each publication is given the title, a brief description of scope and contents, the publisher's name, address and telephone number, and the price. A subject index based on titles is also provided.

Handbooks and Tables in Science and Technology, edited by R. H. Powell. Phoenix, AZ, Oryx Press, 1979.
○ Over 2000 entries are arranged alphabetically by title, each giving bibliographic details of the reference work concerned, sometimes accompanied by descriptive notes. Medical handbooks and US National Bureau of Standards selected data compilations are covered in separate appendices. Indexes of authors/editors and subjects (using narrow subject headings) are provided, as is a list of publishers with their addresses.

Current British Directories: a guide to directories published in the British Isles, edited by C. A. P. Henderson. 10th ed. Beckenham, CBD Research, 1985.
○ Part 1 lists local directories by town/county, and part 2 specialized directories by title. Entries include details of publisher, date first issued, date and price of latest edition, and a brief description of contents. Part 3 consists of an index of publishers, listing their directories, whilst part 4 provides a detailed subject index. *Current European Directories* (2nd ed., 1981) is a companion publication also produced by CBD Research.

Directory of Scientific Directories: a world guide to scientific directories including medicine, agriculture, engineering, manufacturing and industrial directories, compiled by J. Burkett. 3rd ed. Harlow, Essex, Francis Hodgson, 1979.
○ Entries, which are often annotated, include details of publisher and price, the arrangement being by subject field within country. Indexes of authors/compilers/editors and titles are provided. A reorganized 4th ed. has been announced for publication by Longman, Harlow, early in 1986.

The Top 3000 Directories & Annuals 1985/6: a guide to the major titles used in British libraries, edited by A. Wood. 6th ed. London, Alan Armstrong & Associates, 1985.
○ Previously published as *The Top 1000* [or *2000*] *Directories & Annuals*. Entries are arranged alphabetically by title, including details of current edition (with price and ISBN), next edition, publisher, editor, and subject coverage. In a separate section, titles are displayed according to the month and/or year of publication of the next edition. An alphabetical list of publishers with their directories and annuals (giving addresses from which these may be obtained), and a subject index, are also provided. Between issues of this guide, an updating service is available in the form of a monthly bulletin entitled *New Editions*.

An Index of Mathematical Tables, by A. Fletcher, J. C. P. Miller, L. Rosenhead and L. J. Comrie. 2nd ed. 2 vols. Oxford, Blackwell Scientific Publications (for Scientific Computing Service), 1962.
○ Volume 1 gives details of the wide range of mathematical functions covered,

including definitions (and noting variations in definition), referring to published tables by author and date. The bibliography in volume 2, with entries arranged alphabetically by author and then chronologically, provides full details of the appropriate reference: this volume also contains a substantial list of known errors in some published tables (arranged alphabetically by author), and there is an index to the first volume.

Guide to Tables in Mathematical Statistics, by J. A. Greenwood and H. O. Hartley. Princeton, NJ, Princeton University Press, 1962.
○ The general arrangement of this work is similar to that of the preceding item.

Remember that **Subject guides** (p. 42) may also cover the reference material appropriate to their subjects. Finally, *The Search for Data in the Physical and Chemical Sciences* by L. R. Arny (New York, Special Libraries Association, 1984) provides a fairly broad guide to sources of physical and chemical data, with special emphasis on retrieving data from publications of the National Bureau of Standards.

Reports

A scientific/technical report is an account of the work done on a particular project, usually prepared by the research workers themselves, which conveys information to an employer, a sponsor, or to others interested in the same field.

For many years reports were usually reproduced from typescript as hard copy, but latterly increasing use has been made of microfiche—which is an excellent medium for storing this kind of information.

Organizations sponsoring research generally require frequent, up-to-date progress reports, often to justify expenditure or determine whether a project should continue. Reports have the advantages over journal articles of rapid publication and dissemination: they are an ideal means of communication where the material is of limited interest, or where there is need for strict control over distribution, involving security classification both military and commercial.

The majority of reports are not 'published' in the usual sense of the word; neither are they edited to the standard of a journal article, nor, in general, are they refereed (vetted for errors and omissions by an independent expert in the field). Information of lasting value first appearing in a report usually achieves later publication in a journal (subject to security considerations); for this reason the useful life of most reports is limited to a few years and, clearly, progress reports are apt to be superseded. However, journal articles tend to describe only successful research; thus reports are one of the few sources of information on 'failed' work. Again, journal articles normally deal with work which has been completed, so reports can be used to find out about research still in progress.

Reports are invariably produced by, or for, an organization. Most of the work described falls into one of three main categories:

(1) *Government-sponsored research.* Reports in this area can be subject to security classification (military is the most usual), but if this is not the case, they are the easiest to locate and obtain.

(2) *Industrial-sponsored research*. These reports are nearly always subject to commercial security classification and often not available outside the originating company: the work may well lead to a patent application (see p. 78). Reports issued by research associations are usually confidential to members, at least initially.

(3) *Academic research*. Reports produced by university departments, etc., are not normally subject to security classification; however, they can be difficult to locate and obtain, as individual departments are often not very efficient in handling their own reports!

Some reports are produced as single (isolated) items, others as parts of a continuing report series. Organizations may be eager to make their reports freely available, but can be very reluctant to do so. Some arrange distribution of reports themselves, while others leave this to an outside body or clearing house. Reports can become available soon after completion; on the other hand, they may be released only after considerable delay.

Organizations of similar character working in the same field may have widely differing policies, so the acquisition of report literature can be extremely difficult, often presenting libraries with considerable problems. Sometimes a library tries to acquire all reports in particular series; otherwise only reports on given topics are sought.

The BLLD and some similar national centres elsewhere are especially anxious to acquire both reports and semi-published literature of all kinds. Readers of this book in the UK who generate such material, assuming it may be reproduced and circulated freely (from the copyright and security points of view), should send one copy of each relevant document to: Special Acquisitions Section, British Library Lending Division, Boston Spa, Wetherby, West Yorkshire LS23 7BQ. These will be gratefully received, made available for loan at least nationally, announced in *British Reports, Translations and Theses* (see p. 124), and entered in the *SIGLE* database (see p. 148).

An essential feature of almost every report is a serial code number (or numbers), generally enabling any body associated with it to be identified. Serial codes consist of a combination of letters and/or numbers followed by a serial number. For instance, AD-A129045, UWA/DME/TR-83/47 and N83-35417 all represent the same document in systems belonging to different organizations. These codes may be allocated by the body responsible for the work, or by an organization distributing/abstracting the report. Serial codes must *always* be noted or quoted when you encounter a report.

This book contains a section on **Report writing** (pp. 274–279).

Uses

● To back up primary journals; especially to find out about research in progress, 'failed' work and the most recent developments in certain fields.
● To obtain a detailed description of the work carried out by a given organization.
● To get full details of the work done at each stage on a particular project.

Access

● *Via libraries*. The BLLD (see p. 259) is probably the best place to apply for a loan copy or microfiche of any British or American report: this library should be able to supply a copy (generally microfiche) of almost every item mentioned in the guides to reports, below. The Defence Research Information Centre (DRIC, part of the Procurement Executive, Ministry of Defence) holds reports relating to defence-oriented science and technology. Its services are restricted to the Ministry of Defence and UK firms and organizations working on UK government defence contracts or research programmes.

● *Purchase*. Reports may sometimes be acquired from the organizations originally producing them. Alternatively a distributing body or clearinghouse can be approached. Many American reports are marketed by NTIS (in the USA) and its agents elsewhere, including Microinfo Ltd in the UK; further details are given under *Government Reports Announcements & Index* below. It might be worth consulting *How to Get It: a guide to defense-related information resources* (p. 129) in connection with US documents of a military nature. For British reports which can additionally be regarded as government/official publications, see **Government publications—access** (p. 70) and *Catalogue of British Official Publications not published by HMSO* (p. 73). It may be possible to keep reports (from any country) obtained via the BLLD in the form of microfiche or even photocopies, copyright and distribution agreements permitting. Individuals wishing to explore this option should contact a library entitled to borrow from BLLD, as applications must be made on official BLLD loan/photocopy request forms. Permanent retention of the microfiche or photocopy is, of course, also a matter for negotiation with the other library concerned.

● *In case of difficulty*. If the courses of action suggested above fail to produce the required report, try contacting its author(s), assuming an address can be found (see **Guides to people**, p. 15). As a final resort, in cases of extreme difficulty, the BLLD (which is basically a lending library and does not normally accept bibliographical queries) may be able to help identify, and offer advice on obtaining, reports which fall outside the scope of its own holdings: enquiries should be addressed to the Reports Section, BLLD, Boston Spa, Wetherby, West Yorkshire LS23 7BQ.

Caution

● Reports are generally uniquely identified by serial code numbers, which should always be quoted for loans, purchases, etc.
● They may be very detailed, an advantage if similar work is contemplated, but a hindrance when you have to extract small amounts of information quickly.
● Reports may be biased towards an end, particularly where the authors are trying to justify continued expenditure on themselves or their projects.
● Remember that most reports are neither refereed nor edited to journal standard, so bear in mind the increased possibility of errors and omissions.

Guides to reports
See also **Information services**, p. 208

Reports tend to be inadequately covered by most abstracting and indexing journals (though these should always be checked), but are well covered in a few important examples given below. In addition, they may be found in some bibliographies (both subject and general) and guides to government publications (see, especially, *Monthly Catalog of United States Government Publications* described on p. 73); they may also be retrieved by mechanized information services. In some cases it may be helpful to contact organizations working in your field for information about reports they may have prepared or acquired (see **Guides to organizations**, pp. 20, 25, 29 and 33).

Uses

- To indicate the scope, nature and availability of reports.
- To back-up the use of other abstracting/indexing journals.
- To find out about the work done by a particular organization.
- To find out which organizations are working in a particular field.
- To check bibliographic references (non-current guides can still be of value for investigating older reports).

Access

Via libraries.

Caution

- Reports are not covered very well by most abstracting and indexing journals.
- Titles given in indexing journals can be deceptive: it is best to consult an abstract before arranging to borrow or buy a report.
- The coverage for reports of American origin is considerably better than for those of British origin.
- The titles of some of the guides below are misleading, as the subject areas covered are far wider than might at first be supposed.
- Libraries may not have the resources to catalogue individual reports in a series as fully as they catalogue books, etc., though some record of the serial code number will usually be made: in such cases the published guides to reports must be used to identify items by author or subject, and to determine the relevant serial codes.

Examples

Energy Research Abstracts. Oak Ridge, Tennessee, US Department of Energy (DOE), 1977–   .
○ This publication began as *ERDA Reports Abstracts* (1975), and became *ERDA Research Abstracts* (1975–76) before changing its title to *ERDA Energy Research Abstracts* (1976–77). *Energy Research Abstracts* (*ERA*) covers all scientific and technical reports and other kinds of literature originated by the US Department of Energy, its laboratories, energy centres and contractors; *ERA* additionally includes energy reports generated by US federal and state government organizations, foreign governments, and domestic or foreign universities and research institutions. Entries, which are arranged under narrow subject headings, usually supply an abstract number, report number, title, author(s), associated organization(s), date, contract number, page count, note on availability, and an order number followed by the abstract. The following indexes appear in each semi-monthly issue: corporate author (organization), personal author, subject, contract number, and report

number; an order-number correlation section links report numbers with order numbers (listed numerically). Semi-annual and annual cumulative indexes are also provided. The *DOE ENERGY* database may be searched online, but at the time of writing access is easier for users in the US than it is for those in the UK.

Government Reports Announcements & Index. Springfield, VA, National Technical Information Service (NTIS), 1975–   .
○ This publication began as *Bibliography of Scientific and Industrial Reports* in 1946, was known as *Bibliography of Technical Reports* between 1949 and 1954, then changed its name to *U.S. Government Research Reports* (1954–1964), becoming *U.S. Government Research and Development Reports* from 1965 until 1971, and appearing as *Government Reports Announcements* during the period 1971–75. The related index commenced publication as *Government-Wide Index to Federal Research and Development Reports* in 1965 (when it covered *US Government Research and Development Reports*, *Nuclear Science Abstracts*, and *Scientific and Technical Aerospace Reports*), becoming *US Government Research and Development Reports Index* (when it dropped coverage of *NSA* and *STAR*) from 1968 until 1971, and finally appeared as *Government Reports Index* (1971–75) before merging with *Government Reports Announcements*.

Government Reports Announcements & Index is an abstracting journal issued every two weeks which covers US Government-sponsored research, development and engineering reports, as well as some foreign technical reports and other analyses prepared by national and local government agencies, over a very wide range of subjects. Entries include NTIS abstract number (from 1984–   , which must not be used for document ordering purposes), order number and price codes (where the item is available from NTIS), name of the organization associated with the work; report title, author(s), date, page count, serial code number(s), and contract number(s); followed by the abstract. Entries are arranged by abstract number (1984–   ) or alphanumerically by order number (–1983) in broad subject categories according to the COSATI (Committee on Scientific and Technical Information) classification scheme. Each issue contains the following indexes: keyword (subject), personal author, corporate author, contract/grant number, and NTIS order/report number. These cumulate every year and are published as *Government Reports Annual Index*: all entries give the report title, except those in the contract/grant number index. The equivalent *NTIS* database may be searched online; for further details see p. 147.

Also available is the *NTIS Title Index* on microfiche, which divides into two parts. The first (retrospective index) covers reports entering the database between July 1964 and December 1978, providing keyword-out-of-context, author, and report/accession number sequences; a report can be retrieved through any keyword in its title. The second (current index) is a quarterly cumulation of new publications, commencing with those issued in January 1979, editions being cumulated and merged on a two-year cycle.

Although most documents listed can be obtained in the USA from NTIS (National Technical Information Service, US Department of Commerce, Springfield, VA 22161), orders from other countries are dealt with by the appropriate local agent; in the UK, for example, this is Microinfo Limited, PO Box 3, Newman Lane, Alton, Hampshire GU34 2PG.

INIS Atomindex. Vienna, International Atomic Energy Agency (IAEA), 1970–   .

○ INIS (International Nuclear Information System) was set up by the IAEA and its member states to produce a database identifying publications concerned with nuclear science and its peaceful applications. Most kinds of literature are included, but since reports are fairly prominent in this subject field, the guide is dealt with as an example here rather than under abstracting and indexing journals in general. Though it is currently an abstracting journal, issued every two weeks, before 1976 abstracts for most items were not printed in *INIS Atomindex*, but published on microfiche. Entries, which are arranged according to narrow subject headings, include titles, authors, associated organizations, full bibliographic details, notes on availability, and abstracts or descriptors (subject keywords). Five indexes appear in each issue: personal author, corporate entry, subject, conference (in two sequences, date and place), and report, standard and patent number. These indexes now cumulate semi-annually and annually (semi-annually only before 1980), and there is also a multi-annual report, standard and patent number index covering 1972–77. The *INIS-ATOMINDEX* database can be searched online. INIS non-conventional literature on microfiche is another service offered by this system, which embraces all literature other than journal articles and commercially published books. Further information may be obtained from INIS Clearinghouse, International Atomic Energy Agency, PO Box 100, A-1400 Vienna, Austria.

Nuclear Science Abstracts. Oak Ridge, Tennessee, US Atomic Energy Commission (USAEC), or US Energy Research and Development Administration (USERDA), 1948–76.

○ Although no longer published, this example is worth mentioning as a guide to its field in the 1950s and 1960s. *NSA* aimed to provide comprehensive coverage of international nuclear science literature, much of which appears as reports, though it did not exclude alternative types of literature. Reports included those of the USAEC and its contractors, other US government agencies, other governments, universities, and industrial and research organizations. The variety of subjects covered was very wide, ranging from physical sciences, through earth and life sciences to technology. Entries provide an abstract number; report serial code number, title, author(s) and date; name of the organization associated with the work, contract number (where applicable), a note on availability and the abstract. The following indexes were produced: corporate author, personal author, subject and report number. Various annual and multi-annual cumulations of these appeared. For coverage of this field after *NSA* ceased publication, see *INIS Atomindex* and *Energy Research Abstracts* above.

Scientific and Technical Aerospace Reports. Washington DC, US National Aeronautics and Space Administration (NASA), 1963–   .

○ *STAR* aims to provide comprehensive coverage of report literature on the science and technology of space and aeronautics. Reports include those of NASA and its contractors, other US government agencies, and institutions, universities and private firms throughout the world. Coverage also extends to dissertations and theses, translations in report form, and NASA-owned patents and patent applications. The field of subjects covered is very wide, ranging from aeronautics and astronautics through materials technology and electronics to the earth, life, mathematical, physical and social sciences. Entries include a NASA accession number (used for ordering NASA publications) and code indicating whether microfiche can be supplied; name of the organization associated with the work; report title, author(s) and date; contract number (where applicable), and a note on availability/purchase price (where appropriate), followed by the abstract. Entries are arranged by accession number in broad subject categories. Each fortnightly issue contains the following indexes: subject, personal author, corporate source, contract number and report/accession number. In addition, there are annual cumulations. Computerized information retrieval services are associated with *STAR* via the *NASA* database (see p. 147). See also *NASA Thesaurus* (p. 114). A companion guide, *International Aerospace Abstracts*, covers the same subject areas, dealing with journal, book and conference literature, cover-to-cover journal translations, and certain foreign dissertations. For further details see p. 51.

Reports issued by NASA's predecessor NACA (National Advisory Committee for Aeronautics) may be identified using *Index of NACA Technical Publications*, which spans the period 1915–58.

The next examples are mainly concerned with material of European origin.

Euro Abstracts. Luxembourg, Commission of the European Communities, 1970–   .

○ This publication began as *Euratom Information* (1963–69), and since 1975 has appeared in two parts: section 1, Euratom and EEC research (scientific and technical publications, patents, training courses and seminars—conferences and symposia in preparation); section 2, coal and steel (research programmes, research agreements, scientific and technical publications, patents, training courses and seminars—conferences and symposia in preparation). Many reports, especially EUR reports, are covered, hence the inclusion of *Euro Abstracts* as an example here. Entries, arranged by subject, usually provide an abstract number, report number and language code, title, author(s), associated organization(s), and a note on availability (section 1) followed by the abstract. Indexes of authors (for publications) and inventors (for patents, when required) feature in the monthly issues, which also contain information about document ordering (all EUR reports listed can be obtained on microfiche, and a number as hard copy). Some annual indexes have also been produced.

Catalogue EUR Documents, 1968–1979. Luxembourg, Commission of the European Communities, 1983. EUR 7500.

○ This lists all EUR reports published by the Commission during the period under consideration, most of which are available in microform (and some as hard copy) from either the Office for Official Publications of the European Communities, L-2985, Luxembourg, or one of its national sales offices (in the UK, for instance, HMSO Publications Centre, 51 Nine Elms Lane, London SW8 5DR, tel. 01-211 8595). Certain other reports issued by private publishers from 1975 are included as well.

Entries, arranged under narrow subject headings, consist of title (original language, followed by an English translation where appropriate), names of author(s) and institution(s) involved, note on microform availability (except when the report can *also* be supplied as hard copy), number of pages, language code, series reference, EUR report number (essential for ordering purposes) and price (usually in Belgian francs). Title, EUR (report) number, and author indexes are provided. The title index features Danish, German, English, French, Italian and Dutch sections, each arranged in two sequences: alphabetical and series. The English alphabetical sequence contains *all* titles, including many which have been translated, whereas the other language sections list only those titles published in the language concerned.

R & D Abstracts. Orpington, Kent, Technology Reports Centre, 1968–81.
○ Though no longer produced, this example must be mentioned as one of the few British guides to reports. It began as a classified publication in 1948, but became generally available from 1968. Based mainly on material received from UK government R & D establishments, government-supported R & D activities, and other sources in the UK and overseas, the subject field covered was very wide, embracing physical and life sciences as well as technology. Entries include TRC accession numbers, report serial code numbers, names of associated organizations, titles, authors, dates, and abstracts. Annual or semi-annual indexes provide the following sequences (with minor variations depending on the date): subject, title, author, corporate author, conference, report number, and TRC accession number. Most documents listed here are available from BLLD (see p. 259).

British Reports, Translations and Theses. Boston Spa, British Library Lending Division, 1981–   .
○ This has superseded a current-awareness publication, *BLLD Announcement Bulletin*. The material covered is drawn from reports and translations produced by British government organizations, industry, universities and learned institutions, doctoral theses accepted at British universities and polytechnics since 1970, reports and unpublished translations from the Republic of Ireland, and selected British official publications of a report nature not published by HMSO. Entries, which are arranged in subject groups according to the COSATI scheme, include the usual bibliographical details and a BLLD reference number; these must be quoted when items are requested from the British Library Lending Division. A keyterm (subject) index appears in each monthly issue; author, report number and keyterm indexes, which cumulate during the year, are issued quarterly, and hard-copy versions of the annual cumulations have been produced as well. Coverage extends beyond science and technology to social sciences and the humanities.

Dictionary of Report Series Codes, edited by L. E. Godfrey and H. F. Redman. 2nd ed. New York, Special Libraries Association, 1973.
○ This invaluable work lists the relationship between report series codes and the organizations assigning them. The dictionary has two sequences: entries arranged by series code, and entries arranged alphabetically by organization. A useful introductory section explains the history, intricacies and pitfalls of report series coding.

Energy Information Data Base; Reports Number Codes, edited by G. G. Wallace. Oak Ridge, Tennessee, US Department of Energy, 1979. DOE/TIC-85 (Rev. 13).

○ This contains all codes which have been used by the DOE in cataloguing reports. It has two sections. Part 1, arranged alphabetically by code, identifies the corresponding organizations; part 2 is an alphabetical list of organizations, displaying the associated codes.

The term 'grey literature' is sometimes used in connection with reports and certain other documents such as theses. It denotes material which is (probably) not formally published, generally issued in only small quantities, and not readily available for purchase through normal booktrade channels. An online service dealing with this area, *SIGLE* (system for information on grey literature produced in Europe), is described on p. 148. Mention must also be made of papers presented at a conference on the availability and bibliographic control of non-conventional literature, which appeared in *Aslib Proceedings*, Vol. 34, No. 11/12, November/December 1982.

Further information about reports may be found in **General guides** (see p. 40); **Subject guides** (see p. 42); in an article by J. P. Chillag (of the BLLD) entitled 'Don't be afraid of reports' published in *BLL Review*, Vol. 1, No. 2, October 1973, pp. 39–51; and in a paper compiled by Aslib Information Department entitled 'Report literature in the UK' published in *Aslib Proceedings*, Vol. 25, No. 8, August 1973, pp. 320–324.

Reviews and review serials

Reviews, or review articles (not to be confused with book reviews), critically summarise the progress made over a certain time in a given subject field (which may be broad or narrow), including references to the primary literature.

Generally written by experts, they fall into two main categories:

(1) those which survey their topic at an advanced level for the benefit of specialists, backing up abstracting/indexing journal coverage of the literature;
(2) those written for the more general reader, which enable the specialist to be aware of development in fields related to, but outside, his own speciality.

Review articles appear in primary journals, conference proceedings, books and review serials.

Review serials are periodicals entirely devoted to the publication of reviews. They bridge the gap between textbooks or monographs and primary journals. Titles of such publications may include the words *Review* or *Reviews*, though not all periodicals with these words in their titles are review serials. (For example, *Mathematical Reviews* is an abstracting journal; *Physical Review* is a primary journal.)

Many review serials have titles beginning with, or containing, the words *Advances in* or *Progress in*, though again titles can be misleading (e.g. *Advances in the Astronautical Sciences* is a series devoted to the proceedings of conferences held by the American Astronautical Society).

Most review serials of this type are published annually, but some appear less frequently (every two or three years), others more often (quarterly or bimonthly).

Uses

● To give an account of the 'state of the art'.
● To enable specialists to keep informed of developments in their own, or related, subject fields.
● To supplement subject bibliographies or to back up abstracting/indexing journals.
● To bridge the gap between books, etc., and primary journals.

Access

Usually via libraries.

Caution

● Titles of periodicals can be misleading.
● Reviews may not contain the very latest information, especially if published in a review serial (secondary literature).
● Reviews vary in their breadth and depth of subject coverage: some are comprehensive in their coverage of the sources, others not.

Guides to reviews and review serials

The principal guides to review serials are described below. There are very few guides just covering reviews, and these mostly deal with specific fields such as medicine or organic chemistry, so are not general enough to be mentioned here—further details can be obtained from the various **Subject guides** (see p. 42). It should be remembered that review articles published in primary journals, conference proceedings or books must normally be sought by using the guides to these forms of literature, e.g. **Abstracting and indexing journals** (p. 46) and **Subject bibliographies** (p. 59) and certain online databases. It is worth noting that review articles covered by *Science Citation Index* (p. 52) can quickly be identified as such, and also appear in the separate companion publication *Index to Scientific Reviews* described below.

Review serials are included in most **Guides to periodicals** (p. 97) but not separately indexed, so are difficult to retrieve from these sources. One guide to bibliographies, *Bibliographic Index* (p. 60), can be used to identify some review serials and reviews. Book reviews may be found in *Technical Book Review Index* (Pittsburgh, JAAD Publishing Company, 1935–   ), *Book Review Digest* (New York, H. W. Wilson, 1905–   ) and *Book Review Index* (Detroit, Gale, 1965–   ).

Uses

To find reviews in a given subject field, or the titles of review serials.

Access

Via libraries.

Caution

Several of these guides are not very up to date.

Examples

KWIC Index to some of the Review Serials in the English Language held at the NLL. Boston Spa, National Lending Library for Science and Technology, 1969.
○ Produced by the British Library Lending Division's predecessor, this handy list is based on the titles of review serials, the more important keywords being chosen as index entries. Apart from shelfmarks (which do not correspond exactly to BLLD equivalents), no other information is given.

List of Annual Reviews of Progress in Science and Technology. 3rd ed. Paris, Unesco, 1981. PGI/81/WS/27.
○ Nearly 500 annual reviews are listed under fairly broad subject headings, including the following details: title, publisher, date of first issue; date, number and price of latest volume. There is a list of publishers' addresses, but no index.

Directory of Review Serials in Science and Technology, 1970–1973, compiled by A. M. Woodward. London, Aslib, 1974.
○ Entries, which are arranged alphabetically by title, normally supply the publisher's name, frequency of appearance (if more often than annually), volume number(s), and how many review articles were printed, for one calendar year (1972 if issued then, otherwise the latest year of publication). A keyword subject index is also provided.

Index to Scientific Reviews. Philadelphia, Institute for Scientific Information, 1974–   .
○ This semi-annual publication is based on material selected from the *Science Citation Index* database (*SCISEARCH*). The Source Index lists review articles alphabetically by author, whereas in the Permuterm Subject Index significant keywords from titles of reviews are displayed in pairs. From 1982, a Research Front Speciality Index (which may be entered via either the Source Index or the Permuterm Subject Index) links authors of reviews in narrow subject fields of current research importance, the latter being chosen after cluster analysing *Science Citation Index* data. Before 1982, *Index to Scientific Reviews* had a Citation Index section instead.

Further information about reviews and review serials may be found in **General guides** (see p. 40) and **Subject guides** (see p. 42).

Standards

The word 'standards' is used in two ways:

(1) in connection with measurements of length, mass, time, temperature, etc., where a standard unit is maintained, generally in a national laboratory (such as the National Physical Laboratory in Britain, or the National Bureau of Standards in America);
(2) to describe printed standard specifications concerned with quality, fitness for purpose and performance of manufactured articles and materials.

It is the second kind of standard which is dealt with below.
Standards specify how materials and products should be manufactured, defined, measured or tested according to proven and accepted methods. They

may be issued by companies or by other organizations both national and international.

In the UK the British Standards Institution (BSI—an independent national body, partially supported by government grant) is by far the largest originator of standards, although government departments, trade associations and individual companies do produce standards related to their own fields of interest.

In the USA the American National Standards Institute (ANSI—called the American Standards Association until 1966, then being known as the USA Standards Institute until 1969) is the equivalent of BSI, but it does not issue the highest proportion of that country's standards—this is achieved by another national body, the American Society for Testing and Materials (ASTM). It should be remembered that trade association and company standards are more numerous (and important) in America than in Britain.

Most European countries have a national body responsible for standardization. For example, the Deutsches Institut für Normung issues DIN (Deutsche Industrie Normen) standards in Germany, many of which are also available in an English translation.

The two main bodies responsible for international standards are ISO (International Organization for Standardization) and IEC (International Electrotechnical Commission).

Uses

Generally, to find the rules by which something is made, or tested, or the data which are associated with it.

Access

● British Standards are not readily obtainable on inter-library loan in the UK, but they may be purchased from BSI. A counter sales service is available in London (at 195 Pentonville Road, N1, and Hampden House, 61 Green Street, W1, for personal callers, 09.00–17.00, Monday–Friday) and at regional offices in Birmingham, Bristol, Dundee, Glasgow, Leeds, Liverpool, Manchester and Norwich (addresses are given in the *BSI Catalogue* described on p. 130). The Sales Department, BSI, Linford Wood, Milton Keynes MK14 6LE, accepts orders by post or telex (telex number 825777), and telephone orders from subscribing members only, which should be directed to the BSI Enquiry Section, tel. Milton Keynes (0908) 320066.
● Publications of ISO, IEC, national standards bodies in overseas countries and certain other foreign standards can be ordered through BSI Sales Department at Milton Keynes by post, telex or telephone as above, though not every such publication is held in stock. Intending purchasers in the UK should *not* approach foreign standards bodies direct; if they do, they will simply be referred back to BSI. Similarly, intending purchasers of British Standards from abroad should approach the agent in their own country if there is one (a list of these appears in the *BSI Catalogue*); for instance, in the USA contact the American National Standards Institute, 1430 Broadway, New York, NY 10018.
● BSI Library and Database Section at Milton Keynes manage the most comprehensive collection of British and foreign standards in the world. BSI

Library, tel. (0908) 320033, is open for visitors 09.00–17.00, Monday–Friday; complete sets of British, international and overseas national standards, plus the more important overseas technical requirements, and related catalogues/indexes may be consulted. The Database Section uses online databases containing information on standards to answer enquiries, when appropriate. Subscribing members of BSI can borrow items (with the exception of British Standards and certain reference material) from BSI Library in exchange for loan tokens, which should be purchased in advance (in books of 10; for prices see the current *BSI Catalogue*). BSI Library staff can advise on sources for loan and purchase of standards which do not fall within the scope of their collection. A British Standards database, *BSI STANDARDLINE*, has become available for public online access. It is compatible with ISONET (a common system for the bibliographic description of standards and regulations organized by ISO), and a future option exists to make other standards databases more readily available in the UK through merging with the BSI file. Complete sets of British Standards may also be consulted at many public libraries, both in the UK and overseas; they are listed in *BSI Catalogue*.

● British Standards in the field of engineering may be obtained as part of the services offered by Technical Indexes Ltd of Bracknell, Berkshire (see p. 137). Other UK firms specializing in the supply of standards and specifications, particularly of government, military or industrial origin, include: London Information (Rowse Muir) Ltd, Index House, Ascot, Berkshire SL5 7EU, tel. Ascot (0990) 23377, telex 849426—documents obtained from, and supplied to, countries all over the world; and AIRS (American Information Retrieval Service) Ltd, 8–12 Rickett Street, London SW6 1RU, tel. 01-381 5155, telex 22758—delivery service for publications from both the US government and any trade, professional or official body in the USA and Canada. Purchasers of similar material in the USA may find the following work useful.

How to Get It: a guide to defense-related information resources. Revised ed. Arlington, VA, Institute for Defense Analyses, 1982. IDA Paper P-1500 AD/A-110 000.
○ This is intended for all who have to identify or acquire US government published or sponsored documents, maps, patents, specifications or standards, and other resources of interest to the defence community. Entries are arranged alphabetically, in a single sequence, by document type, source, acronym, series designation or short title (cross-references being included). Each main entry provides an identification of the item and detailed acquisition information, such as source, order forms to use, cost, where indexed, and telephone numbers for additional information if required.

Caution

● Standards are subject to constant revision and are apt to become out of date quickly. Always make sure you have the latest standard or amendment to it.
● Standards are usually identified (and quoted) by their serial code and number: it is sometimes difficult to identify the country of origin from this information and there is no single guide to be recommended here. *Technical Information Sources*, by B. Houghton (2nd ed., London, Clive Bingley, 1972, pp. 83–5), lists a few of the more important codes, but when in doubt it is best to consult the BSI experts.

Guides to standards
See also **Information services**, p. 208

Some major guides to standards of British, international and American origin are described below. In addition, information about standards can be sought from organizations, trade literature, bibliographies and (to some extent) periodicals and online databases. One example of the latter, covering material from the USA, is *STANDARDS & SPECIFICATIONS*, which may be searched using the DIALOG service (see p. 153).

Uses

- To indicate the scope, nature and availability of standards.
- To find which standards cover a particular subject or product.
- To check which is the most recent issue of a standard (including its amendments) or whether a particular standard is still current.

Access

Usually via libraries.

Caution

Be sure that you have the latest issue of the guide, as standards are subject to constant revision.

Examples

BSI Catalogue. London, British Standards Institution, annually.
○ At the time of writing, this publication lists British Standards current at 30 September of the year preceding its cover date. Entries are arranged by serial number in several sequences. The main 'general series' is followed by a number of shorter sequences, covering material such as codes of practice, automobile, marine and aerospace series, European (EN) standards, handbooks and special issues, drafts for development, and public authority standards. Further sections indicate withdrawn standards and the degree of correspondence between British Standards and their international (ISO or IEC) equivalents. General-series entries give each standard's serial number and date, the code number(s) of any equivalent (or partially equivalent) international standard(s), title, an abstract, details of subsequent amendments (if relevant), page count and size (if other than A4), price code, and reference number of the BSI Technical Committee responsible for the standard. Less information may be given for items in the other sequences. An alphabetical subject index is provided.
Additionally, *BSI Catalogue* contains lists of libraries which maintain complete reference sets of British Standards, both in the UK and overseas (in the latter case, local sales agents are also identified). Various services available from BSI are briefly described, too; these include quality assurance, certification and assessment, the test-house facility, and technical help to exporters (THE).

BSI News. London, British Standards Institution, monthly.
○ Apart from articles on various aspects of standards, and announcements of forthcoming meetings concerned with standardization, *BSI News* lists the new

British Standards (including revisions and amendments) published each month. Withdrawn standards are also mentioned, together with notes of new work started and draft standards circulated for comment. New international standards are included. *BSI News* updates the *BSI Catalogue*, so should be used in conjunction with it where appropriate.

ISO Catalogue. Geneva, International Organization for Standardization, annually.
○ ISO standards occur in all fields except electrical and electronic engineering, which are the responsibility of the IEC (see next example). This guide features all ISO standards published up to the end of the year preceding its cover date. Standards are recorded numerically under the associated Technical Committee (TC) numbers, the entries consisting of ISO serial number and date, edition, page count, and title in English and French. Other sections include: withdrawals (indicating when withdrawn and if replaced), list in numerical order (with a code representing the page count and the TC number for each standard), UDC/TC index (arranged numerically by Universal Decimal Classification and supplying only the corresponding TC numbers), and subject index (providing TC and standards numbers, in two sequences—English and French).
ISO member bodies are the national standards organizations of participating countries. Most of these act as sales agents for ISO publications in their own countries, and are listed in the *ISO Catalogue*. In the UK and the USA, orders must be addressed to the British Standards Institution and the American National Standards Institute respectively (see **Standards—access**, p. 128, for further details). In countries with no local agent, orders should be sent to ISO Central Secretariat, 1 rue de Varembé, Case postale 56, CH-1211 Genève 20, Switzerland/Suisse.

Catalogue of IEC Publications. Geneva, International Electrotechnical Commission, annually.
○ IEC is responsible for international standardization in the fields of electrical and electronic engineering. This guide records IEC standards published up to the end of the year preceding its cover date. Entries consist of IEC serial number, date, title, page count, price, edition, Technical Committee number, an abstract, and a note concerning amendments. International Special Committee on Radio Interference (CISPR) publications are also included, but listed separately. A subject index is provided, as is a numerical list of IEC Technical Committees. Document orders are handled by the appropriate national standards body (BSI in the UK, ANSI in America) or, in the absence of a local agent, they should be directed to the Sales Department, International Electrotechnical Commission, 3 rue de Varembé, CH-1211 Genève 20, Switzerland.

KWIC Index of International Standards. Geneva, International Organization for Standardization (ISO), 1985.
○ Covers the standardization activities of ISO, IEC, and 27 other international organizations. Single-line entries, which are arranged according to keywords selected from titles, include three-letter codes identifying the organization responsible for issuing each standard followed by the document's serial number. A list of these organizations, with addresses, has been provided,

but there is no numerical index of standards. It is planned to update this work every two years.

Annual Book of ASTM Standards, volume 00.01: subject index; alphanumeric list. Philadelphia, PA, American Society for Testing and Materials, annually.
○ The *1984 Annual Book of ASTM Standards* consists of 66 volumes divided among 16 sections, each containing the standards in a particular subject area (which are updated annually). The subject index has entries under fairly narrow headings, giving the title of each standard, a serial code number, and the volume in which it is to be found. The alphanumeric listing by serial code number also identifies standards which have been discontinued or replaced. The monthly *ASTM Standardization News* carries details of new or revised standards and forthcoming meetings, together with general-interest articles and advertisements for products/services in the field of standardization.

Catalog of American National Standards. New York, American National Standards Institute, annually.
○ ANSI acts as coordinator for America's voluntary standards system, approves standards as American National Standards, and serves as clearinghouse/information centre for national, international and foreign standards. It is the official US member of ISO and IEC.
Entries in the catalogue are made under narrow subject headings, including the title, abbreviation for issuing organization and publication number, date, and price of each standard. A 'listing by designation' section, arranged alphanumerically by serial code number, gives subjects and prices. An index of subject headings is also provided. This work is updated by supplements between editions.

Brief mention must also be made of two guides to American military standards. The US Department of Defense *Index of Specifications and Standards* (*DODISS*), published in looseleaf binder form with cumulative supplements, consists of two parts: an alphabetical listing of products and devices with the corresponding specification and issuing authority; and a numerical listing of specifications with details of the product or device and issuing authority. A weekly current-awareness bulletin *New US Specs & Standards*, issued by London Information (Rowse Muir) Ltd of Ascot, Berkshire, lists new or revised Department of Defense specifications and standards, giving in each case the number, subject, date, and page count; it also covers similar material issued by other bodies (such as ANSI and ASTM) which has been adopted by the US Department of Defense.

Finally, one example of a guide to the national standards of another European country. In addition to comprehensive German-language catalogues of its own publications, the Deutsches Institut für Normung produces a booklet entitled *English Translations of German Standards* (18th ed., Berlin, DIN, 1982), with entries arranged in subject groups based on UDC, subject and numerical indexes also being provided.

Further information about standards may be found in **General guides** (see p. 40) and **Subject guides** (see p. 42).

Theses and dissertations

Theses and dissertations (in practice the terms are interchangeable) give an account of research done at a university or similar institution, generally as part of the requirement for a doctor's or master's degree. (To find out about research work in progress see **Guides to educational organizations**, pp. 25–28.)

Theses are intended to embody a certain amount of original work—indeed they often form the basis for one or more journal articles or monographs. If this work is later published in a journal, it must be greatly condensed (order of magnitude 10:1 reduction) and may also undergo modification.

In the UK a British Standard (BS4821 : 1972) offers guidance on the form, layout and bibliographical presentation of theses.

British theses are produced in a small number of copies, two of which will normally be deposited in the library of the university awarding the degree. However, microfilm copies of many UK doctoral theses, and a few masters' theses (those thought likely to be in heavy demand), are held by the British Library Lending Division. Continental theses are often set up in type and published, whereas microform or xerox copies of American theses can frequently be made to order.

Uses

* To supplement subject information found from other sources.
* To obtain a deep theoretical approach to a subject.
* To get more detail where the work was later published in a journal article.

Access

* British and western European theses may sometimes be borrowed (using the inter-library loan service) from the university library where they are deposited. However, such borrowing is normally restricted to a library (rather than an individual)—so the thesis is only available for consultation within the library concerned.
* The British Library Lending Division is the main source for most post-1970 UK doctoral theses. Microform copies are loaned, and retention copies may also be available provided that the item is required for the purposes of an individual's research or private study (copies for library stock cannot be supplied). A special copyright declaration form signed by the person wishing to consult the thesis must be submitted in advance.
* Many North American doctoral theses, including the majority of those listed in *Dissertation Abstracts International* (see p. 135), can be purchased as microfilm, microfiche (if the order number carries a 76– or later prefix), or xerox copies from University Microfilms International, Dissertation Copies, PO Box 1764, Ann Arbor, Michigan 48106, USA. Prices are given in the current issues of *Dissertation Abstracts International*. UK customers should contact University Microfilms International, 30/2 Mortimer Street, London W1N 7RA, tel. 01-631 5030.
* The BLLD can supply loan copies (only) of theses covered by the University Microfilms International scheme; all dating from 1970–78 should be readily available; those outside this period are obtained on request if not

already held. As far as other doctoral theses are concerned, the BLLD has a small stock and will buy on demand items which can be purchased, or borrow from abroad if appropriate.
● Those wanting to investigate for themselves the accessibility of theses from various institutions all over the world should know about the following publication.

Guide to the Availability of Theses, compiled by D. H. Borchardt and J. D. Thawley. Munich, K. G. Saur, 1981. IFLA Publications, 17.
○ Entries, which are arranged by organization within country, include details such as: name and address of institution, name of relevant library, type of thesis deposited automatically (and number of copies), who holds copyright, where catalogued, conditions for internal consultation, whether personal borrowing is permitted, restrictions on photocopying, availability for inter-library loan, existence of abstracts, and a note of other libraries or institutions usually receiving copies.

Caution

● Theses are often rather academic and the very reason for their production may limit their usefulness.
● Masters' theses are generally far less readily available than doctoral theses, and undergraduate dissertations may be virtually unobtainable.
● Individual theses are usually not well indexed, and specific information may be difficult to locate.
● Work described may be modified before publication in a journal article or monograph.
● Only 'successful' research tends to get written up.

Guides to theses and dissertations
See also **Information services**, p. 209

The principal guides to English-language theses are described below. In addition, theses may be covered by broadly based indexing/abstracting journals (for instance *Chemical Abstracts*, or *Scientific and Technical Aerospace Reports*), certain online databases and, occasionally, bibliographies. Some universities publish, annually, a list of their own theses, often including abstracts, but these are of limited practical use.

Uses

● To find out if a particular subject has been covered by university research.
● To find what work has been done by a particular researcher.
● To back up the use of abstracting/indexing journals.

Access

Usually via libraries.

Caution

● Abstracts are not provided for all titles listed by the major British guide, and those which are available must be sought in a separate microfiche publication.

● The main American abstracting guide only includes material from cooperating institutions.

Examples

Index to Theses accepted for higher degrees by the universities of Great Britain and Ireland and the Council for National Academic Awards. London, Aslib, 1950/1–   .
○ The *Aslib Index to Theses*, as it is generally known, appears twice-yearly. Entries, which are arranged under fairly narrow subject headings, include the title, author, awarding institution, degree and date of each thesis. Subject and author indexes are provided, as are a list of subject headings and notes on the availability of theses. The same publisher also produces *Abstracts of Theses* (from 1977 onwards, microfiche only), although abstracts are given for by no means all theses covered in the former publication. Enquiries regarding a magnetic-tape version of Aslib's *Index to Theses* (1978–   ) should be addressed to Learned Information of Oxford, England. Older British theses may be identified using *Retrospective Index to Theses of Great Britain and Ireland, 1716–1950*, edited by R. R. Bilboul (5 vols. Santa Barbara, California, ABC-Clio, 1975–76); v. 2 Applied sciences and technology, v. 3 Life sciences, v. 4 Physical sciences, v. 5 Chemical sciences; each volume has both subject and author sections.

Dissertation Abstracts International: abstracts of dissertations available on microfilm or as xerographic reproductions. Ann Arbor, Michigan, University Microfilms International, 1969–   .
○ This publication began as *Microfilm Abstracts* (1938–51), then became *Dissertation Abstracts* (1952–69) and was issued in two sections (A: Humanities and social sciences, B: Sciences and engineering) from 1967 until 1976, when they were augmented by a third (C: European abstracts). Sections A and B are produced monthly, whereas C appears quarterly. Coverage is limited to cooperating institutions; that is, those which allow reproduction of their doctoral theses (restricted to the USA and Canada in the case of sections A and B). Copies of most items listed can be purchased from University Microfilms International; see p. 133.
Entries, which are arranged under fairly broad subject headings, include the title, order number (if available for purchase through University Microfilms International), author, degree, awarding institution, date, and page count of each thesis, followed by an abstract. Keyword title and author indexes are provided in every issue, but at present only the author index cumulates at the end of the volume. A retrospective index covering the period 1938–68 (9 vols. in 11) came out in 1970.
The same publisher is also responsible for *Masters Abstracts* (1962–   ) and the next two examples. The University Microfilms International database may be searched as *DISSERTATION ABSTRACTS ONLINE* via the DIALOG system (see p. 153). Another computerized service, DATRIX II, is available from University Microfilms International, 300 North Zeeb Road, Ann Arbor, Michigan 48106, USA.

American Doctoral Dissertations. Ann Arbor, Michigan, University Microfilms International, 1964/5–   .

○ This publication began as *Doctoral Dissertations accepted by American Universities* (1933/4–54/5) and then became *Index to American Doctoral Dissertations* (1955/6–63/4). Issued annually with entries arranged by institution within fairly broad subject divisions, an author index being provided. Aims to give a complete listing of all doctoral dissertations accepted by American and Canadian universities, so includes material omitted from *Dissertation Abstracts International*, but does not give abstracts. A table summarizing the publication, preservation and lending/selling practices of North American institutions regarding doctoral theses is also featured. Computer searches are possible using the *DISSERTATION ABSTRACTS ONLINE* database (see previous example).

Comprehensive Dissertation Index. 37 vols. Ann Arbor, Michigan, University Microfilms International, 1973.
○ Covers American doctoral dissertations over the period 1861–1972, with entries arranged by subject keyword under broad subject headings. The following details are included: title, author, degree and date, awarding institution, page count, citation to *Dissertation Abstracts International*, *American Doctoral Dissertations* or other source, and University Microfilms International order number (where appropriate). An author index is provided. There are annual supplements for 1973– , a 19-volume cumulation (1973–77), and a 38-volume cumulation (1973–82). This publication is also available on microfiche. Computer searches are possible using *DISSERTATION ABSTRACTS ONLINE* database (see above).

British Reports, Translations and Theses (see p. 124) is essentially a current-awareness guide which includes those UK doctoral theses which are deposited at the BLLD. Also worth a brief mention is the series of annual volumes recording *Masters Theses in the Pure and Applied Sciences* accepted by colleges and universities of the United States and Canada (1956–), published in New York by Plenum.
Further information about theses may be found in **General guides** (see p. 40) and **Subject guides** (see p. 42).

Trade literature

We take this to mean literature which an organization produces in connection with its products or services. Its function may be to advertise or instruct the reader in the use, exploitation or maintenance of these products. Some of this literature is more information oriented than sales oriented, and some even encourages the reader to find new applications for the products or materials available.

The literature which we cover in this section may take the form of sales brochures, catalogues, data-sheets, maintenance manuals or even specially produced books or reports. House or trade journals perform similar functions, but are dealt with under **Periodicals** (p. 95).

Uses

● To help you choose suitable products, equipment, materials, services, etc., for your requirements.

- To give you detailed information about products, equipment, materials, etc., which are already in your possession and which you wish to use or maintain.
- To give you details about rival products or services.

Access

- By direct request to the supplier or manufacturer. Many people keep a small personal collection of trade literature for regular reference and so it is often worth checking with likely colleagues.
- Your own organization may keep a 'local' collection, arranged in alphabetical company order. Such collections are usually kept in company libraries, design offices, purchasing offices or stores.
- Some trade directories contain trade literature, such as the *Thomas Register* (p. 24).
- Some public and national libraries maintain collections for reference only, such as the Science Reference Library (p. 259), which also subscribes to catalogue files supplied by Technical Indexes Ltd (see below).
- Commercially available 'product data services'. These take the form of collections of trade literature with specially compiled product and company indexes. These collections are hired out and maintained by service organizations. The rental charges are quite high, so that you have to rely on your own organization or another 'friendly' organization to take the services. Two examples are as follows.

ti *The Technical Indexes System*, available from Technical Indexes Ltd, Willoughby Road, Bracknell, Berkshire RG12 4DW. Tel. Bracknell (0344) 426311.

○ This is a system providing a large number of microfilm/microfiche files with detailed, cross-referenced (between files) hard-copy indexes, all regularly and automatically updated. There are two main types of file, covering products and standards, and the different files are available separately or in groups. The product information is contained in catalogue files which, in the case of the UK, cover: construction and civil engineering, electronic engineering, electronic components, engineering components and materials, laboratory equipment, manufacturing and materials handling, and process engineering; and there are also some files for North America and West Germany. All regular users of an installed system may have their own copy of a 'product data book' which gives access to the product information via suppliers' names, trade names or product names, and also gives suppliers' addresses, telephone and telex numbers. The standards and regulatory document files cover several national, international, military, and US federal and US industry standards. These include UK BSI and Defence Standards; German DIN; ISO; EEC; IEC; American DOD, ACI, ANSI, API, ASCE, ASME, ASTM, IEEE, Mil, and SAE documents. Location of documents is via 'index books' and most files also have detailed subject indexing and cross-references. Microfilm reader/printers can provide instant paper copies; an 'extension 99' telephone enquiry service will provide occasional answers from files not in the user's collection, and even the occasional copy of urgently required documents.

Barbour Index. The Barbour Library information services are aimed at the construction industry, and are available from Barbour Microfiles, New Lodge, Drift Road, Windsor, Berkshire SL4 4RQ. Tel. Winkfield Row (0344) 884121.
○ The information package for architects consists of a customer-tailored set of most-used product catalogues and technical publications, specially indexed by product and manufacturer. This is backed up by a microfiche file covering almost every type of product used in building; a compendium giving all details of manufacturers and trade names with illustrated product entries; and a technical 'microfile' covering legislation, standards and codes, and a wealth of information from official sources. The index contains subject sections and a directory of organizations.

Caution

- It is common for collections, especially 'private' ones, to become out of date.
- Prices are not always quoted because of frequent changes.
- It is wise to check the price and details with the supplier before placing a firm order.
- Manuals are easily lost if they are not kept with the corresponding equipment.

Guides to trade literature

If the name of the supplier or manufacturer is known, but their literature is not available in your organization or locally, the problem is usually to find their address or telephone number, so that you can approach them direct. The directories mentioned under the guides to the various types of organizations or their database equivalents will help you here: for example, telephone directories or *KOMPASS : United Kingdom* for British commercial organizations. The Science Reference Library publishes *Trade Literature in British Libraries* (a brief directory).

If you only know the type of product you want, the first problem is to identify the name of a suitable supplier or manufacturer. You may do this via the product or subject index of your local collection of trade literature, if you have one, or via the indexes in the directories or databases mentioned above.

If you wish to keep up to date with new products in a particular field, you have to rely on the various periodicals for advertisements, or ask appropriate suppliers or manufacturers to keep you informed of their new products. There are some journals which consist almost entirely of advertisements, and there are those which actually list details of new trade catalogues.

Information services

By these, we mean organizations that will supply you directly or indirectly with information, or references to information which could be helpful. The services vary from answering telephone queries to supplying collections of relevant documents, and from broadcast television information to direct connection with computer databases. The organizations vary from small 'one-man bands' to large multinationals.

Most libraries and information offices offer access to a range of external services as well as to their own facilities, but they cannot be described in detail here, owing to their number and variety. Local services should always be considered first (if you have access to them, or can find a suitable one to approach), especially those available within your own organization.

With many present services, the end product is a bibliography, that is a list of references or citations to other documents or literature, such as journal articles and books. Some services offer direct, factual or 'end' information, such as numerical data, statistics, or details of people, organizations, products or services. There is a marked trend towards more sophisticated facilities such as online systems, electronic publication, electronic mail and document ordering, and the computer manipulation, 'massaging' or reprocessing of information.

Services may be retrospective, which means information is supplied that originated over a number of years, or current, where only the latest information is provided.

Computerized information services

About twenty or more years ago, publishers of abstracting and indexing journals started to use computers to help them produce their hard-copy products. Once the information in all the references, abstracts and indexes (records) was available in computer or machine-readable form, people soon realized that it could be searched using a computer and suitable programs or software. After each issue of the abstracting or indexing journal was produced, the tape or other machine-readable equivalent was fed into a computer and searched using a number of different subject strategies or profiles, corresponding to peoples' individual interests. The resulting lists of references ('hits') were then separately printed out on computer paper or cards, and posted to the people concerned. This was known as current awareness or SDI (selective dissemination of information). By the late 1960s these individual tapes were cumulated into single files, which could then be searched to provide retrospective search services. The cumulated files, corresponding to a run of abstracting or indexing journals, are known as bibliographic databases. The first major services were what is known as 'batch mode' or offline, with batches of searches taking their turn with other work to be processed by the computer, and usually with no facility to interrupt the process once started. The first services were also known as serial or sequential because the computer searched all the records in a given database, one after another without stopping, trying to match each one with the search profiles. Before long the searching facilities became more sophisticated, with the searchable parts of records being stored in inverted files, similar, in principle, to coordinate indexes (see p. 270). The databases were also stored on discs rather than tapes, which allowed very rapid random access, as opposed to slow sequential searching. Databases could then be searched online; that is, it became possible to sit at a terminal, interrogate the computer, and get almost immediate answers.

Online information services

For most people, online information services mean organizations like DIALOG, SDC, BRS, DIALTECH, ESA-IRS, DATA-STAR, PERGAMON INFOLINE and BLAISE. For others, the meaning is much wider, covering almost any online use of computers to retrieve information, including the interrogation of one's own company computer or even a personal microcomputer. In theory, there are normally certain essential components. There must be a computer terminal or an equivalent device that can act as a terminal (terminal emulation), and allow you to transmit and receive information or data. There must be a computer of some sort which effectively contains the database to be searched. Finally the terminal and the computer must be linked or connected by leads, cables, wires, or telecommunications systems. In practice, the whole system may be very much more complicated (see p. 160). The cost of searching online varies from as little as £20 per hour to over £100 per hour with royalty payments ranging from zero to several pounds per reference or record printed out. A typical search currently costs between £10 and £50, including telecommunications charges. The significance of the word 'online' is that there is a direct communication path maintained between the terminal and the computer, with resulting 'live', almost immediate communication in both directions. This permits interactive searching, where a search strategy can be changed as the search proceeds, according to what is found, and iterative searching, where a search strategy may be repeated or cycled with slight modifications or refinements until an acceptable result is obtained. Online services are proving to be increasingly popular because of this immediacy. The searcher enters a search term and sees the results within seconds, that is, the numbers of references retrieved (or 'hits') and perhaps a display of some of the titles to see if they are relevant. With the old offline services, the search strategy was compiled without this feedback; it was then sent to the computer, which could be at a remote site, and the search results

141

Chart 7. Online information services in outline

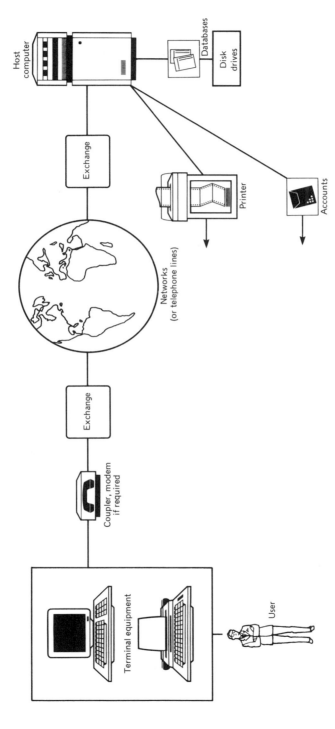

would not be available for some hours or even days. This slow process, which could take weeks if mistakes were made, probably put a lot of people off the computerized information services. The main advantages and disadvantages are described on p. 216, and the importance of online is covered on p. 159.

This section concentrates on the databases which contain the information (in the form of records), the hosts or services supplying online access to the databases, and the way to 'get online' in terms of equipment, telecommunications, procedures with hosts, advice and guidance. Searching is dealt with elsewhere (see p. 211).

Databases, databanks, files and their producers

Currently these are still often thought of as the computerized or machine-readable equivalents of abstracting/indexing journals, directories, handbooks, journals, data compilations, etc., which are searchable online via hosts' computers. However, databases have begun to evolve away from hard-copy forms in that they often contain extra information to enhance online searching, and already some have no hard-copy equivalent. This trend will inevitably continue, perhaps with many of the traditional hard-copy reference works ceasing to exist. Despite the trend, the separation in this book of databases from guides, such as abstracting journals, is somewhat artificial, and future works of the same kind may well include online aspects with each type of information source and guide.

The producers are traditionally the publishers of the various hard-copy equivalents, although this is already changing with the introduction of online databases with no previous hard-copy equivalent. The producers may offer their databases direct from their own facilities instead of, or as well as, via the hosts. They may also offer manuals, thesauri, training courses, help-desks, newsletters and other service and educational facilities. Any conditions of use may be covered by direct contracts or via hosts' contracts.

The term 'database' is somewhat vague. It is frequently used as a general name for all types of collections of machine-readable items or files. Sometimes it is used more specifically as a name for collections of references (a bibliographic database), or for a referral database which does not give you the actual information you want, but only tells you where you are likely to find it. In this book, the term database will be used as a generic or family name for all the various types. The term 'file' is associated with storage on a computer rather than the type of material stored. A file is the name for any collection of items which are effectively stored in one location in the computer and which can be accessed or retrieved by a single address or name. A database may be stored as one file or split up, as sometimes happens for very large databases, into several files. The term 'databank' is most frequently used for databases containing 'source', 'end', or factual or numeric data (as opposed to referral data), e.g. a collection of spectral values or a set of company records.

The unit of information stored in a database is usually called a record. This consists of various fields, such as author, title, journal reference, abstract, classification, indexing terms, etc., in the case of a bibliographic record.

In some areas it is difficult and even artificial to deal with databases separately from hosts, and there will therefore be some repetition in the two appropriate sections. It is important to remember that a particular database may be available from a number of different hosts.

Databases may be seen as having one or more features which must be considered when choosing appropriate types or individual databases. The main features are listed below with some examples, but it must be understood that most of them overlap to a certain extent with each other. The examples are taken mainly from the DIALOG service and give file names and associated hard-copy publications where the name is different.

- Subject. There are databases covering most subjects either specifically or as one of a range of subjects, and they often correspond to the traditional hard-copy journals.
  e.g. Chemistry—*CA SEARCH* (*Chemical Abstracts*)
      Mathematics—*MATHFILE* (*Mathematical Reviews*)
      Multidisciplinary—*NTIS* (*Government Research Announcements and Index*)
- Application. These databases may draw information from many subject areas and different types of document.
  e.g. Space exploration—*NASA* (*Scientific and Technical Aerospace Reports* and *International Aerospace Abstracts*)
      Coffee—*COFFEELINE* (ICO Library monthly entries)
- Document type. Although many databases cover a range of types of documents, some concentrate on one specific type.
  e.g. Books—*BOOKS IN PRINT*
      Conference Papers—*CONFERENCE PAPERS INDEX*
      Multidocument—*COMPENDEX* (*Engineering Index*)
- Referral. These databases refer you elsewhere or give a reference to follow up before you can obtain the information you finally want.
  e.g. Bibliographic (all the above examples)
      Research—*SSIE CURRENT RESEARCH*
- Directory and reference. There are directories and reference books to cover most things, especially people, organizations, products, services and materials, and an increasing number are becoming available online.
  e.g. Biographic—*AMERICAN MEN AND WOMEN OF SCIENCE*
      Chemicals—*TSCA INITIAL INVENTORY*
      Associations—*ENCYCLOPEDIA OF ASSOCIATIONS*
- Full text. A particularly exciting development where journals, groups of journals, encyclopaedias and news items are beginning to be available online or electronically published.
  e.g. Journals—*HARVARD BUSINESS REVIEW*
      Multijournal—*AMERICAN CHEMICAL SOCIETY JOURNALS*
                      IRCS MEDICAL SCIENCE JOURNALS
                        TRADE & INDUSTRY ASAP
- Numeric. These mainly consist of statistics and time series (data versus time).
  e.g. Exports—*US EXPORTS*
      Employment—*BLS LABOR FORCE*
      Time Series ⎱ *PREDICASTS PTS FILES*
      Forecasts   ⎰
- Special purpose.
  e.g. Training—*ONTAP COMPENDEX*

Jobs—*CAREER PLACEMENT REGISTRY*
Citation—*SCISEARCH* (*Science Citation Index*)
Current awareness—*PSYCALERT* (pre-*Psychological Abstracts*)
Multifile—*DIALINDEX* (a cross-file index)

Uses

- As guides to the immense amount of information appearing in the published literature, which are more sophisticated than conventional hard-copy indexes, frequently permitting searching of titles, abstracts and other fields.
- As direct sources of information on people, organizations, products, services, subjects, theories, methods, data, facts, etc.
- To keep in touch with the work of a particular person, organization or articles published in a specific journal.
- To check the accuracy or completeness of references.
- In the case of the producers, occasionally to provide special direct services; education and training; manuals, thesauri, newsletters and guides; and help-desks.

Access

- Directly on your own organization's computer if it can afford to buy in databases and run its own system.
- Via the hosts or service suppliers, such as DIALOG, using your own or your organization's facilities and equipment.
- Occasionally via the database producer as an online, offline, magnetic tape or other special service.
- Indirectly via information brokers or library/information services offering online searching, such as some institute/society or public libraries.

Caution

- Always check on access fees and print royalties, as these can be very high in some cases.
- The reliability of information or data in databases may be variable, both in terms of the original information and its processing.
- Many databases have special guides such as thesauri, manuals, chapters in hosts' manuals, classification schemes, lists of journals covered, etc.
- If possible, check the data sheet for a database to see what period of time is covered, if the database is split into separate files, what fields exist in the records and/or are searchable (such as abstracts).
- If you rely heavily on any database, check with its producer to see if there are any special direct services, facilities, manuals or guides to help you.
- Unless a database is specifically produced for current awareness, it may take anything up to several years for some items to appear.
- Bear in mind that some databases give references to 'end' documents that may not be available in your area.
- Do not rely on scanning the names of databases for potential usefulness, as some have meaningless names or acronyms such as *COMPENDEX* or *ERIC*.
- Some possess unique terminology or classifications for searching more

efficiently (which may change from time to time), making the use of special guides such as thesauri, classification schedules, concordances, etc., essential.
● A few contain special saved searches ('hedges') to cover a particular concept that has not been allocated a classification or thesaurus term, such as 'developing countries' and all equivalent phrases.
● Some databases may be linked together via keywords or codes either directly or indirectly depending on the host, e.g. *CA SEARCH* and *CHEMNAME*, so that results from searching in one database can be transferred to another using a simple command; this could, for example, save typing in a list of Registry Number codes.
● They are not usually comprehensive, and there is rarely an overlap of more than 50 per cent between two databases covering the same subject area.
● Some information may be abbreviated, e.g. journal titles and frequently used words such as prep for preparation.
● Some (exclusives) are only available on one host.
● A database may have different names in different hosts, different fields for searching/printing, different costs, and be available for different periods of time.
● Databases may be split chronologically into separate files by some hosts. This can be an advantage if you wish to limit a search to a particular period of time, or a disadvantage if you wish to search the whole database.
● Some databases are more restricted in their availability and use than others. Some may not be available in your country at all, others may not be searched for external clients or may not permit downloading of electronic copies of records. At least one has different levels of access for subscribers to its hard-copy services.

Examples

Only 12 of the more popular databases are included here, together with *SIGLE* which is a special case; see the appropriate guides for a wider selection (p. 186) and Chart 10 (pp. 192–210).

BIOSIS PREVIEWS (1969–   )
Corresponds to *Biological Abstracts* and *Biological Abstracts/Reports, Reviews, Meetings* (the latter superseding *Bioresearch Index*), with abstracts for recent years and including systematic biological names and codes to enhance searching. All life-science subjects are included, taken from journals, books, reports, theses, meetings, etc., and the coverage is international. Some hosts split the database into several files because of its size (more than 3.5 million records). An average-cost database with online print royalties. It is mounted on most of the larger hosts and the producer, BioSciences Information Service, supplies various facilities direct, such as B-I-T-S, a system supplying references on disc or tape for storing on your own computer, plus BioSuperfile, a program for storing, altering, and searching references on a microcomputer. (Other databases in this subject area include *IRL LIFE SCIENCES COLLECTION* and *ZOOLOGICAL RECORD*.)

CAB ABSTRACTS (1973–   )
Corresponds to a family of 31 journals with abstracts and abstracting journals. Every branch of agricultural sciences is included, taken from journals, books,

reviews, reports, theses, meetings, patents, etc., and the coverage is international (more than 1.5 million records). An average-cost database but with high online print royalties. It is mounted on several large hosts. (Other databases in this subject area include *AGRICOLA*, *CRIS/USDA* and *AGRIS*.)

CA SEARCH (1967–)

Corresponds to *Chemical Abstracts*, normally without the abstracts and systematic chemical compound names, but with Registry Numbers which may be identified via directory files such as *CHEMNAME*. Most aspects of chemistry, biochemistry and chemical engineering are included, and the coverage is international. Some hosts split the database into several files because of its size (more than 5 million records). An average-cost database with online print royalties, even for abstract numbers! It is mounted on most of the larger hosts; the Chemical Abstracts Service mount their own version, in a service called CAS ONLINE, with printable abstracts for more recent years, and a sophisticated chemical compound directory file which permits structural searching. CAS ONLINE is available via the STN network (p. 157), and there may be special discounts for academic users. People interested in searching for chemical structures should also investigate the DARC system and *INDEX CHEMICUS ONLINE* on the French host TELESYSTEMES QUESTEL, and the US-based Chemical Information System Inc CIS service. Industrial chemistry is also covered by *DECHEMA*, *CHEMICAL INDUSTRY NOTES* (*CIN*), *CHEMICAL ENGINEERING ABSTRACTS* (*CEA*) and *CHEMICAL BUSINESS NEWSBASE* (*CBNB*).

COMPENDEX (1970–)

Corresponds to *The Engineering Index* with abstracts and CAL (CARD-A-LERT) subject codes, and is a sister database to *EI ENGINEERING MEETINGS* which relates to *Ei Engineering Conference Index*. Records in the two databases may be linked via meetings numbers where appropriate. Most aspects of engineering are included, taken from journals, publications of the relevant societies, meetings, reports, books, etc., and the coverage is international (about 1.5 million records). An expensive database with high online print royalties. It is mounted on most of the larger hosts. (Other databases in this subject area include *CETIM*, *DECHEMA*, *DOMA* and *ISMEC*.)

EMBASE (1974–)

Corresponds to the Excerpta Medica publications with additional material, with abstracts for over half the records, and various subject codes. All aspects of medicine are included, taken from journals, books, theses and meetings, and the coverage is international. Some hosts split the database into several files because of its size (about 2.5 million records). A fairly expensive database with online print royalties. It is mounted on several large hosts. There is a service called EMSCOPES which is similar to B-I-T-S. (Other databases in this subject area include *MEDLINE*, *PHARMACEUTICAL NEWS INDEX*, *PRE-MED*, and various databases relating to cancer.)

INSPEC (1969–)

Corresponds to *Physics Abstracts*, *Electrical and Electronics Abstracts*, and *Computer and Control Abstracts*, with abstracts and classification codes. Most aspects of physics, electrotechnology, computers and control are included, taken from journals, reports, books, meetings, theses, and patents before 1976, and the coverage is international. Some hosts split the database into separate files because of its size (over 2 million records). An average-cost database with online print royalties. It is mounted on most large hosts. (Other databases in these subject areas include *SPIN*, *INKA-PHYS*, *DRE*, *ZDE*, *MICROCOMPUTER INDEX*, *COMPUTER DATABASE*, and *.MENU— THE INTERNATIONAL SOFTWARE DATABASE.*)

MEDLINE (1966–)
Corresponds to *Index Medicus*, *Index to Dental Literature* and *International Nursing Index* and some additional material, with abstracts for nearly half of the records since 1975, CAS Registry Numbers from 1980, and special subject codes. Virtually all aspects of medicine are included, taken principally from journals, and the coverage is international. A cheap database which is mounted on most of the larger hosts. Some hosts split the database into several files because of its size (about 3 million records). (Other databases in this subject area include *EMBASE*, *PRE-MED* and *PHARMACEUTICAL NEWS INDEX.*)

NASA (1964–)
Corresponds to *Scientific and Technical Aerospace Reports* (*STAR*) and *International Aerospace Abstracts* (*IAA*), with abstracts for many of the more recent records. The name *NASA* database is something of a misnomer as it is significantly smaller than NASA's own database, which also contains research resumés, citations to *NASA Tech Briefs*, software descriptions, etc. Most aspects of space exploration and applications, aeronautics and astronautics are included (nearly all subject areas!) taken from reports, journals, meetings, etc., and the coverage is international. A cheap database available outside US on ESA-IRS (with over 1 million records), but subject to certain contractual agreements with NASA. The *AEROSPACE* database mounted by DIALOG and MEAD DATA CENTRAL is equivalent to *NASA*.

NTIS (1964–)
Corresponds to *Government Reports and Announcements & Index* (*GRA & I*) and various abstract newsletters, with abstracts, and CAS Registry Numbers from 1979. Most aspects of US government sponsored research, development, engineering and associated analyses and reports are included, so that coverage is US biased but multidisciplinary. A cheap database mounted on most of the larger hosts (over 1 million records).

PSYCINFO (1967–)
Corresponds to *Psychological Abstracts* and additional material, with abstracts and various subject codes. Includes many aspects of psychology taken from journals, reports, books and theses, and the coverage is international. An average-cost database but with high online print royalties. It is mounted on many of the larger hosts (about 0.5 million records). (Another database in this subject area is *MENTAL HEALTH ABSTRACTS.*)

SCISEARCH (1974–)
Corresponds to *Science Citation Index* with additional material from the *Current Contents* series, and with access via cited references. Most scientific and technical subjects are included taken from major journals in all these areas and some books, and the coverage is international. An expensive database but with

a substantial discount for subscribers to the hard-copy equivalent. It is mounted on several of the larger hosts. Some hosts split the database into a number of separate files because of its size (about 5 million records).

WPI/WPIL (1963–   depending on subject of patent)
A patents database (see p. 90) which is expensive but available with substantial discounts and special search facilities to certain subscribers of the other Derwent services. It is mounted on several large hosts (about 3.5 million records).

SIGLE (1981–   )
A European solution to the problem of grey literature, that is documents which are issued informally in limited amounts and which are not available through the normal bookselling channels, such as reports, technical notes, etc. The system consists of an online database mounted on European hosts starting with INKA and BLAISE, with national 'authority centres' responsible for data input and document delivery for material originating in their own country. There are no hard-copy equivalents and no abstracts. The UK centre is the British Library Lending Division.

Choosing databases

The following methods assume that money is no object and that the searcher is equally familiar with all the hosts and with all the databases. In practice, cost and familiarity may play a big part in both database and host selection.

(1) *Adapting the traditional approach.* As the vast majority of databases currently have hard-copy equivalents, much of what has been said in **Choosing sources of information and their guides** (p. 6) and in **Abstracting and indexing journals** (p. 46) applies equally well here. So one method is to choose a traditional guide and see whether it has a machine-readable database equivalent.

(2) *Considering the types of database available.* A second method is to consider the various features of databases listed previously (p. 143) and see which are appropriate pointers for suitable databases.
For example, is the information you require likely to be found in a particular subject area such as physics (*INSPEC*), or it is to be found between two separate ones such as chemistry (*CA SEARCH*) and physics, or is it really vague or truly multidisciplinary or interdisciplinary so that you would need to try *NTIS* or *SCISEARCH*?
Does it have associations with any application areas such as space (*NASA*) or the paper industry (*PAPERCHEM*)?
Would the information be particularly associated with a specific type of document such as patents (*WPI/WPIL*) or in a whole range of types of, say, engineering publication (*COMPENDEX*)?
Could the information be so recent that it is still only the subject of current research (*SSIE CURRENT RESEARCH*)?
Would the information have been processed into directory or reference book format (*EMIS*) or into encyclopaedias (*KIRK-OTHMER*) or as statistics (*PREDICASTS PTS* files)?
Is the information likely to be deeply embedded inside the full text of a

document, and unlikely to have been important enough to be indexed in an abstracting or indexing service (*AMERICAN CHEMICAL SOCIETY JOURNALS*)?

(3) *Using cross-files.* A third method is to use files specially created by some hosts from, effectively, all the searchable fields of all the databases mounted on that host (with some exceptions). For example, in DIALOG you can begin in the *DIALINDEX* database, select a group of databases by their file numbers and/or by broad subject group names, then effectively search all of them, one by one, for say 'zinc air batteries' in the title (searches are performed using single-line search statements which yield results that cannot be further combined or manipulated). This would provide you with a listing of the 'hits' for each database so that you can decide which ones are worth searching in depth. The main point about cross-file searching is that it is cheap, although it cannot supply details of records. By doing a pilot search in a cross-file, you can usually see if a longer deeper search is worthwhile in a whole range of databases that you might otherwise have ignored. If you get no hits or very few, you may have saved yourself a lot of time and money searching in vain in an expensive database.

(4) *Using guides* (see also **Subject specialities of selected databases**, p. 192). There are various types of guide to databases, varying from those published by the hosts for their own services to the international guides and reference works (see **Guides to information services and suppliers**, p. 186).

Hosts, service suppliers, vendors and spinners

These are traditionally thought of as names for organizations which buy in tapes from the database producers and mount them on their own mainframe computers using appropriate computer programs, and probably storing the information on large disk drives. They then offer the information for their customers to search and print out records via a further suite of computer programs which involve command languages and search procedures, and some form of telecommunications link. The host normally charges the user for access and prints, passing back any royalties to the database producers. The term 'spinner' is sometimes used to denote a host that relies on offering a variety of other organizations' databases, such as DIALOG, as opposed to a specialist service supplier that concentrates on its own products, such as the Chemical Abstracts Service CAS ONLINE. There is no sharp dividing line between hosts and database producers. Some hosts are database producers, and some database producers provide services direct to the user.

It is tempting for the library and information profession to see DIALOG as 'the' online service, but DIALOG's market area is small compared to that of the business and financial or commercial area. Also, the current number of Prestel subscribers in UK is supposed to be significantly larger than the number of DIALOG password holders worldwide. However, for scientists, engineers and their intermediaries, there are various other aspects of the hosts to consider, and DIALOG certainly excels in some of them. These aspects are as follows.

(1) *Range of databases.* Some hosts, like DIALOG, mount a very large variety of databases, which is helpful to organizations that need this kind of access but would experience difficulties in administration and in maintaining expertise with a number of different hosts. Some hosts mount a number of databases to give good coverage of certain subject areas, such as DATA-STAR's coverage of medicine. Some hosts have a limited number of databases or facilities, but offer very specialized access, such as in the case of CAS ONLINE. Others appear to specialize in collections of exclusive databases, such as PERGAMON INFOLINE and SDC. Many hosts split large databases into a number of files, which makes searching the whole span of the database more tedious and costly. A few hosts such as ESA/IRS and DIALOG offer a gateway service to databases mounted on other organizations' computers. See also **Subject specialities of selected databases**, p. 192.

(2) *Command languages/search sophistication.* For most searchers, the first command language or set of search procedures that they learn appears to be the most popular and easy to use, but there are significant differences between the hosts that may override such familiarity.

Some hosts offer more sophisticated search facilities than others, such as proximity searching (e.g. searching for one term adjacent to another), cross-file searching (see p. 149), after-the-fact field qualification (post-qualification or retrospective limiting of a particular search statement result to a specified field, such as title; see p. 241), distributed proximity searching (proximity within groups of search terms, such as either word A or word B adjacent to either word C or word D), temporary saved search strategies or even complete searches, search-term transfer between files, frequency reports on search terms associated with a given set of records (see p. 231), automatic SDIs, special downloading facilities (see p. 243), and merging and sorting options. Other hosts have a certain lack of sophistication, such as no permitted use of clustering (brackets or parentheses, see p. 238), no online cost reports, and slow response or answer time for highly used (highly posted) terms.

Some command languages are capable of abbreviation; some give a compact search record or history on the screen or printer, such as DIALOG or ESA-IRS. Others are longer, more user-friendly or conversational but less compact, so that the amount of information displayed on one screen page is limited. In the case of end-user services aimed at the private/home sector and practitioners (lawyers, doctors, engineers, scientists, etc.), the user-friendly commands can be maddeningly slow and drawn-out once familiarity has been built up with the system (see BRS After Dark service, p. 153, and DIALOG Knowledge Index, p. 154).

Some hosts offer terminal profile facilities, which means that you can set the system to your own design to a certain extent. For example, you can set it to automatically give online costs, to respond to different hosts' commands, or to print a non-standard line length.

(3) *Fees.* An area of vital concern to many organizations. The effective charges for a specific database can vary from host to host. This may be as a result of international currency exchange rates or due to different contractual arrangements involving hosts, database producers and/or users themselves. Various discount schemes are offered by hosts. Some give an increasing discount with increasing connect hours. Some have minimum-use

agreements, others have prepayment schemes or budget accounts, and some may give special discounts to academic institutions.

(4) *Communications*. This can be a nightmare area for some users with poor telecommunications facilities. Hosts may offer a variety of telecommunications access at different speeds, such as direct-dial national or international networks, or their own private networks. Whatever system is used, it is vital that it is fast, trouble-free, and not too expensive.

(5) *Back-up services*. Different hosts provide widely differing levels of back-up in such areas as manuals; guides (including quick-reference cards or leaflets); newsletters; education and training; national help-desks; a search service if you cannot do it yourself; online ordering of original documents; and special offline print services.

(6) *Special facilities*. Some hosts provide their customers with access to electronic mail; personal database construction, storage, searching and printing; and personal time-series databases which permit entry of data and data manipulation and output.

Uses

● To access databases, some of which may not have any hard-copy equivalent.
● To perform retrospective searches; produce printed bibliographies, lists of references or records, possibly sorted or merged; or current awareness.
● SDIs on given people, organizations, products or topics.
● To perform faster, more thorough, more sophisticated and efficient searches than would be possible by traditional hard-copy methods.
● To provide help (via telephone, letter or online), guidance, education and training, manuals, newsletters, and other guides.
● To download data electronically for editing, manipulation or storage.
● To obtain access to electronic mail/message services, online ordering of original documents or photocopies, and possibly electronically published documents.

Access

● Directly though telephone dial-up or networks.
● Indirectly through libraries, information bureaux or brokers.
● Offline via telephone/letter to some hosts or agents for a remote service (if your equipment fails or your searcher is indisposed).

Caution

● Online searching is still only one of the methods for retrieving information; although it is growing in power, sophistication and efficiency, there are still occasions when other methods, such as traditional hard-copy searching or subscribing to current-awareness journals, are preferable.
● Considerable searcher expertise may be required, especially when using a number of different hosts and databases for one search, when these possess different structures and require differing search commands.

● Watch out for lack of search sophistication, such as proximity searching, or lack of back-up services, such as automatic SDIs or online document ordering.
● Regular searcher practice is required to maintain expertise.
● Overloaded computers or telecommunications networks can lead to delays, frustration and vastly increased costs.
● Different hosts may load the same database differently, may call it by different names, may not have the same back-runs, may charge different rates, may split it up into several files which have to be searched separately, and may differ in speed of updating.
● Watch out for minimum-billing clauses in host service contracts.
● Maintain security of your passwords, but keep a record somewhere, as they are easily forgotten after a long holiday.
● Beware of infringing host or database contracts regarding downloading or searching for third parties.
● Browsing may be difficult and expensive.
● Incoming manuals, updates and newsletters can be difficult to keep up with, in terms of maintaining expertise and keeping a well-ordered reference collection.
● Hosts may occasionally reload databases in a different way (and databases themselves may change), requiring different search techniques and modified SDI profiles.
● Be prepared for VAT charges if they apply, and the need for special arrangements when paying for some foreign services.
● Although coming out of their infancy, these services may still undergo substantial changes from time to time.

Examples

Note that services are undergoing occasional changes and improvements, and these details relate to the 1983/84 situation.

BLAISE

Part of the British Library, this host has two services: BLAISE-LINE, which uses a UK computer and concentrates on coverage of books, conference proceedings, grey literature, eighteenth-century materials, incunabula, audiovisual materials, and periodical articles on education; and BLAISE-LINK, which provides access from the UK to the databases held on the National Library of Medicine (USA computer), which concentrates on medicine, biomedical and toxicological subject fields. Special features include automatic SDI, document request service, cheap offsearch facilities for searching more than one database offline, some cheap training files, a UK help-desk, and training courses. BLAISE also offers special cataloguing services for libraries.

The main problem in using BLAISE is its limited search sophistication in some areas, with no direct word proximity searching available or permitted use of clustering. This problem should disappear with the introduction of the BRS/SEARCH system.
● Conclusion: a good host for medical or book information.
● Contact: The British Library, Bibliographic Services Division, 2 Sheraton Street, London W1V 4BH (tel. 01-636 1544, ext. 242/284).

BRS

Based in the US, it is one of the largest services with over 70 databases and 50 million records. It has a wide subject coverage including science, engineering, medicine, business, finance, education, social sciences and general reference. Special features include sophisticated search facilities, left-hand truncation on *CA SEARCH* and *MEDLINE*, left-hand prefix display (like a left-hand version of EXPAND/ROOT), a highlight feature in displayed or printed records which picks out the terms entered in the search strategy, automatic SDIs with online editing, cross-file searching, merged offline output from multiple databases, after-the-fact field qualification (retrospective limiting), cheap public end-user-friendly services (AFTER DARK and BRKTHRU), and a service for medical professionals called COLLEAGUE. BRS also mount a number of full-text databases including the *KIRK-OTHMER ENCYCLOPEDIA OF CHEMICAL TECHNOLOGY, AMERICAN CHEMICAL SOCIETY JOURNALS*, and *IRCS MEDICAL SCIENCE JOURNALS*. BRS and DATA-STAR have similar command languages, but other links no longer exist.

The main problems for the European users of BRS are telecommunications costs, the dollar exchange rate, and the splitting up of some of the larger databases into separate files.

● Conclusion: an important host for people with varied interests, or those already familiar with DATA-STAR. Its AFTER DARK, BRKTHRU, COLLEAGUE and full-text services offer very attractive facilities to many users.

● Contact: BRS, 1200 Route 7, Latham, New York 12110, USA.

DATA-STAR

An essentially European host located in Switzerland, with a range of larger databases but specializing in business and biomedical/medical subjects. Special features include a discount for academic users, automatic SDIs, a purge command for eliminating unwanted sets, after-the-fact field qualification (retrospective limiting), distributed proximity searching, cross-file searching, large databases in single files, a useful quick-reference guide, a UK help-desk available every day with free access from France and Germany, and good training programmes.

The current problems with using DATA-STAR are a certain lack of information in the screen/printer displays, such as online charges, and long response times for some terms in the larger chemical files. As the service is fairly new and expanding rapidly, these problems will hopefully disappear. DATA-STAR's rates are among the lowest and very competitive. (See also BRS.)

● Conclusion: an essential host for European users in medical subjects or in academic organizations, and a useful additional service to those already familiar with BRS.

● Contact: DATA-STAR, Plaza Suite, 114 Jermyn Street, London SW1Y 6HJ (tel. 01-930 5503) in the UK, or Radio-Suisse SA, Laupenstrasse 18, 3008 Berne, Switzerland, for outside the UK.

DIALOG Information Services

Based in the USA, it is the largest service of its kind, with about 200 databases and 100 million records. It has a very wide subject coverage, in addition to science and engineering, which includes business, economics, law, social sciences and humanities, and current affairs. There is also a wide source coverage, including patents, books, directories, theses, conference papers, news, citations, time series, as well as journal articles and reports. Special features include a sophisticated search facility, distributed proximity searching, after-the-fact field qualification, online document ordering, automatic SDI, cross-file searching, information transfer between files, cheap training files, a 'REPORT' feature enabling the sorting of data from certain databases into tabular format, a gateway service to databases on other computers, such as *OAG ELECTRONIC EDITION* (Official Airline Guides), and a cheap evening/weekend public service called KNOWLEDGE INDEX. DIALOG have a UK help-desk, excellent manuals, a quick-reference guide, newsletter and comprehensive training sessions. Electronic mail is expected to be made available soon.

The main problems in using DIALOG are those affecting the European users, namely the dollar exchange rate, telecommunications costs, and some difficulties in gaining access during peak periods. DIALOG do have their own telecommunications network (DIALNET) which extends to PSS access in the UK, and is cheaper than IPSS. The current practice of splitting some of the larger databases into a number of separate files (five for CA Search) can also be a disadvantage, although this should eventually be stopped.

• Conclusion: a vital host for people with varied database requirements, especially for those already familiar with the ESA-IRS command language, and those who wish to access a minimum number of hosts.

• Contact: DIALOG Information Services, PO Box 8, Abingdon, Oxford OX13 6EG (tel. Oxford 0865 730969) for Europe, or DIALOG Information Services Inc, Marketing Department, 3460 Hillview Avenue, Palo Alto, CA 94304, USA.

ESA-IRS

The European Space Agency Information Retrieval Service. (IRS-DIALTECH is the UK agent with full back-up services, including a local offline printing service, free use of PSS via an IRS NUI (Network User Identity), and alternative access via Prestel sets.) It has a good selection of databases specializing in science and engineering. Special features include automatic SDI, cross-file searching, frequency reports on search terms associated with given sets of records (the ZOOM command, see p. 231), online document ordering, sophisticated and compact command language and displays (similar to DIALOG), large databases searchable in single files, downloading facilities, a gateway service to databases on other computers, private file services (QUESTMENU), a UK help-desk, good training programmes, cheap training files and an online facility for trainers (QUESTLEARN), a microcomputer terminal package with communications and file-handling software (MIKROTEL), and access to the *NASA* database. The main problem with ESA-IRS is currently a delay in manual updates from ESA, which has been aggravated by the phasing out of DIALTECH's own manual.

• Conclusion: an essential host for users involved in aerospace or any of the many associated subject areas, and good for science and engineering in general. DIALOG users might find the similar command language helpful.

● Contact: IRS-DIALTECH, Department of Trade and Industry, Room 392, Ashdown House, 123 Victoria Street, London SW1E 6RB (tel. 01-212 5638/8225), or ESA-IRS, ESRIN, via Galileo Galilei, 00044 Frascati, Italy.

PERGAMON INFOLINE

A London-based service backed up by one of the largest publishing and printing organizations in the world. Its specialities include patents, chemistry, technology and business, and a number of exclusive, specialized databases such as *PIRA*, *RAPRA*, and *ZINC LEAD AND CADMIUM ABSTRACTS*. Special facilities, apart from the exclusive databases, include automatic SDIs, document delivery, a UK help-desk, private file and advanced electronic publishing services, frequency reports for terms in a given set of references retrieved during a search (the GET command, see p. 231), a handy quick-reference guide, and a back-up search service. There are agents with local training and help-desk facilities in France, Scandinavia, the Netherlands, and with new agents coming up in Italy, the Far East, Australia, and South America.

The main problem with using INFOLINE is having to change to other hosts for some of the larger popular databases, although this will not be important if automatic host switching becomes available.

● Conclusion: an essential host for certain exclusive databases.

● Contact: Pergamon Infoline Ltd, 12 Vandy Street, London EC2A 2DE (tel. 01-377 4650; help-desk 01-377 4957). There are offices in McLean VA, USA, and Willowdale, Ontario, Canada.

SDC

Based in the USA, it is one of the original large services with about 80 databases. It has a wide subject coverage including business, science, engineering and the social sciences, with many exclusive databases, and a number of important patents and energy files. Special features include a sophisticated searching facility, online document ordering, automatic SDIs, cross-file searching, information transfer between files, terminal profile facilities, fairly good manuals and a quick-reference guide, newsletter and training programmes, and a UK help-desk.

The main problems with SDC are telecommunications costs, dollar exchange rates, and the splitting of larger databases into separate files. There is also minimum billing for the use of some databases in certain cases.

● Conclusion: an important host for people with varied interests, and essential to those whose interests are covered by the exclusive databases such as *FOREST* and *SAE*.

● Contact: SDC Information Services, Bakers Court, 4th Floor, Bakers Road, Uxbridge, Middlesex UB8 1RG (tel. Uxbridge 0895 37137, ask for SDC), or SDC Information Services, 2500 Colorado Avenue, Santa Monica, CA 90406, USA.

Other services which may be investigated via the various guides include the following.

BELINDIS, specializing in economics, law and energy, and located in Belgium.

BRITISH MARITIME TECHNOLOGY (BMT) provides access to its own database, BMT/SHIP ABSTRACTS, and is located in Wallsend, Tyne & Wear, UK.

CAS ONLINE, specializing in a single file of *Chemical Abstracts* with printable abstracts, sophisticated structural searching facilities in a chemical compound file to obtain Registry Numbers, and a document delivery service. It is located in the USA with various European addresses for contact, including the Royal Society of Chemistry in Nottingham for the UK. There is a direct networking link (STN, see below) between the US computer and the FIZ Karlsruhe computer in West Germany.

CIS (Chemical Information Service), formerly known as CIS NIH/EPA, is now owned by Chemical Information Systems Inc, and located in the USA with European support from Fraser Williams (Scientific Systems) Ltd, of Poynton, Cheshire, UK.

CISI-WHARTON, specializing in worldwide financial and economic data, and offering econometric modelling and data manipulation. This is a Paris-based group with an office in London.

CONTROL DATA B.V. provides access to the database of the Dutch Maritime Information Centre.

DATACENTRALEN, specializing in European time series, agricultural research, various Nordic databases, and located in Denmark.

DATASOLVE, specializing in world news media, international commercial development, and European law. Located in Sunbury-on-Thames, UK.

DATASTREAM provides both online computer-based information and computation services, specializing in international business economics, finance, modelling etc., and located in London.

DIMDI, specializing in medicine, agriculture and biosciences, and ISI databases, located in Cologne, West Germany.

ECHO, the European Commission's own host, offering various free services to mainly European information on research projects, documentation, terminology, databases and centres. It is located in Luxembourg.

EDS DATA LIMITED, covering a wide range of computer services. The World Trade Statistics Database, *TRADSTAT*, and the Royal Society of Chemistry's new *CHEMICAL BUSINESS NEWSBASE* (CBNB) are currently available. EDS DATA is located in Watford.

FINSBURY DATA SERVICES provide TEXTLINE, specializing in international and national business information with particularly good coverage of UK newspapers (also available via ESA-IRS); NEWSLINE a service to back up TEXTLINE with the very latest news; and DATALINE, specializing in company accounts. Finsbury Data is located in London.

FIZ TECHNIK, specializing in many aspects of engineering and located in Frankfurt, West Germany.

GEOSYSTEMS, specializing in geological information in depth and its management, and located in London.

INKA, a large West German service with about 50 databases covering many important national databases as well as foreign ones, located in Karlsruhe, and expecting its databases to become available via STN (see CAS ONLINE above and STN below).

JORDANS, specializing in UK company information and located in London and other cities.

KODA ONLINE, a service recently established by Reed Publishing, offers *EKOL*, a database equivalent to the European editions of *KOMPASS* (pp. 21-22), and is located in East Grinstead, UK.

MEAD DATA CENTRAL (MDC) is currently expanding into popular bibliographic databases, in addition to its full-text 'libraries', with 'The Reference Service'. It is based in the US, with a London office, and with Butterworths Telepublishing in London also acting as an agent for *LEXIS* (the legal database).

SAMSOM, specializing in maritime, offshore and hydraulics databases, but recently subject to a takeover with no details yet available. Located in the Netherlands.

SCICON, specializing in European and national and local UK government information, and located in Milton Keynes, UK.

STN INTERNATIONAL, the Scientific and Technical Information Network. A new service formed by the American Chemical Society and the Fachinformationszentrum Energie, Physik, Mathematik GmbH, linking the Columbus and Karlsruhe computers. STN currently offers direct access from Europe (Karlsruhe), North America (Columbus, Ohio) and Japan (Tokyo) to CAS ONLINE and *PHYSICS BRIEFS*. It is planned that other INKA databases will follow. Database producers may be able to mount their databases on the system, which uses the sophisticated and user-friendly 'MESSENGER' software, and then deal directly with their own users.

TECH DATA, a new end-user service for engineering and technical information with a user-friendly command language, developed with BRS by Information Handling Services of Englewood, Colorado.

TELESYSTEMES QUESTEL, the large French service specializing in national databases, particularly patents, the very large multidisciplinary *PASCAL* database, and the DARC chemical sub-structure search facility derived from Chemical Abstracts Services and located in Paris.

WILSONLINE, a potentially very useful service which should become available eventually in the UK, covering H. W. Wilson Company's large range of indexing services, which includes *Applied Science and Technology Index*. It is already available in part direct from Wilsons in the USA, and access via established hosts may follow in due course.

Choosing hosts

We shall assume that you have already selected one or more databases, and that you can only choose from the hosts to which you have access; other hosts would be available through brokers, etc.

(1) *Database coverage.* You are immediately limited to the host or hosts that mount all or most of the databases you have selected. Most people prefer to stay with one host for searching a number of databases, unless they consider that another host is significantly better for searching in certain databases (see below).

(2) *Database loading.* If the full back-run of a database is to be searched, hosts that have split it into numerous files are normally to be avoided. Some hosts do not cover the maximum back-run available. Where it is required to search particular fields, such as abstracts, it should be remembered that some hosts may not provide this facility, and may be rejected.

(3) *Command languages.* If a search strategy requires certain sophisticated techniques, such as word proximity or zoom-type analysis, then this may rule out some hosts. Otherwise people normally choose the host they know best, especially if the search is at all awkward or has to be done in a rush. The availability of good quick-reference guides and manuals should be considered if it does become necessary to turn to an unfamiliar host.

(4) *Charges.* The same database can vary significantly in cost between different hosts, although in practice this may also be due to the factors mentioned in 5 below. Some hosts charge for services like storing saved search strategies.

Chart 8. Popular large databases versus hosts for online searching

Hosts	BIOSIS PREVIEWS	CAB ABSTRACTS	CA SEARCH	COMPENDEX	EMBASE	INSPEC	MEDLINE	NASA/AEROSPACE	NTIS	PSYCINFO	SCISEARCH	WPI/WPIL
BLAISE							●					
BRS	●		●	●	●	●	●			●	●	
CAS ONLINE			●									
DATA-STAR	●		●	●	●	●	●			●	●	●
DIALOG	●	●	●	●	●	●	●	●	●	●	●	●
DIMDI	●	●			●		●			●	●	
ESA-IRS	●	●	●	●		●		●	●			
INKA			●			●			●			
MEAD DATA CENTRAL	●						●	●	●			
PERGAMON INFOLINE			●									
SDC			●	●		●				●	●	●
STN INTERNATIONAL			●							●		
TELESYSTEMES QUESTEL			●									●

(5) *Availability, speed, and reliability of operation.* This is often a feature of both the host and the telecommunications route. You will find it necessary to avoid certain hosts for particular types of searches and at peak congestion times of the day and week, and also to avoid certain telecommunication routes at busy periods.

(6) *Special facilities.* If these are required, in the form of, say, downloading formats, online document ordering, cross-file searching, or high-quality offline prints, then this will further reduce your options.

Setting up an online information service

Although it has not been our policy to include this kind of information in other parts of the book, we believe that because these services are relatively new and yet are potentially of vital importance to many organizations, we are justified in providing this section. Only details of commercially available services are covered. In-house online services are not included.

Why go online?

For certain people it must seem reasonable and tempting to stay with the traditional, well-tried and apparently reliable methods, and perhaps use commercial online services, via bureaux, if really necessary. The question is, is it really worth all the expense of purchasing special equipment, the training of staff, and running the service? There are four ways of considering this.

(1) *The information.* The services now provide information (or references to information) which may be unavailable from any other source, or which can only be obtained elsewhere with considerable difficulty. In any case, the services provide a key to tens of millions of papers, records and documents, and a previously undreamt-of wealth of information in the vast majority of subject areas. The systems probably already cover about half of the traditional Western scientific and technical literature that has ever been published (over 100 million records, although, so far, probably less than one per cent of the actual 'end' information is available online). Can you afford to continue with a limited access to such information?

(2) *The searching process.* The services can effectively scan ten or twenty million references or data records in a few minutes, using sophisticated searching strategies to see if there is anything relevant to your interest. In many cases, this would be a gigantic task by any other method. Can you afford to rely on the slower traditional methods or on an information bureau?

(3) *The competition.* Quite simply, your competitors may have their own local in-house access to these services, so can you afford the risk of giving them a monopoly on such a powerful information tool?

(4) *The evolutionary process.* We are certainly entering a new information age. The services are growing, getting more sophisticated and efficient, taking over from existing services, and they are here to stay. The way people find and use information is an integral part of the way they work, and new techniques cannot be adopted overnight. So can you afford not to evolve with the services?

Each organization must answer these questions for itself. Some, especially the smaller ones, may continue to thrive without online services, because of the nature of their particular businesses, but for most it is a question of *when* and not *whether* to go online.

How to get online

For some people, especially those opting for a very basic set-up (such as a terminal with no software requirement), a complete package or a 'turnkey' system, the procedure for getting online can be quite fast and simple. For others, the whole process can turn into a nightmare. This section will attempt to outline the basic steps required and some of the pitfalls to avoid. The components of a typical set-up consist of equipment, software, telecommunications, and agreements with hosts. This means that before you can go online you must:

(1) Investigate, choose, order/purchase (or rent), have installed, and learn to operate equipment (terminals or microcomputers, printers, disk drives, and leads) and possibly software (on tapes, disks or microchips/ROMS).
(2) Investigate, select, apply for/lease or rent, have installed or connected, and learn to operate telecommunications systems (telephone lines, special lines, and networks), and possibly purchase/rent acoustic couplers and/or modems.
(3) Investigate, select, apply for contracts with, and learn how to operate online services (hosts).

Equipment

This is the requirement that most people think of first, because of the time it takes to select, order, and wait for delivery and installation. This can be a pitfall if the other components of the set-up are neglected for too long. For example, telephone lines and telecommunications contractual and connection procedures may take more time than you realize. Another pitfall is to ignore any software requirements in the early stages, in the case of the more sophisticated or intelligent equipment, such as microcomputers. The equipment may be capable of doing what you want it to, but only with the help of a compatible software package that may not yet exist.

The main requirement is for a computer terminal or equivalent, normally with some printing facility. This may consist of one or more of the following: a simple teletype terminal, a video display unit (VDU) terminal, a printer terminal, a microcomputer, or in some cases a telex or Prestel TV set. Separate printers are available for most equipment. Some equipment is portable, such as small typewriter-sized printer terminals, small microcomputers and possibly the hand-held six-key Microwriter. Printers can vary in print quality and speed of operation. Dot-matrix printers may be quite fast and relatively cheap, but they tend to be noisy and give poor print quality. Daisywheel printers can give better quality print but are relatively slow, while the more exotic ink-jet and laser printers are still too expensive for most applications. The simplest and cheapest terminals have little or no intelligence, require no software, and are sometimes referred to as being 'dumb'. Terminal equipment may have varying degrees of 'intelligence'. Intelligence varies from such attributes as local editing facilities, and stored passwords or search terms, to full offline compilation, editing and storage of search strategies and downloading, editing, storage and searching of search results.

Intelligence requires software and memory, and memory can take various forms, the main types being: RAM (random-access memory, that is read or write memory) on microchips, which is usually lost when the power is turned off (volatile memory), or maintained for a limited time span by batteries; ROM (read-only memory) on microchips, which is non-volatile but requires special procedures to write items into the memory; and floppy or hard disks, which are effectively memory devices with moderate to very large capacities for information storage. The more exotic facilities require separate software packages (computer programs).

Intelligence gives greater sophistication and possibly ease of use, but there is more to go wrong and, currently, it often results in compatibility problems between different items of equipment. Some equipment has built-in acoustic couplers or modems for connecting (interfacing) with a telephone service, but even here there may be compatibility problems, such as between US equipment/modems and UK telecommunications services. Compatible items of equipment can be connected together, for example a VDU or microcomputer and a printer. The whole problem of compatibility has been the subject of various national and international studies. Probably the major current attempt to encourage compatibility is OSI (Open Systems Interconnection), an international plan which should provide standards for computers and telecommunications networks.

Software

This is the name for the computer programs that drive the equipment (hardware) or tell it what to do, where any intelligent facilities are involved, such as memories and editing. Software is not required for dumb terminals. A software package might include stored programs, documentation and procedures.

The programs may be stored on microchips (as part of the equipment, as plug-in extras, or as part of an add-on device or 'black box') or on tape or disks. There are also intelligent printers, modems, and even cables. Networks also have built-in intelligence, and are usually highly sophisticated.

The software may just allow a terminal to have simple editing and memory facilities, or enable a microcomputer to act as a terminal (terminal emulation). Further sophistication might permit automatic logging-on procedures (auto logon), simple word processing, offline search-strategy compilation, and limited storage of search results. The most sophisticated packages have additional special facilities for online searching, such as downloading, file handling, and automatic command-language conversion; and will also allow interaction with other packages, such as those for word processing and information storage and retrieval. Packages that might be investigated include CONNECT from Learned Information of Oxford; HEADLINE and HEADFORM from Head Computers of Oxted; ASSIST and IT (Information Transfer) from Userlink Systems of Stockport; the Derwent Workstation (mainly for patents) from Derwent Publications of London; and various US packages that may be adapted to work in Europe, such as Sci-Mate Universal Online Searcher from ISI of Philadelphia (Uxbridge in the UK). Most of these software packages cost around £200.

The three main considerations affecting choice of equipment and software are:

(1) *What do you want from the equipment/software, or what sort of service do you hope to provide?* The host services can be accessed at different speeds, normally 30 characters per second (cps), 120 cps, or split-speed with different rates for transmitting and receiving (7.5/120 cps). This split-speed access is sometimes called high speed, which can be misleading because the transmit speed of 7.5 cps is very slow if you are interested in sending (uploading) large amounts of data to a remote computer. Equipment capable of the higher speeds is required if it is intended to do a lot of online printing or displays, or downloading. Even higher speeds are sometimes possible with certain hosts and the use of Datalines (see p. 165).

Intelligence in the form of memory can be very useful, for example to store automatic logging-on procedures, including passwords; for storing search strategies prior to going online; for automatic transmission of data after connection to a host (uploading); and for storing the retrieved records/search results prior to editing and printing (downloading). Intelligence may also be associated with control codes, such as those for turning equipment on or off. The same control code may do different things to different pieces of equipment.

An auto-dial facility may be available in the terminal, modem or separately, to save repeated dialling of long numbers.

Other facilities in addition to online access may be required: for example, word processing or computation. This would normally require a microcomputer.

If a microcomputer is to be used as a terminal, or for sophisticated procedures such as downloading combined with word processing, then appropriate software should be identified first, and compatible equipment found next.

Compatibility is a vital requirement. Equipment must be compatible with the host computer, the telecommunications systems, and any other equipment it may be connected to. Other equipment may just mean printers or disk drives (extra memory-storage devices), but may also mean the company mainframe computer or other microcomputers on a company network.

Ergonomic or comfort factors are also of prime importance, because most operators are not used to working with VDUs or keyboards, and in such cases it is very easy to cause stresses and strains. Noise and poor visibility are frequent problems with printers. Acoustic hoods can improve on the sound insulation, but can make the visibility of the printout worse.

(2) *Cost.* As with most things in life, you tend to get what you pay for. If you are going to spend a lot of time online, get the best equipment you can afford. The cheapest set-ups (under £1000) are basic VDUs, small portable terminals (such as the Texas Instruments 700 series or Digital Equipment's LA12 DECwriter Correspondent) or personal microcomputers (such as the BBC B) with cheap slow-speed printers. The cost can be gradually increased by purchasing accessories such as disk drives, high-speed printers, add-on intelligence (such as USERKIT from Userlink Systems of Stockport), or simple software packages on disk or microchips (ROMs). One package that

must be mentioned is British Telecom's own Merlin 1100 system, which comes with screen, keyboard, modem and choice of printers for under £1300, and is fully compatible with Viewdata, Telecom Gold and PSS. BT also produce the Merlin Tonto system, which is almost identical to ICL's One-per-desk. The next stage in expense (about £2000 to £2500) includes the purpose-designed intelligent terminals and microcomputers with built-in modems (such as the Torch from Torch Computers of Cambridge or the Teleputer 3 from Rediffusion Computers of Crawley). Then the cost could be increased by selecting increasingly more powerful microcomputers and software (such as those in the IBM PC family, which seem likely to set an international standard). Some services require special equipment for certain facilities, such as a graphics capability for CAS ONLINE, and this can add extra expense, especially to printers. Do not forget that equipment costs about ten per cent of its purchase price to maintain per year.

(3) *Support services*. Equipment and software need to be maintained, and users or operators need advice in setting up and dealing with changes in operating procedures. If your parent organization has standardized on a specific make of equipment and associated software and accessories, and can offer advice, back-up and maintenance, then this could be well worth considering, even if the equipment is not quite your first choice. Otherwise look for equipment that is well known, or has an experienced and reliable local distributor/agent.

Telecommunications

Telecommunications is the name for the communications system that can link your terminal to the host computer; it includes telephone, telex, cable, fibre-optics, radio, networks and satellite communications. See Chart 9.

The simplest example is a direct telephone line, with a telephone or telephone socket at your end. This can be a PSTN (Public Switched Telephone Network) line on which you can dial up a telephone number for a host computer in the same way you would dial an ordinary number, or it could be a dedicated leased line or privately leased circuit (PC). In either case, ordinary telephone wires are used (although private circuits are engineered to be noise-free). Telephone lines linked through an organization's own switchboard are not usually recommended because of extra line noise and the possibility of being cut off by the switchboard operator, although the latest automatic electronic switchboards are supposed to be virtually noise-free. Your terminal equipment is linked to the telephone line using an acoustic coupler (normally a device with rubber cups to fit over the ends of a traditional hand-piece, or special universal couplers for fitting most modern telephones), or a modem (a more efficient device), which is plugged into the telephone socket or directly connected to the telephone in the case of British Telecom equipment (Datel 300, 600 or 1200 Duplex services, which all make use of the telephone dialling facility, and will also permit normal telephone use). Your terminal equipment is plugged into the coupler or modem using an appropriate lead with D-type connectors (also referred to as RS232 or V24 interfaces).

Packet networks. With the increasing sophistication of telecommunications, most communications between terminals and computers over any distance make use of packet switching and networking. This system is cheaper to use

Chart 9. Telecommunications links for online access
A simplified guide to access from the UK. Note that teletex terminals
may be connected to all the circled networks but may not access online hosts at this time.

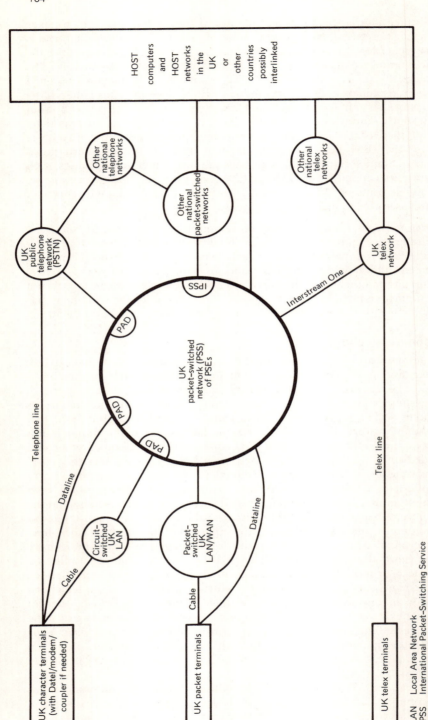

LAN   Local Area Network
IPSS   International Packet-Switching Service
PAD   Packet Assembler/Disassembler
PSE   Packet-Switching Exchange
PSS   Packet SwitchStream
PSTN   Public Switched Telephone Network
WAN   Wide Area Network

and more efficient than the traditional telephone system. Instead of maintaining and dominating a continuous-transmission path between two parties, the messages are broken up into separate packets, labelled with various information including the addresses of the sender and of the destination, and sent through the network to their destination where they are put together again to reform the original message. The route through the network is preset, but if there is a breakdown on that path the packets are automatically rerouted. The devices used for the assembling and disassembling of packets are called PADs (Packet Assembler/ Dissassemblers). The packets from different sources can share the same transmission paths, resulting in a more efficient use of the network and a cheaper service. The user is unaware of the packeting and sharing processes, unless the system becomes overloaded and slows down, or works intermittently. Congestion or overloading usually occurs at the host end, on their datalines into the networks, or in joint networks. A network, at its simplest, is a number of service points connected together by telecommunications, where each service point could be a telephone or other type of exchange or a computer facility of some kind.

Packet SwitchStream. The packet network usually used for online service access in the UK is British Telecom's Packet SwitchStream service (PSS). PSS is separate from the telephone network, and the cost for using it is independent of distance. In the case of the normal character terminals, PSS may be accessed via ordinary dial-up (using a modem or coupler) and the traditional PSTN service, or alternatively via a dedicated British Telecom Dataline link (Dataline 300 or 1200, which do not include a telephone as part of the system) to a special packet-switching exchange (PSE). A growing number of PSEs are coming into operation, and most online users are already within local telephone dialling distance. People with packet terminals (normally intelligent devices, such as computers, that have PAD facilities) use higher-speed Dataline services (Dataline 2400, 9600, or 48K). The PSS network connects with other networks in many overseas countries (such as the various national Post and Telecommunications Administrations (PTTs) networks like Telepak, Datapak and Transpac for Europe; and Telenet, Tymnet and Uninet for the USA), permitting its UK customers to contact online services based in other countries. The International Packet Switching Service (IPSS) is the name given to this international connection or 'gateway'. Other networks in the UK may have gateways into PSS, enabling anyone with access to such a network (such as JANET, the Joint Academic Network, which is known as a Wide Area Network), to gain entry to PSS and overseas networks. Organizations with private networks (operating X25, the same packet-switching protocol used by PSS), and requiring multi-terminal access to PSS, may use a special BT networking switch system called Packet Netmux. Packet Netmux also enables X25 Local Area Networks (LANs) to be created, with or without PSS access. These networks are ideally suited to a large office complex or self-contained site environment such as a hospital, and enable an organization's terminals, computers and microcomputers to communicate with each other and share memory, printing and other facilities. British Telecom has also launched a family of LAN products called NetStream, including 'Baseband' and 'Broadband' networks, fibre-optic cables, and 'Gateway', which integrates the LANs with PSS.

There are also private networks available either as an independent service or as part of a host's services, such as DIALOG's DIALNET. It should be noted that the present European network grew out of Euronet, which was a service provided by Euronet Diane, an organization that promotes European hosts, for example, by publishing directories, encouraging cooperation, especially the use of the European Common Command Language by European hosts.

Networks have built-in error correction, so that if any data is corrupted during transmission by, say, line noise, the error is recognized, and a request to retransmit the data is sent back. The ordinary dial-up PSTN system does not have this facility, but a new British Telecom service called MultiStream should eventually extend error protection to even this service, plus increased user-friendliness and lower costs. MultiStream is also expected to support videotex and teletex terminals (see pp. 182, 185).

British Telecom publish a news magazine called *NETWORK Journal* for their network users, which contains details of the latest products and services, and in-depth articles.

Telecommunications costs for online use are normally composed of various combinations of the following:

- Inland telephone rental and call charges to the nearest PSE or to UK hosts.
- Modem rental.
- Dataline rental and call charges (duration and data-volume dependent).
- PSS/IPSS Network User Identity (NUI) rentals and call charges (duration and data-volume dependent).
- Network charges. In the UK, the charges for these, if applicable, are normally included with the invoices for PSS/IPSS.

Contacts: for telephone, Datel or Dataline services, contact your local telephone manager's office; for PSS, contact Packet SwitchStream, Customer Service Group, G07 Lutyens House, 1–6 Finsbury Circus, London EC2M 7LY (tel. 01-920 0661 or Freephone 2170); for IPSS, contact IPSS Operations and Development, Room 723, Holborn Centre, 120 Holborn, London EC1N 2TE (tel. 01-936 2750).

Agreements with hosts

Potential customers normally obtain and complete an application form for each separate host for which access is required. The applications then form the basis of a contract between the two parties, describing the limits of the service and the conditions for use, including acceptance of the individual database producers' conditions. The contract is normally expected to accommodate any changes that the host may require to make from time to time, which are often at the request of the database producers.

Many of the comments on choosing hosts for searching (p. 157) apply to choosing hosts for service contracts. If it is not obvious which hosts are the best ones to start with, or the best ones to add to existing services, you should seek advice from someone in a similar position who has solved this problem successfully. You should be particularly careful to check on the different methods available for paying for the services, such as deposit accounts, minimum commitments, and academic or volume discounts. Normally start with the service that you expect to use the most.

A typical online session

Do not logon to a host until you know how to logoff! At £1 per minute, it could be expensive. See the example searches in INSPEC for DATA-STAR and DIALOG at the end of this section. Note that the way to send what you have typed in on your terminal to the computer is normally to press the ENTER, RETURN, CARRIAGE RETURN, or CR key.

This example assumes that the searcher has switched on the mains electricity and the equipment, and has worked out which database, host and search strategy to start with. (The example is not specific to any one host.) The searcher:

(1) Dials local PSS exchange (PSE), waits for ringing tone followed by high-pitched computer tone.
(2) Pushes data button on modem telephone, or pushes telephone hand-piece into acoustic coupler cups.
(3) Types in two carriage returns (CRs), a terminal identifier (such as A7) followed by one more CR, resulting in a response from the PSS exchange. A further CR results in a request for the user's identity (NUI, Network User Identity).
(4) Enters NUI (types in NUI plus one CR), resulting in a request for the address of the host computer (ADD).
(5) Enters the address for the selected host (known as an NUA, Network User Address), and hopefully receives confirmation of connection to the host, with a request to enter a password. (The NUI and the NUA could have been entered together in one sequence or string, separated by a hyphen.)
(6) Enters password and receives acknowledgment, perhaps some news messages, and a request to enter details of the selected database.
(7) Enters database name or code and receives a prompt to start searching.
(8) Enters first search statement (see p. 231) and receives a reply giving a running search statement or set number, and the number of 'hits' that the search statement has retrieved.
(9) Enters further search statements, if appropriate, and receives corresponding replies.
(10) Enters a command to print out some titles to see if the search is producing relevant results.
(11) Refines search strategy by making various combinations and permutations, and by observing the results.
(12) Enters a command to print out the full records of the final selection, and waits for the printer to work its way through the references.
(13) Enters logoff command and receives confirmation and estimate of costs.
(14) Records details of search, including name, date, subject and cost.

Sample searches

The following searches were performed in the *INSPEC* database on both DIALOG and DATA-STAR, and show how a search might be performed to find some references on the problems of radiation from colour TV sets, especially X-rays. The search starts by using fairly obvious *INSPEC* thesaurus terms, which produce only one reference. The search is then continued by using that reference to provide additional search terms for enriching the search strategy. Explanations of the commands may be found from p. 236 onwards. Note that, in DIALOG, S is the abbreviation for SELECT and SS is the abbreviation for SELECT STEPS. The DIALOG search is partly a projection, as *INSPEC* was not available on DIALOG version 2 at the time that this was written. Typed commands are in **bold** type.

DIALOG search in INSPEC

POR\A01-7050040019   (*response from Portsmouth PSS Exchange/PSE*)
NUI?
NDIALOG-A21230012011   (*part of the NUI is automatically suppressed for security*)

23421230012011+COM
DIALOG INFORMATION SERVICES
PLEASE LOGON:
?■■■■■■■■
Welcome to DIALOG

Dialog version 2, level 3.10.5
LOGON File001 04jan85 10:08:27
New DIALOG Version 2 search features now available!

File 1:ERIC—66-84/DEC   (*automatic entry into ERIC*)

Set	Items	Description

?B13
   04jan85 10:09:12 User006649
   $0.30   0.011 Hrs File1
   $0.12   DialnetE
   $0.41   Estimated cost this file

File13:INSPEC—77-84/ISS24
(Copr. IEE)
See File 12(1969 Thru 1976)

Set	Items	Description

?S COLOUR TELEVISION RECEIVERS AND X-RAYS

	802	COLOUR TELEVISION RECEIVERS
	577	X-RAYS
S1	1	COLOUR TELEVISION RECEIVERS AND X-RAYS

?T1/8/1
 1/8/1
446657
   RADIATION PROTECTION CONSIDERATIONS ON THE USE OF COLOUR TV SETS
   Descriptors: COLOUR TELEVISION RECEIVERS; X-RAYS; RADIATION PROTECTION; ELECTRONIC EQUIPMENT MANUFACTURE; COLOUR TELEVISION PICTURE TUBES; ELECTRON TUBE MANUFACTURE
   Identifiers: X-RAY RADIATIONS; RADIATION ABSORPTION CHARACTERISTICS; RADIATION PROTECTION; TELEVISION KINESCOPE; COLOUR TELEVISION RECEIVER; TELEVISION PICTURE TUBE; MANUFACTURE; X-RAY EFFECTS

Class Codes: B6420D; B0170E; B2360
?S (COLOUR OR COLOR)(W)(TELEVISION OR TV)

	15611	COLOUR
	6519	COLOR
	11954	TELEVISION
	9049	TV
S2	2534	(COLOUR OR COLOR)(W)(TELEVISION OR TV)

?SS X(W)RAYS OR RADIATION PROTECTION

S3	91070	X
S4	13730	RAYS
S5	5695	X(W)RAYS
S6	2417	RADIATION PROTECTION
S7	8076	X(W)RAYS OR RADIATION PROTECTION

?S S2 AND S7

	2534	S2
	8076	S7
S8	8	S2 AND S7

?T8/5/1-8
8/5/1
1008066   C83011164
IS YOUR TV A RADIATION HAZARD?
HALFHILL, T.R.
COMPUTE. J. PROG. COMPUT. (USA) VOL. 4, NO. 12 108-112 DEC. 1982
Coden: COMPER   ISSN: 0194-357X
    Treatment: PRACTICAL
    Document Type: JOURNAL PAPER
    Languages: ENGLISH
HOME COMPUTERS AND VIDEO GAME MACHINES USED WITH OLD COLOR
TV SETS COULD EXPOSE PEOPLE TO POTENTIALLY HAZARDOUS DOSES OF
RADIATION. THE HAZARDS ARE EXAMINED AND ADVICE IS OFFERED
    Descriptors: TELEVISION RECEIVERS; RADIATION PROTECTION; PERSONAL
COMPUTING
    Identifiers: RADIATION HAZARD; COMPUTERS; VIDEO GAME MACHINES;
COLOR TV SETS; HAZARDOUS DOSES OF RADIATION
    Class Codes: C7830
8/5/2

. . . .
. . . .
?LOGOFF
    04jan85 10:13:38 User006649
    $6.24   0.065 Hrs File13
    $0.65   DialnetE
    $1.44   8 Types
    $8.33   Estimated cost this file
    $8.74   Estimated total session cost 0.076 Hrs.
Logoff: level 3.10.5 10:13:39   (*sign off from DIALOG*)
CLR PAD (00) 00:00:05:30 134 14   (*sign off from PSE giving time duration and
                                    numbers of segments received and
                                    transmitted*)

DATA-STAR search in INSPEC

LO1\A02-1920040104   (*response from London PSE*)
NUI?
NPARKER   (*next 6 characters are suppressed automatically for security*)
ADD?

A9228464110115
228464110115+COM
DATA-STAR, PLEASE ENTER YOUR USERID: **ABCDEF**
ENTER YOUR A-M-I-S PASSWORD
■■■■■■■■■■■
ENTER YES IF BROADCAST MSG IS DESIRED__: **N**
ENTER DATA BASE NAME__: **INSP**

*SIGN-ON   14.49.40   04.01.85
D-S/INSP/1980-V85, I02/1985   SESSION 359
COPYRIGHT BY INSTITUTION OF ELECTRICAL ENGINEERS, LONDON, GB.

D-S—SEARCH MODE—ENTER QUERY
  1__: **COLOUR-TELEVISION-RECEIVERS AND X-RAYS**
  RESULT   1
  2__: **. . P 1 TI, DE, ID/ALL**
    1
TI   Radiation protection considerations on the use of color TV sets.
DE   colour-television-picture-tubes, colour-television-receivers, electron-tube-
     manufacture,   electronic-equipment-manufacture,   radiation-protection,
     X-rays.
ID   X-ray-radiations, radiation-absorption-characteristics, radiation-protection,
     television-kinescope,   colour-television-receiver,   television-picture-tube,
     manufacture, X-ray-effects.
R0601 * END OF DOCUMENTS IN LIST

D-S—SEARCH MODE—ENTER QUERY
  2__: **(COLOUR OR COLOR) ADJ (TELEVISION OR TV)**
  RESULT   1616
  3__: **(X ADJ RAYS) OR RADIATION-PROTECTION**
  RESULT   5495
  4__: **2 AND 3**
  RESULT   6
  5__: **. . P 4 ALL/1-6**
    1
AN   C83011164.
AU   Halfhill-T-R.
TI   Is your TV a radiation hazard?
SO   Compute. J. Prog. Comput. (USA), vol. 4, no. 12 P: 108–112, 0 REFS.
CD   COMPE.
LG   EN.
RN   ISSN: 0194-357X.
YR   82.
PT   JOURNAL PAPER.
TC   PRACTICAL.
DE   television-receivers, radiation-protection, personal-computing.
ID   radiation-hazard,   computers,   video-game-machines,   color-TV-sets,
     hazardous, doses-of-radiation.
AB   Home computers and video game machines used with old color TV sets could
     expose people to potentially hazardous doses of radiation. The hazards are
     examined and advice is offered.
CC   C7830.
      2
  . . . .
  . . . .
R0601 * END OF DOCUMENTS IN LIST

D-S—SEARCH MODE—ENTER QUERY
  5_:. . **OFF**

CONNECT TIME INSP: 0:05:25 HH:MM:SS   0.090 DEC HRS.   SESSION 359
*SIGN-OFF 14.54.44   04.01.85   (*sign off from DATA-STAR*)
CLR PAD (00) 00:00:05:34 79 14   (*sign off from PSE giving time duration and
                                  numbers of segments received and
                                  transmitted*)

Publicizing the service

It is relatively easy to create a demand by offering free trial searches, but this interest can fade rapidly, especially if any form of charging is then introduced. Many online operators or intermediaries find that healthy growth is maintained by word of mouth, provided that the service is useful. Although notices, leaflets, talks and demonstrations all help, they may not be remembered by people whose work does not require this type of service in the immediate future. The aim should be to catch people at the appropriate stage in their work, so that online will be actually tried out when it is most needed, and then remembered as a tool to be used again in the future.

Running the service

The main problems are finance, records, staffing and maintaining expertise. Annual costs are bound to increase in most cases, as the service becomes known. If there is a central budget, there will be a need to justify growth. If an organization accepts research or development contracts, future proposals or applications should include a cost element for online searching when appropriate. There will normally be a flow of invoices each month from the hosts, some requiring foreign-currency payment. Records may be required to check expenditure against invoices, to provide usage statistics for management, or to store search details in case of repeat requests, or as a means of putting people involved in similar work in touch with each other. Search forms for filling out by the requestor prior to searching, and by the searcher during searching, can be of help. Search staff need to use the services regularly to maintain expertise, and the service should be available when required, or the use will fall off. A system for handling the large volume of newsletters, publicity leaflets, manuals, updates etc. has to be organized.

Advice, guidance and training

Nearly everyone needs help, both at the initial stages in setting up and starting a system, and then in maintaining expertise while operating it. There are a number of potential sources of help:

● Online information, computer, and telecommunications experts in your own organization or its branches or subsidiaries.
● The Online Information Centre offers various services to its subscribers, including a newsletter; guides to databases, databanks, equipment and *Going Online* 1984 (see p. 189); register of subscribers and their equipment etc.; and an enquiry service. First-time equirers will normally be given some initial advice free of charge. The Online Information Centre is at Aslib, The

Association for Information Management, Information House, 26/27 Boswell Street, London WC1N 3JZ (tel. 01-430 2502).
● Local or national groups, associations or bodies with appropriate expertise, such as online user groups (the Online Information Centre can help with contacts) or local/regional information cooperatives (see p. 181). In the UK there is a national online user group (UKOLUG) which is a special interest group of the Institute of Information Scientists, but open to non-members of the Institute. UKOLUG has a newsletter, runs training courses, meetings and conferences, and acts as a pressure and consumer group on behalf of its members. UKOLUG may be contacted through the Institute of Information Scientists, 44 Museum Street, London WC1A 1LY (tel. 01-831 8003). There are also numerous local online user groups scattered throughout the country.
● Other users of online services. Many large organizations, libraries and information services have online facilities and will often be prepared to give advice.
● Published books, guides and journal articles (see p. 186).

The previous examples will be naturally biased, to some extent, towards their own selected systems, but the following examples may be very heavily biased towards their own products because they rely on selling them in order to survive.

● Manufacturers, agents and suppliers of equipment and software. Beware of salesmen who try to sell microcomputers, or any other equipment, without regard to appropriate software or compatibility with the other components involved with online services. Insist on a demonstration under the same conditions that you hope to use, unless you have already seen identical equipment in use elsewhere, operating in the same way as you wish to operate your own. Remember that the equipment specifications in brochures rarely give details of compatibility with other equipment or facilities.
● Hosts and database suppliers. Many hosts and some database suppliers have considerable experience with certain equipment and with telecommunications, and will usually be willing to help with enquiries in this area as well as in their own product and service areas. Their own training courses, guides and help desks are, of course, essential for maintaining expertise. People attending training courses would be well advised to do some basic homework before they go, such as becoming familiar with the contents of any quick-reference guides. It is also a good idea to take along some search topics or search strategies that are of real interest to you, as you may be asked to supply an example for a demonstration, or to try out for yourself during a practical session.
● British Telecom. For most telecommunications enquiries and assistance and for details of some products. Although BT still has a monopoly in some areas, it is beginning to have to compete in others. BT supplies various guides and maintains a number of enquiry services (see p. 166).

Checklist

(1) Check on the type of service, if any, that would be appropriate for your organization, by discussing with potential customers and with online staff in

similar organizations. Most important of all, find out what money will be available for setting up the service, and how much will be available each year for running it.

(2) Check on the facilities already available in your organization, such as equipment, experts and telecommunications.

(3) If no telecommunications links already exist, then make appropriate applications to British Telecom for installation, NUIs etc. Having a line installed can take a long time.

(4) Order selected equipment (do not forget things like leads, printer paper and floppy disks), preferably after having seen an adequate demonstration. This can take a long time, especially for new and popular models.

(5) Sign on with selected hosts and order relevant manuals, guides and thesauri.

(6) Arrange appropriate training with hosts, database producers, or other available sources. This may have to be booked some time in advance.

(7) Organize internal publicity and administrative procedures for running the service.

The future

Although nobody can say for sure, all the signs are that online information retrieval is here to stay and that it will grow. We may expect the following:
• An increase in the number and variety of databases, especially those aimed at the end user and those providing direct information.
• An increase in sophistication for equipment, software, telecommunications, command languages, and back-up and special services.
• An increase, hopefully, in standardization and compatibility between services, systems and facilities.
• An increase in user-friendliness, especially for the end-user services, such as Knowledge Index, After Dark, and the EASYNET service which can switch between hosts (sponsored by NFAIS, the National Federation of Abstracting and Information Services in the USA). It is interesting to note the required use of credit cards for payment for some of these services. Hosts may permit access to other hosts though linking, switching or gateways, reducing the number of contracts logons, command languages to be learnt, and invoices to be paid for the user. The user-friendliness may become available via hosts, networks, or software for local microcomputers. It may be available for different levels of expertise; help with host and database selection, enquiry clarification, search monitoring, and information evaluation; support multiple screen windows; and may even provide tutorial and training facilities.
• An increase in direct access by end users due to the availability of specific end-user services, user-friendliness, availability of popular files such as *MAGAZINE INDEX*, and the growing access to and familiarity with microcomputers.
• An increase in the supply of 'end' information, as opposed to references, via electronic document delivery and publishing systems.

● A general integration of online information retrieval and other electronic information systems, such as word processing, electronic publishing and electronic mail, into our everyday working lives. This might have a significant effect in education, where the concept of 'open learning' is being explored by such organizations as the Council for Educational Technology (CET), in the UK. (With open learning, people study when, where and how it suits them, as far as possible.) Perhaps the greatest impact will occur if really powerful personal microcomputers become both available and popular. If these micros contain the massive memories being predicted for optical disks, we really will be entering a very different era of personal and organizational information management. Cheap and massive storage or memories will encourage local, in-house or decentralized systems, while cheap, rapid, and reliable telecommunications will encourage remote or central systems such as DIALOG. In practice we shall probably end up with a blend of local and remote 'distributed' services.

The main clouds on the horizon are those arising out of the problems of copyright, piracy, and profitability for the database producers and the publishing trade in general; congested computers and telecommunications systems; and restrictive practices by publishers, database producers, national governments and telecommunications organizations. There may well be severe problems for countries which are unable to take advantage of these services, for political or financial reasons.

The information age offers a future sparkling with possibilities, but we all face a growing vulnerability to information technology, finance, politics and crime, and an increasing dependence on equipment, software, telecommunications and electrical power supply.

Information bureaux, brokers and other services

By information bureaux and brokers we mean organizations or departments which will perform searches on your behalf, either online, offline, or using traditional, hard-copy 'manual' methods. They may also offer back-up services, such as document delivery, photocopying, advice/consultancy and training. Remember that many types of organization, libraries and information offices offer this kind of service. Commercial organizations that offer a public service are normally restricted to certain hosts, database producers, or commercial information services, such as Geosystems (London). Examples of other types of organization offering specific services are:

Professional   Institution of Mechanical Engineers (London).
Educational   Biomedical Information Service (University of Sheffield).
Research   PIRA (Research Association for the Paper and Board, Printing & Packaging Industries, Leatherhead).
Official   Industrial Waste Information Bureau (Harwell Laboratory, AERE, Oxford).
Mixed   Transport and Road Research Laboratory, Technical Information and Library Services (Crowthorne).

By other services we mean organizations offering all the alternative facilities that may be provided, ranging from punched-card systems to standard SDI profiles.

Many public, educational, research, professional, and commercial libraries offer information services, although some of the services may only be available for a fee or to members of the library or parent organization. Special libraries probably include the largest number of miscellaneous specialist information services. The significance of the word 'special' is that it is usually taken to mean a restricted area of subject interest or application, such as electrochemistry, textiles, or the interests of one particular company.

Uses

● To obtain access to online search services for people without their own equipment.
● To provide an enquiry service for all kinds of information.
● To supply back-up services such as document delivery and photocopies.
● To give advice and consultations.

Access

Generally via letter, telephone, or personal visit.

Caution

● Find out whether fees are involved, and, if so, what you will be charged. Remember that membership of some organizations attracts free or discounted rates for its services.
● Make sure you say precisely what you want and what you do not want; if possible, say why you want the information and how you hope to use it.
● If large sums of money are involved, keep accurate records.
● In the case of free or subsidized services, do not take advantage of the system or take the staff for granted.

Examples

The following examples include some of the most popular and useful services, and a selection of other services to illustrate the range available. It is essential to use the appropriate guides to identify all or any of the latest relevant services.

Aslib, The Association for Information Management
A professional body providing its members with a comprehensive advisory, consultancy, training, referral and information service. Specialities include information management, information technology and online services. The majority of members' enquiries are handled free of charge. Membership costs depend on the type of organization and the number of employees, and start at about £100 per year. A membership list is published irregularly. Aslib produces its own periodicals, reports and books, and also organizes its own conferences, meetings and training courses. Groups and branches have been formed in certain interest or geographical areas, to promote useful contacts and discussion between their members. The groups include the following areas: audiovisual, biological and agricultural sciences, chemical, computer,

economic and business information, electronics, engineering, informatics, one-man bands, social sciences, technical translation, and transport and planning. The regional branches exist in the Midlands, the North, and Scotland.
● Contact: Aslib, The Association for Information Management, Information House, 26/27 Boswell Street, London WC1N 3JZ (tel. 01-430 2671).

BioSciences Information Services (BIOSIS)
Apart from producing its own database and abstracting journals, BIOSIS offers various current-awareness services including C.L.A.S.S., a custom-tailored weekly SDI service based on *BIOSIS PREVIEWS*; BIOSIS Standard Profiles, a monthly service similar to C.L.A.S.S. but operating on 28 different fixed topics; and *BIOSIS/CAS SELECTS* (see p. 286). See also BIOSIS B-I-T-S, p. 145.
● Contact: Thompson Henry Ltd, London Road, Sunningdale, Berks SL5 0EP (tel. Ascot 0990 24615/22639) or BioSciences Information Service, 2100 Arch Street, Philadelphia, PA 19103-1399, USA.

BSI Information Department (Standards and Foreign Trade)
The British Standards Institution is mentioned in the standards section (p. 127), but since the move of some of the departments to Milton Keynes, there is now one location from which you can obtain information, see, buy, or borrow almost anything to do with standards, technical requirements, UK or overseas certification, approval or similar systems. Note that non-members of BSI may have to pay charges for services that are free to members. A database covering UK and foreign standards is in preparation for online searching.
● Contact: Information Department, BSI, Linford Wood, Milton Keynes MK14 6LE. General enquiries and orders: tel. 0908 320066; Library, Database section, and Technical Help to Exporters section: tel. 0908 320033.

Commonwealth Scientific & Industrial Research Organization (CSIRO)
A service covering all subjects, including retrospective computerized searches and SDIs.
● Contact: Commonwealth Scientific & Industrial Research Organization, Central Information Service, Box 89, 314 Albert Street, East Melbourne, Victoria 3002, Australia.

GEODEX International
Offers two coordinate index-card services: GEODEX SYSTEM/S covering structural engineering back to 1956; and GEODEX SYSTEM/G covering soil and rock mechanics and foundation and geological engineering back to 1925 (and incorporating *Geotechnical Abstracts* since 1970). Both services provide regular volumes of digests and punched cards on a subscription basis. Back-files for both systems may be purchased, and there is a service for supplying copies of original articles for the GEODEX SYSTEM/S only.
● Contact: Geodex International Inc, 669 Broadway, PO Box 279, Sonoma, CA 95476, USA.

Hampshire County Library
An example of a public-library system with major service points in Winchester, Southampton, Portsmouth, Basingstoke and Farnborough,

providing a comprehensive reference and information service within the county by manual and online resources; strong in commercial and technical information. A full back-up service of loans, photocopies and personal attention is available, both as part of the public-library service (HANTSLINE), and as the headquarters of HATRICS, the local, regional information cooperative.

• Contact (for the Hampshire area): Central Reference and Information Library, Hampshire County Library Headquarters, 81 North Walls, Winchester, Hampshire SO23 8BY (tel. Winchester 0962 60644).

IEE/INSPEC Literature Search Service
Via the IEE (Institution of Electrical Engineers) Library, this offers online and manual services backed up by a lending and copy service for documents in the IEE Library, and via the British Library Lending Division.

• Contact: Library Information Service, Institution of Electrical Engineers, Savoy Place, London WC2R 0BL (tel. 01-240 1871).

INSPEC
Apart from producing its own database and abstracting and indexing publications, INSPEC offers various current-awareness services, such as SDI, which is a custom-tailored weekly service covering the *INSPEC* database and providing output on cards; TOPICS, which is a weekly standard profile version of SDI, covering about 90 different topics; and *Key Abstracts*, see p. 287.

• Contact: INSPEC Marketing, Station House, Nightingale Road, Hitchin, Herts SG5 1RJ (tel. 0462 5331).

Institute for Scientific Information (ISI)
Well known for its *Current Contents* and citation indexes, ISI also provides a number of other services. These include the weekly SDI services ASCA, using custom-tailored profiles, and ASCATOPICS, which uses standard profiles on some 300 topics, both covering the citation indexes and *Current Contents*. ISI also provide ANSA (Automatic New Structure Alert), a monthly customized SDI service based on the Index Chemicus Registry System. All these services are supported by The Genuine Article (formerly OATS), ISI's document delivery system, which often supplies original tearsheets from the actual journals. ISI now also supplies software: Sci-Mate offers menu-driven searching and downloading of a wide variety of online databases, and the creation and management of text files. The ISI Search Service is an online searching bureau service covering ISI's own databases and many others. See also **Current awareness** (p. 285) and **Abstracting and indexing journals** (p. 52).

• Contact: ISI European Branch, 132 High Street, Uxbridge, Middlesex UB8 1DP (tel. Uxbridge 0895 70016), or ISI 3501 Market Street, Philadelphia, PA 19104, USA.

London Researchers
An information-broker division of Alan Armstrong and Associates Ltd. The following services are available: online computer searching, enquiries, desk research, referrals, bibliographic verification, and document tracing and supply.

● Contact: Mary Ann Colyer, London Researchers, 76 Park Road, London NW1 4SH (tel. 01-723 8530 or 01-258 3740).

London University, CIS

An online broker service specializing in biomedicine, marketing, social sciences, political sciences, and humanities, and offering back-up advisory and online training courses.
● Contact: CIS, University of London, Senate House, Malet Street, London WC1E 7HU (tel. 01-636 8000, ext. 5089).

Metals Information

A joint service of the Institute of Metals, London, and the American Society for Metals. Online searches can be conducted (set up over the telephone if required), mainly on *METADEX, WORLD ALUMINIUM ABSTRACTS*, and *METALS DATAFILE*, although other databases may be searched if required. The service is backed up by the provision of high-speed original document copies if required. Translations of many of the original articles are also available. A series of specialized subject bibliographies, updated annually, and the current-awareness publications *Steels Alert, Nonferrous Alert, Polymers/ Ceramics/Composites Alert* and *Iron & Steel Industry Profiles* are also available.
● Contact: The Institute of Metals, 1 Carlton House Terrace, London SW1Y 5DB (tel. 01-839 4071), or the American Society for Metals, Metals Park, Ohio 44073, USA.

Microinfo Limited

A publisher, agent and service supplier, and especially good for US documentation. All subjects are covered, with specialities in micrographics, video, energy, pollution, health and safety, biotechnology, computer software, cancer and pesticides. Microinfo markets newsletters, American standards, Canadian government literature and Canadian company financial reports, World Bank, International Monetary Fund, JICST (Japanese), US Government Printing Office, National Technical Information Service (NTIS, USA), publications and services. These include *US Government Weekly Newsletter* (a series of 28 different subject publications), reports, published searches, monthly *Tech Notes* (a series of 10 different subject compilations of technological innovations), foreign translations, international market data, and *Innovations Digest*. Their services include custom searches; retrospective online searches of the *NTIS* database, plus back-up ordering facilities; SRIM (Selected Research in Microfiche), a current-awareness service covering fortnightly mailings of full-text microfiche copies of reports entering NTIS in practically any subject area.
● Contact (except for North America): Microinfo Limited, PO Box 3, Newman Lane, Alton, Hants GU2 PG (tel. 0420 86848).

National Physical Laboratory (NPL)

As the UK's national standards laboratory, the NPL maintains and develops the national measurement standards for all physical quantities, provides essential calibration services for their dissemination, fosters good metrological practice and carries out research programmes necessary to support these activities. It is responsible for ensuring that the UK's national measurement system conforms with other countries with whom the UK trades and collaborates scientifically, and it provides three specific services. (1) The

British Calibration Service (BCS) accredits laboratories as competent to provide authenticated and traceable calibrations of measuring instruments, gauges and reference standards. About 120 laboratories, mainly in the private sector, are covered and may issue official BCS calibration certificates. (2) The National Testing Laboratory Accreditation Scheme (NATLAS) accredits laboratories for general testing in much the same way as BCS does for calibration. There are some 240 accredited laboratories covering a wide range of testing services including mechanical, physical, chemical and electrical tests. (3) The National Corrosion Service (NCS) seeks to help industry reduce its losses due to corrosion, by the provision of information and consultation. The National Corrosion Coordination Centre, operated by the NCS, undertakes projects on special corrosion problems of interest to companies on a shared-cost basis.

In addition, the Laboratory undertakes research on key technologies and promotes their uptake by UK industry. At present, it has programmes on engineering materials and information technology. Free advice is offered where practicable, and consultancies, research investigations and calibration services are available on a repayment basis.

● Contact: National Physical Laboratory, Teddington, Middlesex TW11 0LW (tel. 01-977 3222).

Overseas Technical Information Unit (OTIU)

A service to help UK industry and government departments keep in touch with significant technological developments overseas. The OTIU is part of the Department of Trade and Industry and acts as a focal point for the activities of UK science and technology staff stationed in British Embassies in Bonn, Moscow, Paris, Peking, Tokyo and Washington. All are qualified engineers and scientists able to report on a wide cross-section of topics including industrial automation and robotics, microelectronics, information technology, biotechnology, energy technology and pollution control. Reports on these topics and others are disseminated free of charge. During visits to the UK, the OTIU arranges for counsellors to give seminars and meet senior industrialists for individual briefing.

● Contact: Overseas Technical Information Unit, Department of Trade and Industry, Ashdown House, 123 Victoria Street, London SW1E 6RB (tel. 01-212 0449 or 5678).

PERA

One of Europe's largest independent technology centres, with over 400 staff providing research and development, consultancy, advisory, training and information services to manufacturers and other firms. Department of Trade and Industry schemes of consultancy, advisory and practical aid to industry, operated by PERA, are the Manufacturing Advisory Service, the Technical Enquiry Service for small firms, and the Quality Assurance Advisory Service.

● Contact: PERA, Melton Mowbray, Leicestershire LE13 0PB (tel. Melton Mowbray 0664 64133).

Piper Andrews Consultants

This company offers an enquiry service covering scientific instruments, apparatus, equipment, materials, chemicals, and their suppliers, and specializes in biochemical laboratory requirements.

● Contact: Piper Andrews Consultants Ltd, 37 Mildmay Grove, London N1 4RH (tel. 01-354 2601).

Royal Society of Chemistry

This professional body offers a computer-based postal search service. Searching covers identifying Registry Numbers from given molecular formulae/structures and names; the *TOXLINE* database for toxicology and related literature; the CA file on CAS ONLINE and/or *BIOSIS PREVIEWS* (and possibly other files on request); and chemical substructures in the CAS ONLINE Registry File. In the last case, monthly update searches and a crossover facility for transferring the results from a substructure search in the Registry File to the CA File are also available. An individual customized current awareness service is available, covering *CA SEARCH* and/or *BIOSIS PREVIEWS*, with output on cards or paper, and with or without abstracts. Both services offer more sophisticated search strategies than are normally available with *CA SEARCH* through the usual spinners. See also **Current awareness** (p. 287). The Society also acts as an agent for the Chemical Abstracts Service publications, microforms, tapes and facilities. An online search service is also available from the Society's library in London.
● Contact: Royal Society of Chemistry, The University, Nottingham NG7 2RD (tel. Nottingham 0602 507411).

Rubber and Plastics Research Association (RAPRA)

The International Technical Centre for Rubber and Plastics offers an online search service covering all subjects but specializing in science and technology, and especially RAPRA's own field. The RAPRA library contains the world's most comprehensive collection of data on the rubber and plastics industries, and there is a back-up loan and photocopy service for documents cited in RAPRA Abstracts/database.
● Contact: Rubber and Plastics Research Association of Great Britain, Shawbury, Shrewsbury, Shropshire SY4 4NR (tel. Shawbury 0939 250383).

Science Reference Library's Computer Search Service

This is a wide-ranging online information broker service, supported by a very large library collection and a photocopy service.
● Contact: Science Reference Library, 25 Southampton Buildings, Chancery Lane, London WC2A 1AW (tel. 01-405 8721, ext. 3371).

Scientific Documentation Centre Ltd (SDC)

A service specializing in current awareness in 1400 different subject areas with weekly or monthly sets of cards, and retrospective searches via SDC's own large databank or traditional literature sources. Most subjects and types of literature are covered, and charges are very competitive.
● Contact: Scientific Documentation Centre Ltd, Halbeath House, Dunfermline, Fife, Scotland (tel. 0383 723535).

Techsearch

A service of IRS-DIALTECH (p. 154) which is a bureau service to ESA-IRS, for those people unable to establish their own online connection.

Vital Information

A specialist in handling life-sciences information, this company provides a wide range of services from carrying out hard-copy and online searches and

SDIs, to consultancies and setting up in-house information-retrieval systems and databases.
- Contact: Vital Information Limited, 51 George Street, Berkhamsted, Hertfordshire HP4 2EQ (tel. Berkhamsted 04427 6693).

Regional services, cooperatives and networks

In addition to the usual library networks, there are various regional services operating in the UK which are usually assisted by local councils, such as HERTIS (Hertfordshire Technical Library and Information Service) and HATRICS (Hampshire Technical Research Industrial Commercial Service). The services are usually available to organizations which become members by paying a small annual subscription. Details of these services are often available from your local public library, university, polytechnic or technical college, as well as in the appropriate guides.

Council for Small Industries in Rural Areas (CoSIRA)

This is an agency of the Development Commission, whose objective is to revitalize country areas (towns/villages with a catchment area of less than 10 000 people) by helping small rural firms to become more prosperous. It works closely with the Small Firms Service (see below), but specializes in its own narrower area. The help available includes local information and advice, technical and professional advice, business-management and financial advice, finance, grants and training. There are over 30 country offices throughout the UK.
- Contact: CoSIRA, 141 Castle Street, Salisbury, Wiltshire SP1 3TP (tel. Salisbury 0722 336255).

Departments of Trade and Industry (DTI) and Employment

The government operates a large number of schemes and services for UK companies. These range from providing various kinds of advice and information for companies, through to providing financial assistance with starting up a company, market research, innovation and capital investment. The range of services is wide and may, at first sight, appear complicated. A good starting point for a small company, or someone who is thinking of starting up a business, would be the nearest Small Firms Centre. These are easily reached by dialling 100 and asking the operator for Freephone Enterprise. Medium-sized and larger companies will often find that their best initial point of contact is their nearest regional office. A number of schemes are operated nationally with contact points at headquarters buildings in London. A useful directory is published periodically as a supplement to *British Business*, called *Guide to Industrial Support: British business guide to government schemes and services*, which includes sections on technical advice, information and services, and export services. The two main regional services offered are:

(1) The Regional Network of the Department of Trade and Industry. The regional offices represent the department in their dealings with industry, local authorities, the regional offices of other government departments, and other local bodies and organizations. They are responsible for administering selective assistance under sections 7 and 8 of the Industrial Development Act 1982, and provide export advisory services. They are located in Birmingham, Bristol, Leeds, Liverpool, London, Manchester, Newcastle and Nottingham. The telephone number for the London office is 01-730 4678.

(2) Small Firms Service. This provides information and advice to existing and prospective small businesses. Information is available through a network of 12 regional centres; in addition, confidential and impartial business advice is provided by over 200 Counsellors working throughout the country. By asking for Freephone Enterprise, enquirers will be connected to their nearest centre. These are located at Birmingham, Bristol, Cambridge, Cardiff, Glasgow, Leeds, Liverpool, London, Manchester, Newcastle, Nottingham and Reading. Provision of information is free, as are the first three counselling sessions. Various booklets are available, such as *Starting Your Own Business: the practical steps*, and *How to Make Your Business Grow: a practical guide to Government schemes*.

See also Overseas Technical Information Unit (p. 179), IRS-Dialtech (pp. 154 and 180) and TechAlert (p. 290).

INTERLAB

This is a free-membership cooperation scheme for contact between all kinds of laboratories and research and development departments, which is also organized by the Department of Trade and Industry. Cooperation may include the location of specific experimental facilities; purchase of specialized equipment; guidance and information on research techniques; the location of supplies of materials and equipment; and many other kinds of help and advice. Enquiries are made via a directory issued to all members, or via the regional INTERLAB Secretaries.

● Contact: Department of Trade and Industry, South Eastern Region, Room 219, Ebury Bridge House, Ebury Bridge Road, London SW1W 8QD (tel. 01-730 4678, ext. 517).

University Directors of Industrial Liaison (UDIL)

This group was established to promote understanding and collaboration between industry and the universities. It publishes the *Directory of University/ Industrial Liaison Services*, 6th ed. by Brunel Industrial Services Bureau, Brunel University. A more thorough service will be offered as an online database (*BRITISH EXPERTISE IN SCIENCE AND TECHNOLOGY*) including detailed information about research, expertise and services in universities and polytechnics.

● Contact (for database): Mr M. T. Tobert, Longman Cartermill, PO Box 33, The Technology Centre, North Haugh, St Andrews, Fife KY16 9EA (tel. 0344 77660).

Videotex services

Videotex or videotext is usually taken to mean the information services available through a domestic-type television receiver, and includes teletext and viewdata.

Teletext is the name given to the broadcast services, such as BBC's CEEFAX and ITV's ORACLE (not to be confused with teletex, p. 185). Pages of text (about 800 altogether) are transmitted alongside the normal television programmes, and although the service is free it requires the purchase or renting of a special teletext TV set. CEEFAX and ORACLE provide mainly news, and are especially good for national, international, business and financial news. The main usefulness of teletext is its currency:

pages can be updated every hour or so if necessary. The service may expand to offer more local news, educational services, and transmissions for home computers, such as business or educational computer programmes (BBC's Telesoftware for example).

Viewdata is the name used for videotex services offered via the telephone network, such as Prestel. This system is much more like the usual online information-retrieval terminal, with about 250000 pages of text available in the public Prestel service. Some of these pages or frames are free, but others are subject to a frame charge. The main differences between viewdata and teletext are that viewdata is an interactive system which also contains a greater variety of information, sometimes in greater depth, and that most of the information is input and updated by the information providers, who can therefore offer a truly commercial information or advertising operation. Apart from this public facility, there is also a private Prestel service for setting up closed user groups. Other facilities include Prestel MAILBOX and TELEX LINK. A special receiver is required, different to the teletext equipment; this may be a Prestel TV, an ordinary TV with an adaptor, or a microcomputer with a Prestel interface. The service is always available, and includes a response frame or reply facility (almost a combined information and mailbox service).

The amount of information currently available from both types of public videotex service is limited as far as scientists and engineers are concerned. Possibly the arrival of cable TV and 'intelligent' television receivers and telephones may result in a boosting and an expansion of the services. Prestel, however, may be used as a gateway to other services or computers, via Prestel Gateway, which permits full interactive communication. Information providers may offer a service through public or private Prestel, as well as through Prestel Gateway. Examples of services using Prestel include the following.

● Prestel Microcomputing is a gateway service for microcomputer owners (home or business), and is accessed using Micronet 800 or Viewfax 258, subsidiary services which offer all the necessary hardware and software. Micronet 800 provides its members with ordinary Prestel services plus information, software, discount ordering, news, enquiries, and mailbox facilities. (Micronet even offers an intermediary online bureau service for its members via other established bureaux.)

● LAWTEL is a closed-user-group service providing a current synopsis of all significant legal developments in England and Wales, for solicitors, barristers and other workers requiring legal information.

● The Teleordering Viewdata Service was developed to allow small booksellers cheaper but more limited access to the computer-based book-ordering Teleordering system.

● Farmlink is a comprehensive farm management information service.

Information transmission systems

Although not technically true information services, these are destined to become so much a part of the information supply system that it is essential to be at least aware of their existence.

Electronic mail

This service allows anyone with the type of terminal equipment used for online services to compose, edit, send, receive, store, reply to, and file electronic messages. There need never be paper copies, although it is always possible to produce them if desired. The features may include instant despatch of messages, simultaneous transmission of one message to several destinations, shared information or news files for closed user groups, overseas communications, and mailbox facilities, so that incoming messages are stored by the service until you log on and receive an appropriate alert. Some of the online information service hosts provide this facility, and at least one has introduced an experimental 'voicemail' to cover verbal messages. In the UK, British Telecom's Telecom Gold is a good example, with all the usual facilities and also access to BT's Telex, Radiopaging and Telemessaging services. Connection to Telecom Gold is via direct-dial to London or via PSS. BT's Merlin have just launched their VM600 Voice Mail system.

● Contact: Telecom Gold Limited, Capital House, 42 Weston Street, London SE1 3QD (tel. 01-403 6777; Merlin is on 01-631 2122).

Facsimile transmission (fax, telefax or telefacsimile)

This is a system for transmitting copies of whole pages over the telephone networks, or over private wires. The system works by using a line-scanning method somewhat similar to domestic television transmissions. There must be fax machines at both transmitting and receiving ends, and they must be compatible. Fax machines are known by their group or generation number. Groups 1 and 2 are relatively slow, ranging from 6 to 2 minutes to transmit one page of A4, but the machines in the two groups may be compatible with each other. Group 3 machines take about 20 to 40 seconds to transmit a page, depending on printed line density, and are compatible with Group 2 and, sometimes, Group 1 machines also, by means of a 'handshaking' process automatic in the higher-level machine.

Documents have to be fed into the machines as single sheets, which means that pages from bound volumes must be photocopied first. This could create copyright problems in certain cases. Transmission may be automatic (in the case of machines fitted with automatic reception) or manual, commencing with an ordinary telephone call and then with operators at both ends switching over to the fax machines. Generally, only the sending machine requires the presence of an operator during transmission, and even this is not necessary if the other machine is 'calling off' documents which have been placed ready for it on the sending machine (a technique known as 'polling'). The change from the ordinary telephone (PTT) tone to a digital tone on dialling a number is automatic, as for online work, but at the end of the fax transmission it is possible to revert to PTT for voice conversation without breaking the line.

Perhaps the most important thing to stress about fax is that it is, at present, the only digital medium which will handle pictures, graphics, photographs, handwriting, etc., through the telephone network. It therefore has a unique position for supportive material compared to online or telex.

Users receive one free entry in the national directory, and one free copy of the directory for each of their entries listed. The system may be used for transmission overseas and there are appropriate directories available. There are various bureaux services available for people without equipment, such as

British Telecom's Bureaufax. For further details, see the telephone numbers given near the beginning of your telephone directory, or contact your local telephone offices.

Telex and teletex

Most people are familiar with the traditional slow, large, typewriter-like telex machines, which use transmission paths similar to but separate from the public telephone network. Textual information (includes numeric) can be transmitted as it is typed in, or composed offline and then sent in one batch, to many parts of the world via the telex service. More sophisticated terminals are beginning to appear and the telex network is being modernized. The advantage over the postal service is that it is fast and messages can be acknowledged quickly. The advantage over the telephone service is that the sender can check what is sent before it goes, and both the sender and the recipient have permanent records of what is transmitted. The disadvantage compared to ordinary online systems is that telex is not fully interactive in the 'immediate' sense. National and international telex directories are available in the same way as telephone directories, and many of the directories to organizations give telex numbers.

Teletex is a new high-speed text communications service from British Telecom (BT), which permits high-speed transmission and the use of more sophisticated equipment (special teletex terminals or certain other equipment with appropriate conversions or conversion devices connected). The terminals offer all the advantages of electronic processing and storage of data. Teletex makes use of both the Public Switched Telephone Network (PSTN) and the Packet SwitchStream (PSS) networks. To send a message from a teletex terminal on PSS to one on PSTN, a BT gateway service called Interstream Two is used. To send messages between teletex terminals on either PSTN or PSS to telex installations, another BT gateway service is used called Interstream Three. Interstream One is a further BT gateway service which allows PSS terminals to call inland telex numbers, and telex machines to call PSS dataline customers (see p. 165). Teletex is currently a terminal-to-terminal, non-interactive electronic mail service, although some terminals will be able to use the more traditional communication routes as separate, additional facilities. It may also be possible to adapt existing equipment for use as a teletex terminal by employing appropriate software or 'black box' hardware.

● Contact for teletex: Teletex Marketing, 5th Floor, Cheyne House, Crown Court, Cheapside, London EC2V 6BE (tel. 01-236 5516 or Freephone 2789).

Electronic document delivery

A service allied to electronic publishing which should speed up the delivery of the full text information in documents, for example, after the online retrieval of references. The service will probably use satellite data transmission systems between service centres, which can operate some 40 times faster than land-line networks. The actual delivery to the end user may be via online, fax, or local mail delivery.

Electronic publishing

Currently a somewhat vague term used for any form of production, publication or distribution of information via electronic as opposed to hard-copy forms. These processes are still in their infancy, and are presently taken to include electronic mail, word processing, online information retrieval, videotex and even the sale of software in the High Street. Electronic publishing will almost certainly become a significant if not the principal method of publishing in the future.

Guides to information services and suppliers

As many of these services are still relatively new and undergoing frequent changes and expansions, there are no good guides which cover all the aspects of all the types of service. Online services are covered quite well, but because of constant changes it is the directory type of guide which is published on a regular basis that is the most useful. As far as individual services or databases are concerned, service suppliers or hosts or database producers normally offer the best guidance via their manuals and training sessions. In the UK, Aslib (and especially its Online Information Centre, p. 175), large public, national, regional or academic libraries or information services may be helpful. In the US, the Special Libraries Association (SLA) may be contacted at 1700 18th Street NW, Near Dupont Circle, Washington DC 20009, USA.

Uses
- To find details of services which cover your subject field or interest.
- To find details of databases covering a particular subject, application, or type of literature.
- To find which hosts offer which databases and vice versa.
- To find out how services work and how to use them.

Access

Via libraries or information services.

Caution

- These guides become out of date very quickly, owing to the constant changes in many services and their charges. It is always necessary to obtain the latest details from the appropriate supplier.
- General guides for online services are not usually a substitute for the manuals, guides and training offered by the hosts and database producers.
- Due to the constant changes, the stream of newsletters and manual updates is difficult to keep up with and to organize.

Examples

As already suggested, the guides such as manuals, newsletters, quick-reference guides and thesauri which are supplied by hosts and database producers are the most important for actually using online services.

Directory of Online Databases. Santa Monica, Cuadra Associates Inc, 1979–   .
○ Published quarterly, this guide consists of two directory issues and two update supplements. Coverage includes about 2600 bibliographic, numeric,

referral, and full-text databases. The main section consists of an alphabetical list of databases including details such as type, subject and content, producer, online service supplier, language, time span, and updating information. There is a separate list of service suppliers and database producers giving addresses and telephone numbers. The separate indexes cover subject, database suppliers, online services and telecommunications, and a master index covers all the organizations and their products or services. Also available online as *CUADRA DIRECTORY OF DATABASES*.

Computer-Readable Databases: a directory and data sourcebook, by M. E. Williams, L. Lannom and C. G. Robins. Illinois, American Society for Information Science, 1982.
○ The directory covers some 773 databases which are publicly available, word-orientated and so mainly bibliographic. The information given is similar to that in the previous guide, but goes into considerably more detail; for example, it lists the data elements present in each record and describes the subject analysis and indexing data. It does not appear to be comprehensive in its coverage of the more popular databases in the UK. A new edition covering about 2800 databases is to be published in two volumes by ALA Publishing Services of Chicago. Also available online as *DATABASE OF DATABASES*.

Directory of Periodicals Online: indexed, abstracted & full-text, edited by C. Chung. 3 vols. Washington DC, Federal Document Retrieval Inc., announced for 1985–86.
○ Potentially useful in an area likely to become increasingly important.

Directory of Data Bases and Data Banks. Luxembourg, Euronet Diane, March 1984 (irregularly updated).
○ A guide to European services, giving an alphabetical list of broad subject headings with relevant databases under each one; two lists of hosts, one giving the names of databases offered, the other giving contact details; and one list of databases giving brief details of subject coverage and names of appropriate hosts. An online database version is available on ECHO as *DIANEGUIDE*.

Datapro Directory of Online Services. Delran, USA, Datapro Research Corporation, 1983–   .
○ This directory is composed of two parts: two large looseleaf binders of directory-type information, including company profiles, databases, and online services; and a current-awareness journal which can be filed in the back of one of the binders. The coverage is probably poor for UK or European organizations, but gives a great deal of information about US and US-related organizations.

The North American Online Directory 1985. New York, Bowker, announced for 1985.
○ A potentially useful guide for the online industry and users in North America, including coverage of networks, consortia, centres and brokers.

EUSIDIC Database Guide 1983. Oxford, Learned Information, 1983.
○ The guide is an international directory but with a European emphasis. The main section consists of an alphabetical list of database producers and operators (over 900 organizations). Where available, information given

includes address, telephone number, name of contact and databases covered (including year started and subject coverage). There are separate indexes to organizations and databases (and their abbreviations), broad subject headings, and networks. While this guide appears to be up to date and comprehensive, it is not too easy to find what you want. The subject index is too broad, and it is not possible, for example, to turn to a separate list of information brokers.

Online Bibliographic Databases, by J. L. Hall and M. J. Brown. 3rd ed. London, Aslib, 1983.
○ This work is intended as a quick-reference directory to principal English-language bibliographic databases readily available to the European user. The main section consists of an alphabetical list of databases including details of the supplier, subject fields, time span, printed equivalent, names of hosts supplying access and costs, and details of supporting guides and thesauri. Sample records are included. There are various additional guides and lists, including subject keyword and subject grouping listings. Although limited in its coverage (179 databases and excluding some specialized databases such as those devoted to patents) it is a practical and usable guide.

Data Base Directory. White Plains, NY, Knowledge Industry Publications, annually.
○ This directory covers more than 1800 numeric, full-text, textual/numeric, property, bibliographic and referral databases available in North America, giving full descriptions and details of availability. There are indexes to subjects, producers and vendors. Available online with the same name.

UK Online Search Services, compiled by J. B. Deunette. 2nd ed. London, Aslib, 1982.
○ This directory lists intermediary or broker services (over 100 organizations) in the UK which offer existing or planned online search services. The services are arranged in alphabetical order and include details of availability, coverage, charges, speed of service, number of searches performed per year, number of staff, and back-up services such as loans and photocopying. There are additional indexes covering hosts accessed, general subject specializations, and geographical locations. A very practical guide for people in the UK without their own online facilities (or temporarily without them), or without access to certain hosts.

Online Database Search Services Directory: a reference and referral guide to libraries, information firms, and other sources providing computerized information retrieval and associated services using publicly available online databases, edited by J. Schmittroth and D. M. Maxfield. 2 vols. Detroit, Michigan, Gale Research Company, 1983–84.
○ The directory covers services offered by some public, academic and special libraries, private information firms and other organizations in the US and Canada. It generally excludes bodies whose services are not available in any form to outside users. Over 1100 entries, arranged alphabetically by organizations within each volume, include details such as addresses and telephone numbers, names of key contacts and search personnel, online systems accessed, subject areas searched, databases frequently used, availability of services, fee policies, and search request procedures. There are

indexes to organizations by name, by online systems accessed, by databases searched, and by subject areas searched (under rather broad headings); search personnel and geographic indexes are also provided. Cumulative versions of all indexes appear in the second volume.

Going Online. London, Online Information Centre, 1984.
○ A first-class guide for people considering going online or actually just starting to use online. Apart from trying to answer the most obvious questions, there are also helpful costings, lists of addresses for hosts, networks, user groups (local, UK and European), and reference sources including directories, textbooks, articles, journals and newsletters. The Centre has also published separate subject guides to medical, law and patents databases, and is in the process of producing guides to building/construction/architecture and news databases, and a potentially very useful guide to selecting equipment.

Comparative Cost Chart: for online files, by M. Woodrow. Hertford, Hertfordshire Library Service, quarterly.
○ A computer printout showing the prices for online access and printing, for a large number of databases on over a dozen hosts. A very useful quick-reference guide and costing tool.

Prompt Cards for Online Search Languages, by M. Henry and M. Young. London, Macmillan Press, 1980.
○ A useful quick-reference guide to the command languages of BLAISE, BRS, DIALOG, ESA-IRS, Infoline, SDC and DECO-UCSL.

Online International Command Chart. Weston, CT, Online Inc, announced for 1985.
○ This stand-up flip chart should be particularly valuable for people using unfamiliar command languages, as it permits direct comparisons between equivalent commands in different systems. Forty-one online commands are listed for each of sixteen command languages, covering nineteen hosts in North America and Europe.

Online Information Retrieval Systems: an introductory manual to principles and practices, by B. Houghton and J. Convey. 2nd ed. London, Clive Bingley, 1984.
○ The most recent book for the European searcher, with a lot of practical help for using the limited number of systems that it covers.

Computerized Literature Searching: research strategies and databases, by C. L. Gilreath. Boulder, Colorado, Westview Press, 1984.
○ A modern, chatty book which is quite good at describing databases and aids to searching under very broad subject areas.

Online Searching: an introduction, by W. M. Henry and others. London, Butterworths, 1980.
○ One of the best general works, covering most aspects of the field, but inevitably becoming outdated on some of the service details.

Online Searching: a primer, by C. H. Fenichel and T. H. Hogan. New Jersey, Learned Information, 1981.
○ A good introduction but briefer, with a US bias and with less detail of services and databases than the previous book.

Answers Online: your guide to informational databases, by B. Newlin. New York, McGraw-Hill, announced for 1985.
○ This appears to be a practical beginner's guide for North Americans, especially those using microcomputers.

Online Search Strategies, edited by R. E. Hoover. New York, Knowledge Industry Publications, 1982.
○ An excellent guide with plenty of examples divided into broad subject sections.

Online Bibliographic Searching: a learning manual, by C. Chen and S. Schweizer. New York, Neal-Schuman Publishers, 1981.
○ A fairly good introduction with a US bias.

Basics of Online Searching, by C. T. Meadow and P. A. Cochrane. New York, John Wiley & Sons, 1981.
○ A similar book to the previous one.

Easy Access to DIALOG, ORBIT, and BRS, by P. J. Klingensmith and E. E. Duncan. New York, Marcel Dekker, 1984.
○ A useful guide to the three large North American hosts, which considers each type of command and compares its use on each of the three services.

Online Reference and Information Retrieval, by R. C. Palmer. Littleton, CO, Libraries Unlimited, 1983.
○ A suitable guide for North American students covering BRS, DIALOG and SDC, with some good examples.

An Introduction to Online Searching, by Tze-Chung Li. Westport, CT, Greenwood Press, announced for 1985.
○ This book should also interest North American students as it includes sections on searching major US hosts.

An Introduction to Automated Literature Searching, by E. P. Hartner. New York, Marcel Dekker, 1981.
○ A less practical approach that goes deeper into the theory and background of the subject.

Online Searching: the basics, settings & management, by J. H. Lee. Littleton, Colorado, Libraries Unlimited, 1984.
○ This guide gives good coverage of the philosophy, administration and management of online services, especially those in corporations, and public and academic libraries.

Online Terminal/Microcomputer Guide & Directory 1982–83. 3rd ed. Weston, CT, Online, 1982.
○ Potentially a very useful and informative guide despite a US bias, provided it is kept up to date.

British Information Services Not Available Online: a select list, compiled by A. Dewe and M. A. Colyer. London, Aslib, 1980.
○ The list includes commercially available services providing information in response to direct enquiries, but excludes publishers of abstracting and indexing services and suppliers of collections of trade literature, unless they provide some form of back-up service. Research and trade associations are included only if there is no restriction on access apart from preferential rates to

members. Each entry includes address, telephone number, subject or scope, date of first operation, nature of service, sources of information and cost. There are organization and subject indexes. Nearly 60 parent organizations are covered, although the number of different services is nearer 100.

British Scientific Documentation Services. London, Longman (for the British Council), 1976.
○ A small but wide-ranging guide covering nearly 100 organizations and several hundred titles, which still contains useful information despite its age.

Industrial and Related Library and Information Services in the United Kingdom, by J. Burkett. 3rd ed. London, Library Association, 1972.
○ This is a study of the various services rather than a directory, but it is still a useful reference work which includes details of a number of services.

Encyclopedia of Information Systems and Services, by A. T. Kruzas and J. Schmittroth. 5th ed. Detroit, Gale Research Company, 1983.
○ The body of this directory consists of entries for organizations involved with all kinds of information services, in alphabetical order, giving full descriptions of the service. It includes international coverage (with a US bias) of databases; hosts; commercial, educational and governmental organizations; software; telecommunications; brokers; networks; videotex and teletex; and many other products, services and types of organizations. There are over 20 indexes covering such things as names, subjects and locations. A good guide for tracking down the more obscure services, especially in the USA. The enlarged 6th ed. for 1985–86 has been announced; it will be in two volumes (International and United States).

European Sources of Scientific and Technical Information, edited by A. P. Harvey. 6th ed. Harlow, Longman, 1984.
○ This guide covers the key sources of information including national offices, patents and standards offices, and organizations active in science and technology, with identifiable library and information services available to the public. Coverage includes eastern Europe. Entries for each organization are given in 25 subject sections, sub-divided by country. Each entry gives details of the organization and its services. The organizations are indexed by title, keyword and subject.

Information Trade Directory 1983: an international directory of information products and services. Oxford, Learned Information, 1982.
○ A comprehensive guide including coverage of about 40 countries, giving details of database producers, hosts, library and telecommunications networks, information brokers, terminal manufacturers, etc. There are subject classified, name and geographic indexes. (Also published as *Information Industry Market Place* in North and South America by R. R. Bowker Company, New York.)

See also *Document Retrieval: sources and services* (p. 255), *Register of Consulting Scientists: contract organizations and other scientific and technical services* (p. 18), *Directory of Special Libraries and Information Centers* (p. 262), and the various guides to organizations (p. 20, 25, 29, 33), abstracting and indexing services (p. 54), periodicals (p. 97) and libraries (p. 260).

Chart 10. Subject specialities of selected online databases

Appropriate hosts are shown immediately after each database and/or, in the case of eight larger hosts, in the separate display grid. Training files have not been included. The grid also allows you to see which hosts provide good coverage of a specific topic. Some details of hosts are given on pages 149–159. Scope notes are provided where the subject coverage is not too obvious. When you are looking for a suitable heading, remember to try applications and document types as well as the usual subjects. Also check the 'Multidisciplines' heading and any appropriate multidisciplinary topics, such as environmental sciences. An asterisk (*) denotes specific types of information source. The following simplifications have been used:

BLAISE for BLAISE-LINE
BLAISE/NLM for BLAISE-LINK
DATASTAR for DATA-STAR

ESA for ESA-IRS/IRS-DIALTECH
INFOLINE for PERGAMON INFOLINE
TELESYSTEMES for TELESYSTEMES QUESTEL

Summary of headings

Aeronautics/astronautics
Agriculture
Biosciences/pharmacology/biotechnology
*Books
Business/management/finance/economics
Chemical engineering
Chemistry
Civil engineering
Computer science/technology and software
*Conferences/papers/proceedings
Control/robotics
Education
Electrical engineering
Electronics
Energy
Environmental sciences
Food science and technology
Geology/earth sciences and technologies
*Governments/official publications
Health and safety
Information/library science
*Information services
Law
Leisure/recreation/tourism
Marine science/engineering

Materials science/technology
Mathematics
Mechanical engineering
Medicine
Metallurgy
MULTIDISCIPLINES/INTERDISCIPLINES
*News/newspapers
*Organizations
*Patents
*People
*Periodicals
Petroleum industry
Physics
Psychology/psychiatry
*Reference works
*Reports
Research
Social/behavioural sciences
*Standards
Telecommunications
Textiles
*Theses/dissertations
*Translations
Transportation
Water science and technology

Aeronautics/astronautics

Database	BRS	DATASTAR	DIALOG	DIMDI	ESA	INFOLINE	INKA	SDC
AEROSPACE (restricted access) on MEAD DATA CENTRAL &			DIALOG					
AEROSPACE DAILY	BRS	DATASTAR	DIALOG		ESA			
COMPENDEX see p. 146 (engineering)		DATASTAR	DIALOG		ESA	INFOLINE	INKA	
FLUIDEX					ESA			
NASA see p. 147					ESA			
SATELDATA (satellites)					ESA			
SPACECOMPS (spacecraft components)					ESA			

Agriculture see also Biosciences

Database	BRS	DATASTAR	DIALOG	DIMDI	ESA	INFOLINE	INKA	SDC
AGREP (European res.) on DATACENTRALEN &	BRS			DIMDI				
AGRICOLA	BRS		DIALOG					
AGRIS			DIALOG	DIMDI	ESA			
AGROCHEMICAL DATABANK		DATASTAR						
BIOLOGICAL & AGRIC. INDEX on WILSONLINE								
CAB ABSTRACTS see p. 145 (agriculture)			DIALOG	DIMDI	ESA			
COMPENDEX see p. 146 (engineering)	BRS	DATASTAR	DIALOG		ESA	INFOLINE	INKA	SDC
CRIS/USDA (agricultural research)			DIALOG					
FOREST (wood products)								SDC
IALINE (agric./food ind.) on TELESYSTEMES								
TROPAG (tropical agriculture)								SDC

Astronomy/space science see Physics and Aeronautics/astronautics

Automotive engineering see Mechanical engineering

Biosciences/pharmacology/biotechnology

Database	BRS	DATASTAR	DIALOG	DIMDI	ESA	INFOLINE	INKA	SDC
ABDA (drugs)				DIMDI				
ABSTRACTS IN BIOCOMMERCE		DATASTAR						
AQUACULTURE			DIALOG					
AQUATIC SCIENCE & FISHERIES ABS.			DIALOG	DIMDI				
BIOBUSINESS		DATASTAR	DIALOG					
BIOLOGICAL & AGRIC. INDEX on WILSONLINE								
BIOSIS PREVIEWS see p. 145 (biology) on MEAD DATA CENTRAL &	BRS	DATASTAR	DIALOG	DIMDI	ESA			SDC
BIOTECHNOLOGY ABSTRACTS (subscribers only)								
CAB ABSTRACTS see p. 145 (agriculture)			DIALOG	DIMDI				
CHEM. ABS. (p. 146) on STN, TELESYSTEMES &	BRS	DATASTAR	DIALOG		ESA			SDC
CHEMLINE (biomed cherns) on BLAISE/NLM &				DIMDI				
COMPENDEX see p. 146 (engineering)	BRS	DATASTAR	DIALOG		ESA	INFOLINE	INKA	SDC
CONSUMER DRUG INFORMATION FULLTEXT			DIALOG					

Biosciences/pharmacology/biotechnology—continued

	BRS	DATASTAR	DIALOG	DIMDI	ESA	INFOLINE	INKA	SDC
CURRENT AWARENESS IN BIOLOGICAL SCIENCES		DATASTAR				INFOLINE		
CURRENT BIOTECHNOLOGY ABSTRACTS	BRS	DATASTAR	DIALOG		ESA	INFOLINE		
DRUG INFORMATION FULL TEXT				DIMDI				
EUROPEAN BIOTECH. INFO. PROJ. on PRESTEL & INSPEC see p. 146 (physics/elec./comput.)	BRS	DATASTAR	DIALOG	DIMDI	ESA		INKA	SDC
INTERNATIONAL PHARMACEUTICAL ABSTRACTS	BRS	DATASTAR	DIALOG	DIMDI	ESA			
LIFE SCIENCES COLLECTION			DIALOG					
MARTINDALE (drugs)		DATASTAR						
MEDITEC (biomed. eng.) on FIZ TECHNIC & MERCK INDEX (chemicals & drugs) on CIS &	BRS			DIMDI				
PESTDOC (pestic./herbic., subscribers only)								SDC
PHARMACEUTICAL NEWS INDEX			DIALOG					
PHARMAPROJECTS		DATASTAR						
PHARMLINE		DATASTAR						
PHYTOMED				DIMDI				
PIE (Pacific Islands Ecosystems)								SDC
RINGDOC (pharmaceuticals, subscribers only)								SDC
SDF (drugs, subscribers only)								SDC
TELEGEN (genetic engineering)			DIALOG	DIMDI	ESA			
VETDOC (veterinary, subscribers only)								SDC
ZOOLOGICAL RECORD			DIALOG					

*Books

	BRS	DIALOG	DIMDI	WILSONLINE	BLAISE	SDC
BOOK REVIEW DIGEST on WILSONLINE						
BOOK REVIEW INDEX		DIALOG				
BOOKS IN PRINT	BRS	DIALOG				
CUMULATIVE BOOK INDEX on WILSONLINE						
DPB (British Library) on BLAISE						
ISI/STP&B (science & technology)			DIMDI			
LC MARC on BLAISE, WILSONLINE & LC/LINE (US and Canada)		DIALOG				
SUPERINDEX (back of book indexes)	BRS					SDC
UKMARC on BLAISE						
WHITAKER on BLAISE						
WILEY CATALOG/ONLINE		DIALOG				

Botany see Biosciences

Business/management/finance/economics see also Organizations

Database	BRS	DATASTAR	DIALOG	DIMDI	ESA	INFOLINE	INKA	SDC
ABI/INFORM (business) on MEAD DATA CENTRAL &	BRS	DATASTAR	DIALOG		ESA			SDC
ACCOUNTANTS (finance)			DIALOG					SDC
AP NEWS (business)			DIALOG					
BUSINESS PERIODICALS INDEX on WILSONLINE								
BUSINESS RESEARCH SERVICES on DATASTREAM								
ECONOMIC LITERATURE INDEX			DIALOG					
EPCA (chem. ind. trade statistics) on EDS								
FINANCIAL TIMES ABSTRACTS (UK)	BRS	DATASTAR						
FIND/SVP REPORTS & STUDIES (market research)			DIALOG					
FOREIGN TRADE & ECON ABS on BELINDIS &		DATASTAR	DIALOG					
HARVARD BUSINESS REVIEW	BRS	DATASTAR	DIALOG					
INFORMAT BUSINESS DATABASE (news) on MEAD DATA CENTRAL &						INFOLINE		
JORDANWATCH on JORDANS &						INFOLINE		
LABORDOC (labour)								SDC
MANAGEMENT AND MARKETING ABSTRACTS						INFOLINE		
MANAGEMENT CONTENTS on MEAD DATA CENTRAL &	BRS	DATASTAR	DIALOG					SDC
PAIS INTERNATIONAL (public affairs)	BRS	DATASTAR	DIALOG					
PTS PROMPT (marketing and technology)	BRS	DATASTAR	DIALOG					
SOCIAL SCISEARCH (citation index)	BRS		DIALOG	DIMDI				
TEXTLINE and NEWSLINE on FINSBURY DATA &					ESA			
TRANSDOC (transport economics)					ESA			
TRADSTAT (world trade statistics) on EDS								
WORLD EXPORTER/REPORTER on DATASOLVE								

Chemical engineering

Database	BRS	DATASTAR	DIALOG	ESA	INFOLINE	INKA	SDC
CHEMICAL ABSTRACTS on STN, QUESTEL &	BRS	DATASTAR	DIALOG	ESA			SDC
CHEMICAL BUSINESS NEWSBASE on EDS &		DATASTAR	DIALOG				
CHEMICAL ENGINEERING ABSTRACTS		DATASTAR		ESA	INFOLINE		
CHEMICAL HAZARDS IN INDUSTRY		DATASTAR			INFOLINE		
CHEMICAL INDUSTRY NOTES (CIN)		DATASTAR	DIALOG				
COMPENDEX see p. 146 (engineering)	BRS	DATASTAR	DIALOG	ESA	INFOLINE	INKA	SDC
CORROSION							SDC
DECHEMA (chemical eng.) on FIZ TECHNIK							
EAST EUROPEAN CHEMICAL MONITOR		DATASTAR					
KIRK-OTHMER ENCYCLOPEDIA OF CHEM. TECH.	BRS	DATASTAR					

Chemistry

Database	BRS	DATASTAR	DIALOG	DIMDI	ESA	INFOLINE	INKA	SDC
AGROCHEMICAL DATABANK		DATASTAR						
AMERICAN CHEM. SOC. JOURNALS	BRS							
ANALYTICAL ABSTRACTS		DATASTAR				INFOLINE		
CAS REGISTRY FILE on **STN** and **TELESYSTEMES**								
CHEM. ABS. (p. 146) on **STN, TELESYSTEMES &**	BRS	DATASTAR	DIALOG		ESA			SDC
CHEMICAL HAZARDS IN INDUSTRY		DATASTAR				INFOLINE		
CHEMICAL REGULATIONS & GUIDELINES (US)			DIALOG					
CHEMLINE (biomed chems) on **BLAISE/NLM &**				DIMDI				
CHEMNAME/CHEMDEX (CAS chem. nomenclature)		DATASTAR	DIALOG					SDC
CORROSION								SDC
CRDS (chemical reactions/syntheses)								SDC
ECDIN (Europ. chem. data) on **DATACENTRALEN**								
FINE CHEMICALS DIRECTORY						INFOLINE		
IFP-TH (physico-chem. data) on **TELESYSTEMES**								
INDEX CHEMICUS ONLINE on **TELESYSTEMES**								
KIRK-OTHMER ENCYCLOPEDIA OF CHEM. TECH.	BRS	DATASTAR			ESA	INFOLINE		
LABORATORY HAZARDS BULLETIN		DATASTAR			ESA	INFOLINE		
MASS SPECTROMETRY BULLETIN/MASSLIT					ESA	INFOLINE		
MERCK INDEX (chemicals and drugs) on **CIS &**	BRS				ESA	INFOLINE		
PESTICIDES MANUAL								
TSCA INITIAL INVENTORY (chemicals—toxicity)			DIALOG		ESA			SDC
TSCA PLUS (chemicals—toxicity)								
WORLD SURFACE COATINGS ABSTRACTS					ESA	INFOLINE		
ZINC LEAD CADMIUM ABSTRACTS					ESA	INFOLINE		

Civil engineering

Database	BRS	DATASTAR	DIALOG	DIMDI	ESA	INFOLINE	INKA	SDC
ACOMPLINE (GLC Library) on **SCICON &**					ESA			
BRIX (building research)					ESA			
COLD (regions)								SDC
COMPENDEX see p. 146 (engineering)	BRS	DATASTAR	DIALOG		ESA	INFOLINE	INKA	SDC
DELFT HYDRO (fluid mechanics)					ESA			
FLUIDEX			DIALOG		ESA	INFOLINE		
IBSEDEX (building services)					ESA	INFOLINE		
IRRD (road research)					ESA			
PICA (construction and architecture)					ESA	INFOLINE		
URBALINE (GLC news) on **SCICON**					ESA	INFOLINE		

Companies see Organizations and Business

Computer science/technology and software

Database	BRS	DATASTAR	DIALOG	DIMDI	ESA	INFOLINE	INKA	SDC
ABI SOFT (business)								
BUSINESS SOFTWARE	BRS	DATASTAR	DIALOG		ESA			
COMPENDEX see p. 146 (engineering)	BRS	DATASTAR	DIALOG		ESA	INFOLINE	INKA	SDC
COMPUTER DATABASE on MEAD DATA CENTRAL &		DATASTAR	DIALOG		ESA			
COMPUTERPAT						INFOLINE		
COSMIC (aerospace software)	BRS				ESA			
DATA PROCESSING & INFORMATION SCIENCE	BRS							
INSPEC see p. 146 (physics/elec./comput.)	BRS	DATASTAR	DIALOG		ESA		INKA	SDC
.MENU (international software)			DIALOG					
MICROCOMPUTER INDEX			DIALOG					
MICROSEARCH								SDC
RESOURCES IN COMPUTER EDUCATION	BRS							
SOFT (micro software)					ESA			
ZDE (like INSPEC) on FIZ TECHNIK								

*Conferences/papers/proceedings

Database	BRS	DATASTAR	DIALOG	DIMDI	ESA	INFOLINE	INKA	SDC
COMPENDEX see p. 146 (engineering)	BRS	DATASTAR	DIALOG		ESA	INFOLINE	INKA	SDC
CONF (conferences)							INKA	
CONFERENCE PAPERS INDEX			DIALOG		ESA			
CONFERENCE PROCEEDINGS INDEX on BLAISE								
EI ENGINEERING MEETINGS/EI MET		DATASTAR	DIALOG		ESA			SDC
ISI/ISTP&B (science & technology)				DIMDI				
SIGLE (grey literature) on BLAISE &							INKA	

Control/robotics

Database	BRS	DATASTAR	DIALOG	DIMDI	ESA	INFOLINE	INKA	SDC
COMPENDEX see p. 146 (engineering)	BRS	DATASTAR	DIALOG		ESA	INFOLINE	INKA	SDC
INSPEC see p. 146 (physics/elec./comput.)	BRS	DATASTAR	DIALOG		ESA		INKA	SDC
ISMEC (mechanical)			DIALOG		ESA			
ROBODATA on DATASOLVE								
ROBOMATIX (robotics)								
ROBOTICS INFORMATION	BRS				ESA			
ZDE (like INSPEC) on FIZ TECHNIK								

Dissertations see Theses

Earth sciences see Geology

Economics/finance see Business

Education see also Psychology and Social sciences

BRITISH EDUCATION INDEX on **BLAISE**
EDUCATION INDEX on **WILSONLINE**

Database	BRS	DATASTAR	DIALOG	DIMDI	ESA	INFOLINE	INKA	SDC
ERIC (education)	BRS		DIALOG					SDC
EUDISED (European education R&D)	BRS				ESA			
EXCEPTIONAL CHILD EDUCATION RESOURCES			DIALOG					
LANGUAGE AND LANGUAGE BEHAVIOR ABS			DIALOG					
RESOURCES IN COMPUTER EDUCATION	BRS							
PSYC/INFO see p. 147	BRS	DATASTAR	DIALOG	DIMDI				SDC

Electrical engineering see also Energy

Database	BRS	DATASTAR	DIALOG	DIMDI	ESA	INFOLINE	INKA	SDC
AMPEREDOC (electrical engineering)					ESA			
COMPENDEX see p. 146 (engineering)	BRS	DATASTAR	DIALOG		ESA	INFOLINE	INKA	SDC
EDF-DOC (elec. power) on **TELESYSTEMES &**					ESA			
ENEL (energy)					ESA			
INSPEC see p. 146 (physics/elec./comput.)	BRS	DATASTAR	DIALOG		ESA		INKA	SDC
MERLIN-TECH (electrical/electronic)					ESA			
ZDE (like INSPEC) on **FIZ TECHNIK**					ESA			

Electronics

Database	BRS	DATASTAR	DIALOG	DIMDI	ESA	INFOLINE	INKA	SDC
COMPENDEX see p. 146 (engineering)	BRS	DATASTAR	DIALOG		ESA	INFOLINE	INKA	SDC
ELECTRONICS & COMMUNICATIONS ABSTRACTS					ESA			
EMIS (electronic materials data)					ESA			
INSPEC see p. 146 (physics/elec./comput.)	BRS	DATASTAR	DIALOG		ESA		INKA	SDC
MERLIN-TECH (electrical/electronic)					ESA			
SPACE COMPS (space components)					ESA			
ZDE (like INSPEC) on **FIZ TECHNIK**					ESA			

Energy see also Electrical engineering and Physics

Database	BRS	DATASTAR	DIALOG	DIMDI	ESA	INFOLINE	INKA	SDC
COAL (IEA) on **BELINDIS &**							INKA	
COALRIP (IEA) (coal research projects)							INKA	
COMPENDEX see p. 146 (engineering)	BRS	DATASTAR	DIALOG		ESA	INFOLINE	INKA	SDC
DOE ENERGY (US DoE.) on **MEAD DATA CENTRAL &**			DIALOG				INKA	SDC
EBIB (energy bibliography)			DIALOG					
EDF-DOC (elec. power) on **TELESYSTEMES &**					ESA			
ELECTRIC POWER DATABASE			DIALOG					
ELECTRIC POWER INDUSTRY ABSTRACTS								SDC

Database	BRS	DATASTAR	DIALOG	DIMDI	ESA	INFOLINE	INKA	SDC
ENEL (energy)					ESA			
ENERGIE							INKA	
ENERGIRAP (esp. nuclear) on **TELESYSTEMES**								
ENERGY DATABASE							INKA	SDC
ENERGYLINE			DIALOG		ESA		INKA	SDC
ENERGYNET (people/organizations/work)			DIALOG		ESA		INKA	
INIS-ATOMINDEX (Atom-nucl.), on **BELINDIS** &					ESA		INKA	
POWER (US DoE Library)								SDC

Engineering (general) see MULTIDISCIPLINES

Environmental sciences see also Health

Database	BRS	DATASTAR	DIALOG	DIMDI	ESA	INFOLINE	INKA	SDC
ACOMPLINE (GLC Library) on **SCICON** &					ESA			
APTIC (air pollution)	BRS	DATASTAR	DIALOG					
COMPENDEX see p. 146 (engineering)			DIALOG		ESA	INFOLINE	INKA	SDC
ECDIN (Europ. chem. data) on **DATACENTRALEN**								
ENDOC (information centres) on **ECHO**								
ENREP (European Research) on **ECHO**								
ENVIROLINE			DIALOG		ESA			SDC
ENVIRONMENTAL BIBLIOGRAPHY			DIALOG		ESA			
POLLUTION ABSTRACTS	BRS		DIALOG		ESA			
URBALINE (GLC news) on **SCICON**								

Food science and technology see also Biosciences and Agriculture

Database	BRS	DATASTAR	DIALOG	DIMDI	ESA	INFOLINE	INKA	SDC
AQUATIC SCIENCE & FISHERIES ABS			DIALOG	DIMDI				
COFFEELINE			DIALOG					
COMPENDEX see p. 146 (engineering)	BRS	DATASTAR	DIALOG		ESA	INFOLINE	INKA	SDC
CORALIE (food additives) on **BELINDIS**								
FOOD SCIENCE & TECHNOLOGY ABSTRACTS			DIALOG	DIMDI	ESA			SDC
FOODS ADLIBRA (food industry)			DIALOG					
IALINE (agric./food ind.) on **TELESYSTEMES**								

Geology/earth sciences and technologies

Database	BRS	DATASTAR	DIALOG	DIMDI	ESA	INFOLINE	INKA	SDC
ASIAN GEO					ESA			
COLD (regions)								SDC
COMPENDEX see p. 146 (engineering)	BRS	DATASTAR	DIALOG		ESA	INFOLINE	INKA	SDC
GEOARCHIIVE								
GEODE (geology) on **TELESYSTEMES**								

Geology/earth sciences and technologies—*continued*

	DATASTAR	BRS	DIALOG	DIMDI	ESA	INFOLINE	SDC	INKA
GEOLINE						INFOLINE		INKA
GEOMECHANICS ABSTRACTS			DIALOG			INFOLINE	SDC	
GEOREF			DIALOG			INFOLINE	SDC	
METEOROLOGICAL & GEOASTROPHYSICAL ABS						INFOLINE		
MINSEARCH (minerals/economics)					ESA			
MOLARS (meteorology)					ESA			
OCEANIC ABSTRACTS			DIALOG					

*Governments/official publications

	DATASTAR	BRS	DIALOG	DIMDI	ESA	INFOLINE	SDC	INKA
ACOMPLINE (GLC Library) on **SCICON** &			DIALOG		ESA		SDC	
AS/(US statistical pubs.)			DIALOG				SDC	
CIS(US Congress)			DIALOG				SDC	
CONGRESSIONAL RECORD ABSTRACTS			DIALOG					
DHSS-DATA (UK health/social) on **SCICON** &	DATASTAR		DIALOG					
FEDERAL INDEX(US)			DIALOG					
FEDERAL REGISTER ABSTRACTS(US)			DIALOG				SDC	
GPO MONTHLY CATALOG(US)		BRS	DIALOG					
GPO PUBLICATIONS REFERENCE FILE(US)			DIALOG					
POLIS (UK parliamentary docs.) on **SCICON**								
URBALINE (GLC news) on **SCICON**								

Health and safety see also Environmental sciences and Medicine

	DATASTAR	BRS	DIALOG	DIMDI	ESA	INFOLINE	SDC	INKA
ACOMPLINE (GLC Library) on **SCICON** &			DIALOG		ESA			
CHEMICAL EXPOSURE			DIALOG					
CHEMICAL HAZARDS IN INDUSTRY	DATASTAR					INFOLINE		
CHEMICAL REGULATIONS & GUIDELINES(US)			DIALOG		ESA			
CISDOC (occupational health)								
DHSS-DATA (UK health/social) on **SCICON** &	DATASTAR	BRS	DIALOG					
EXCERPTA MEDICA/EMBASE see p. 146	DATASTAR	BRS	DIALOG	DIMDI				
HAZARDLINE on **MEAD DATA CENTRAL** &		BRS						
HEALTH PLANNING AND ADMINISTRATION		BRS	DIALOG	DIMDI				
HECLINET (health care)				DIMDI				
HSELINE (health and safety)					ESA	INFOLINE		
LABORATORY HAZARDS BULLETIN		BRS			ESA	INFOLINE		
MEDLINE (p. 147) on **BLAISE/NLM, MEAD DATA CENTRAL** &	DATASTAR		DIALOG	DIMDI				
MENTAL HEALTH ABSTRACTS	DATASTAR		DIALOG					
OCCUPATIONAL HEALTH & SAFETY (NIOSH)			DIALOG			INFOLINE		

PHARMACEUTICAL NEWS INDEX ——————————— DIALOG
RTECS (chemical toxicity) on BLAISE/NLM & ———————————————— DIMDI
TBD (toxicology) ———————————————— DIMDI
TOXLINE on BLAISE/NLM & ———————————————— DIMDI
TSCA INITIAL INVENTORY (chemicals—toxicity) ——————— DIALOG
TSCA PLUS (chemicals—toxicity) ——————— DIALOG ——— SDC
URBALINE (GLC news) on SCICON

Industry see Organizations and subject fields

Information/library science
DATA PROCESSING & INFORMATION SCIENCE ——— BRS
ELECTRONIC MAGAZINE ——— ESA
ELECTRONIC PUBLISHING ABSTRACTS ——— INFOLINE
INFORMATION SCIENCE ABSTRACTS ——— BRS DATASTAR DIALOG
INSPEC see p. 146 (info. tech.) ——— BRS DATASTAR DIALOG ESA INKA SDC
LIBRARY LITERATURE on WILSONLINE
LISA (library and information science) ——— DIALOG
ONLINE CHRONICLE (online news) ——— DIALOG
SOCIAL SCISEARCH (citation index) ——— BRS DIALOG DIMDI ——— SDC
ZDE (like INSPEC) on FIZ TECHNIK

*Information services
CUADRA DIRECTORY OF DATABASES ——— DATASTAR
DATABASE OF DATABASES ——— DIALOG
DIANEGUIDE (databases) on ECHO
DUNDIS (UN information) on ECHO
ENDOC (environmental) on ECHO
KIPD—DATABASE DIRECTORY ——— BRS

*Journal articles see subjects

Law
CRIMINAL JUSTICE PERIODICAL INDEX ——— DIALOG
INDEX TO LEGAL PERIODICALS on WILSONLINE
LAWTEL on PRESTEL
LEGAL RESOURCE INDEX (law literature) ——— BRS DIALOG
LEXIS on MEAD DATA CENTRAL/BUTTERWORTHS
PATLAW ——— DIALOG INFOLINE
POLIS (UK and European legislation) on SCICON
SOCIAL SCISEARCH (citation index) ——— BRS DIALOG DIMDI

Leisure/recreation/tourism

Database	BRS	DATASTAR	DIALOG	DIMDI	ESA	INFOLINE	INKA	SDC
ACOMPLINE (GLC Library) on SCICON &					ESA			SDC
SPORT								
URBALINE (GLC news) on SCICON	BRS							

Library science see Information

Management see Business

Marine science/engineering see also Water and Transportation

Database	BRS	DATASTAR	DIALOG	DIMDI	ESA	INFOLINE	INKA	SDC
AQUACULTURE			DIALOG					
AQUATIC SCIENCE & FISHERIES ABS			DIALOG	DIMDI				
BMT/SHIP ABSTRACTS on BRITISH MARITIME TECHNOLOGY								
FLUIDEX			DIALOG		ESA			
MARNA (maritime inf.) on CONTROL DATA B.V.						INFOLINE		
NAVAL RECORD								
OCEANIC ABSTRACTS			DIALOG		ESA			
SHIPDES on CONTROL DATA B.V.			DIALOG					
TRIS (inc. maritime research)			DIALOG					

Materials science/technology see also Chemical engineering and Mechanical engineering

Database	BRS	DATASTAR	DIALOG	DIMDI	ESA	INFOLINE	INKA	SDC
AMERICAN CERAMICS ABSTRACTS						INFOLINE		
CETIM (mechanical engineering)								
CHEMICAL ABSTRACTS on STN, QUESTEL &	BRS	DATASTAR	DIALOG		ESA			SDC
COLD (regions)					ESA			SDC
COMPENDEX see p. 146 (engineering)	BRS	DATASTAR	DIALOG		ESA	INFOLINE	INKA	SDC
CORROSION			DIALOG		ESA	INFOLINE		SDC
DKI (plastics/rubber/fibres) on FIZ TECHNIK								
EMIS (electronic materials data)								
INSPEC see p. 146 (physics/elec./comput.)	BRS	DATASTAR	DIALOG		ESA		INKA	SDC
ISMEC (mechanical)			DIALOG		ESA			
METADEX (metals)			DIALOG		ESA		INKA	SDC
NONFERROUS METALS ABSTRACTS			DIALOG		ESA			
PAPERCHEM			DIALOG		ESA			SDC
PSTA/PACKABS (packaging sci./tech.)				DIMDI	ESA	INFOLINE		
RAPRA (rubber and plastics)			DIALOG				INKA	
RHEO (rheology)						INFOLINE		
WELDASEARCH			DIALOG				INKA	
WORLD SURFACE COATINGS ABSTRACTS			DIALOG			INFOLINE		

Mathematics

Database	Online hosts
INSPEC see p. 146	BRS · DATASTAR · DIALOG · ESA · INKA · SDC
MATHEMATICS ABSTRACTS (Zentralblatt) on STN &	BRS · DATASTAR · DIALOG · ESA · INKA
MATHFILE/MATHSCI (Mathematical Reviews)	

Mechanical engineering

Database	Online hosts
CETIM (mechanical engineering)	ESA
COMPENDEX see p. 146 (engineering)	BRS · DATASTAR · DIALOG · ESA · INFOLINE · INKA · SDC
DKF (automotive eng.) on FIZ TECHNIK	
DOMA (mechanical eng.) on FIZ TECHNIK	
FLUIDEX	DIALOG · ESA
ISMEC (mechanical)	DIALOG · ESA
SAE ABSTRACTS (automotive engineering)	
TRIBOLOGY INDEX	SDC
VOLKSWAGONWERK	DATASTAR · INKA

Medicine see also Biosciences and Health and safety

Database	Online hosts
ABLEDATA (rehabilitation)	BRS
BIOETHICS on BLAISE/NLM	
BRITISH MEDICAL ASSOCIATION PRESS CUTTINGS	
CANCERLIT on BLAISE/NLM &	DATASTAR · DIMDI
CANCERNET on TELESYSTEMES	DATASTAR
CANCERPROJ on BLAISE/NLM &	DIMDI
CHEMLINE (biomed chems) on BLAISE/NLM &	DIMDI
CHID (health)	BRS
CLINICAL ABSTRACTS	DIALOG
CLINICAL NOTES ONLINE	DATASTAR
CLINPROT (anti-cancer) on BLAISE/NLM &	BRS · DIMDI
COMPREHENSIVE CORE MEDICAL LIBRARY	BRS
DHSS-DATA (UK health/social) on SCICON &	DATASTAR
DRUGINFO/ALCOHOL USE & ABUSE	BRS
EPILEPSYLINE	BRS
EXCERPTA MEDICA/EMBASE see p. 146	BRS · DATASTAR · DIALOG · DIMDI
HECLINET (health care)	DIMDI
INTERNATIONAL PHARMACEUTICAL ABSTRACTS	DIALOG · DIMDI · ESA
IRCS MEDICAL JOURNALS	BRS · DATASTAR · DIMDI
ISI/BIOMED	BRS · DIMDI
LIFE SCIENCES COLLECTION	DIALOG

204

Medicine—*continued*

Database	BRS	DATASTAR	DIALOG	DIMDI	ESA	INFOLINE	INKA	SDC
MEDIS on **MEAD DATA CENTRAL**								
MEDLINE (p. 147) on **BLAISE/NLM, MEAD DATA CENTRAL &**	BRS	DATASTAR	DIALOG	DIMDI				
MENTAL HEALTH ABSTRACTS	BRS	DATASTAR	DIALOG					
NURSING AND ALLIED HEALTH	BRS		DIALOG					
PHARMACEUTICAL NEWS INDEX			DIALOG					
PRE-MED (current awareness)	BRS	DATASTAR						
PSYCINFO see p. 147	BRS	DATASTAR	DIALOG	DIMDI				SDC
WILEY MEDICAL RESEARCH DIRECTORY		DATASTAR						

Metallurgy

Database	BRS	DATASTAR	DIALOG	DIMDI	ESA	INFOLINE	INKA	SDC
BIIPAM CTIF (foundry work)					ESA			
COMPENDEX see p. 146 (engineering)	BRS	DATASTAR	DIALOG		ESA	INFOLINE	INKA	SDC
CORROSION								SDC
MDF/I (metals information)								SDC
METADEX (metals)			DIALOG		ESA		INKA	SDC
NONFERROUS METALS ABSTRACTS/BNF			DIALOG		ESA			
SDIM (metallurgy)							INKA	
WELDASEARCH			DIALOG		ESA			
WORLD ALUMINIUM ABSTRACTS			DIALOG			INFOLINE		
ZINC LEAD CADMIUM ABSTRACTS						INFOLINE		

Meteorology see Geology

MULTIDISCIPLINES/INTERDISCIPLINES see also Research and the different types of literature such as books, conference papers, reports and theses

Database	BRS	DATASTAR	DIALOG	DIMDI	ESA	INFOLINE	INKA	SDC
ABSTRAX 400 (popular US periodicals)	BRS							
ACADEMIC AMERICAN ENCYCLOPEDIA	BRS	DATASTAR	DIALOG					
APPLIED SCI & TECH INDEX on **WILSONLINE**								
BIBLAT (Latin America) on **TELESYSTEMES**								
BIBLIOGRAPHIC INDEX on **WILSONLINE**								
COMPENDEX see p. 146 (engineering)	BRS	DATASTAR	DIALOG		ESA	INFOLINE	INKA	SDC
EI ENGINEERING MEETINGS/EI MET		DATASTAR	DIALOG		ESA			SDC
FEDERAL RESEARCH IN PROGRESS (US)			DIALOG					
FLUIDEX			DIALOG		ESA			
GENERAL SCIENCE INDEX on **WILSONLINE**								
INDUSTRY DATA SOURCES	BRS	DATASTAR	DIALOG	DIMDI				
ISI/MULTISCI								
MAGAZINE ASAP/INDEX (popular American)	BRS		DIALOG					

Database	BRS	DATASTAR	DIALOG	ESA	Other
MIDDLE EAST: ABSTRACTS & INDEX			DIALOG		
MIDEASTFILE			DIALOG		SDC
MONITOR (Christian Science Monitor)					
NASA AEROSPACE (p. 147)				ESA	
NTIS (US Govt. Res. see p. 147) on MEAD DATA CENTRAL &	BRS	DATASTAR	DIALOG	ESA	INKA SDC
PASCAL (Bull. Signal.) on TELESYSTEMES &		DATASTAR	DIALOG	ESA	INKA
PHYSCOMP (data guides)					
PTS DEFENSE MARKETS & TECHNOLOGY		DATASTAR	DIALOG		
PTS FORECASTS (statistical)	BRS	DATASTAR	DIALOG		
PTS TIME SERIES (statistical)	BRS	DATASTAR	DIALOG		
SCISEARCH (citation index)		DATASTAR	DIALOG	DIMDI	
SIGLE (grey literature) on BLAISE &			DIALOG		INKA
SOVIET SCIENCE AND TECHNOLOGY			DIALOG		
SUPERINDEX (back of book indexes)	BRS				
TRADE & INDUSTRY ASAP/INDEX	BRS		DIALOG		
US EXPORTS			DIALOG		
WORLD AFFAIRS REPORT			DIALOG		

*News/newspapers

Database	BRS	DATASTAR	DIALOG	ESA	Other
AP NEWS (business)			DIALOG		
NATIONAL NEWSPAPER INDEX (US)	BRS		DIALOG		
NDEX (Newspaper Index—mainly US)	BRS				SDC
NEWSEARCH (US—current awareness)			DIALOG		
NEXIS on MEAD DATA CENTRAL					
TEXTLINE and NEWSLINE on FINSBURY DATA &				ESA	
UPI NEWS (United Press International)			DIALOG		
WASHINGTON POST INDEX			DIALOG		
WORLD REPORTER on DATASOLVE &				ESA	

Nuclear science/technology see Physics and Energy

Oceanography/oceanology see Marine science and Water science

*Official publications see Governments

*Organizations/companies/associations/institutions

Database	BRS	DATASTAR	DIALOG	ESA	Other
ASSOCIATIONS PUBLICATIONS IN PRINT	BRS				
DATALINE (EXSTAT) on FINSBURY DATA					
D&B—DUN'S MARKET IDENTIFIERS (US)			DIALOG		

Organizations/companies/associations/institutions—continued

	BRS	DATASTAR	DIALOG	ESA	INFOLINE	INKA	SDC
D&B—MILLION DOLLAR DIRECTORY (US)			DIALOG				
D&B—PRINCIPAL INTERNATIONAL BUSINESSES			DIALOG				
EKOL (European Kompass) on KODA ONLINE			DIALOG				
ELECTRONIC YELLOW PAGES (US/various topics)			DIALOG				
ENCYCLOPEDIA OF ASSOCIATIONS (US—nonprofit)			DIALOG				
FINANCIAL TIMES ABSTRACTS	BRS	DATASTAR					
ICC BRITISH COMPANY DIRECTORY on ICC &		DATASTAR	DIALOG				
INDUSTRIAL MARKET LOCATIONS (UK companies)					INFOLINE		
INDUSTRIAL RESEARCH LABORATORIES					INFOLINE		
JORDANWATCH on JORDANS &					INFOLINE		
KEY BRITISH ENTERPRISES (D&B)			DIALOG		INFOLINE		
PTS ANNUAL REPORTS ABSTRACTS (US/internat.)	BRS	DATASTAR	DIALOG				
PTS F&S INDEXES (general-statistical)	BRS	DATASTAR	DIALOG				
STANDARD & POOR'S CORPORATE DESCRIPTIONS/US			DIALOG				
TEXTLINE and NEWSLINE on FINSBURY DATA &				ESA			
THOMAS REGISTER ONLINE (US manufacturers)			DIALOG				
WHO OWNS WHOM			DIALOG		INFOLINE		

Packaging see Materials

*Patents see p. 90–91

*People

	BRS	DATASTAR	DIALOG	ESA	INFOLINE	INKA	SDC
AMERICAN MEN & WOMEN OF SCIENCE	BRS		DIALOG				
BIOGRAPHY INDEX on WILSONLINE	BRS						
BIOGRAPHY MASTER INDEX			DIALOG				
MARQUIS WHO'S WHO/PRO-FILES			DIALOG				

*Periodicals

	BRS	DATASTAR	DIALOG	ESA	INFOLINE	INKA	SDC
CALIFORNIA UNION LIST OF PERIODICALS	BRS						
CASSI (Chemical Abstracts sources)							SDC
JOURNAL DIRECTORY on WILSONLINE	BRS						
READER'S GUIDE TO PERIODICALS on WILSONLINE	BRS						
ULRICH'S INTERNATIONAL PERIODICALS DIRECTORY	BRS		DIALOG				

Petroleum industry

	BRS	DATASTAR	DIALOG	ESA	INFOLINE	INKA	SDC
APILIT/APIPAT (restricted access)				ESA		INKA	SDC
COMPENDEX see p. 146 (engineering)	BRS	DATASTAR	DIALOG		INFOLINE		SDC

	BRS	DATASTAR	DIALOG	DIMDI	ESA	INKA	SDC
P/E NEWS (petroleum/energy business)			DIALOG				SDC
TULSA (petroleum literature—restricted access)							SDC

Pharmacology see Biosciences and Medicine

Physics see also Energy

	BRS	DATASTAR	DIALOG	DIMDI	ESA	INKA	SDC
INIS-ATOMINDEX (atom/nucl) on **BELINDIS** &					ESA	INKA	
INSPEC see p. 146 (physics/elec./comput.)	BRS	DATASTAR	DIALOG		ESA	INKA	SDC
PHYS (Physics Briefs) on **STN** &						INKA	
SPIN (general physics)			DIALOG				

Physiology see Medicine

Plastics see Materials science

Pollution see Environmental sciences

***Products/services** see Organizations

Psychology/psychiatry see also Medicine

	BRS	DATASTAR	DIALOG	DIMDI	ESA	INKA	SDC
EXCERPTA MEDICA/EMBASE see p. 146	BRS	DATASTAR	DIALOG	DIMDI			
FAMILY RESOURCES/NCFR	BRS		DIALOG				
LANGUAGE AND LANGUAGE BEHAVIOR/OR ABS			DIALOG				
MEDLINE (p. 147) on **BLAISE/NLM, MEAD DATA CENTRAL** &	BRS	DATASTAR	DIALOG	DIMDI			
MENTAL HEALTH ABSTRACTS	BRS	DATASTAR	DIALOG				
MENTAL MEASUREMENTS	BRS						
NATIONAL CLEARINGHOUSE FOR MENTAL HEALTH	BRS						
NATIONAL REHABILITATION INFORMATION CENTER	BRS						
PSYCALERT (PSYCINFO in process)			DIALOG				
PSYCINFO (psychology) see p. 147	BRS	DATASTAR	DIALOG				SDC
PSYN (psychology)				DIMDI			
SOCIAL SCI SEARCH (citation index)	BRS		DIALOG	DIMDI			
SOCIOLOGICAL ABSTRACTS	BRS	DATASTAR	DIALOG	DIMDI			

***Reference works** see also specific topics

	BRS	DATASTAR	DIALOG	DIMDI	ESA	INKA	SDC
ACADEMIC AMERICAN ENCYCLOPEDIA	BRS		DIALOG				
EURODICAUTOM (sci/tech terms/trans) on **ECHO**							
KIRK-OTHMER ENCYCLOPEDIA OF CHEM. TECH.	BRS	DATASTAR					

*Reports

	BRS	DATASTAR	DIALOG	DIMDI	ESA	INFOLINE	INKA	SDC
COMPENDEX see p. 146 (engineering)	BRS	DATASTAR	DIALOG		ESA	INFOLINE	INKA	SDC
NASA AEROSPACE (see p. 147)			DIALOG		ESA		INKA	SDC
NTIS (US Govt. Res. see p. 147)	BRS	DATASTAR	DIALOG		ESA		INKA	SDC
SIGLE (grey literature) on BLAISE &					ESA		INKA	

*Research

	BRS	DATASTAR	DIALOG	DIMDI	ESA	INFOLINE	INKA	SDC
ACS DIRECTORY OF GRADUATE RESEARCH	BRS							
AGREP (European) on DATACENTRALEN &				DIMDI				
COALRIP (IEA) (coal research projects)							INKA	
CRIS/USDA (agricultural research)			DIALOG					
EABS (European research) on ECHO								
FEDERAL RESEARCH IN PROGRESS (US)			DIALOG					
INDUSTRIAL RESEARCH LABORATORIES (US)						INFOLINE		
NTIS (US Govt. Res. see p. 147) on MEAD DATA CENTRAL, STN &	BRS	DATASTAR	DIALOG		ESA		INKA	SDC
SSIE CURRENT RESEARCH			DIALOG					

Science (general) see MULTIDISCIPLINES

Social/behavioural sciences

	BRS	DATASTAR	DIALOG	DIMDI	ESA	INFOLINE	INKA	SDC
ABLEDATA (rehabilitation)	BRS							
ACOMPLINE (GLC Library) on SCICON &					ESA			
AGELINE (retired persons)	BRS							
CHILD ABUSE & NEGLECT			DIALOG					
ARTS AND HUMANITIES SEARCH	BRS							
DHSS-DATA (UK health/social) on SCICON &		DATASTAR						
FAMILY RESOURCES/NCFR	BRS		DIALOG					
HUMANITIES INDEX on WILSONLINE								
LABORDOC (labour) on TELESYSTEMES &					ESA			
LABORINFORMATION (labour legislat./policy)					ESA			
PAIS INTERNATIONAL (public affairs)	BRS	DATASTAR	DIALOG					SDC
PSYCINFO (psychology) see p. 147	BRS	DATASTAR	DIALOG	DIMDI				SDC
SOCIAL SCIENCES INDEX on WILSONLINE								
SOCIAL SCISEARCH (citation index)	BRS		DIALOG	DIMDI				
SOCIAL WORK ABSTRACTS	BRS							
SOCIOLOGICAL ABSTRACTS	BRS	DATASTAR	DIALOG					
URBALINE (GLC news) on SCICON	BRS							
US POLITICAL SCIENCE DOCUMENTS			DIALOG					

*Standards
BSI STANDARDLINE ——————————————— INFOLINE
DITR (Germany) on **FIZ TECHNIK**
INDUSTRY & INTERNATIONAL STANDARDS ——— BRS
MILITARY & FEDERAL SPECS & STANDARDS (US) ——— BRS
STANDARDS & SPECIFICATIONS (US) ——— DIALOG — ESA
STANDARDS SEARCH (SAE & ASTM) ——— SDC

Telecommunications
COMPENDEX see p. 146 (engineering) —— BRS — DATASTAR — DIALOG — ESA — INFOLINE — INKA — SDC
ELECTRONICS & COMMUNICATIONS ABSTRACTS ——— ESA
INSPEC see p. 146 (physics/elec./comput.) —— BRS — DATASTAR — DIALOG — ESA — INKA — SDC
ZDE (like *INSPEC*) on **FIZ TECHNIK**

Textiles
TEXTILE TECHNOLOGY DIGEST ——— DIALOG
TITUS on **TELESYSTEMES**
WORLD TEXTILES ——— DIALOG — INFOLINE

*Theses/dissertations
DISSERTATION ABSTRACTS ONLINE —— BRS — DIALOG
SIGLE (grey literature) on **BLAISE** & ——— INKA

Toxicology see Health and safety

*Translations
WTI (World Transindex) ——— ESA

Transportation
ACOMPLINE (GLC Library) on **SCICON** & —— BRS — DATASTAR — DIALOG — ESA — INFOLINE — INKA — SDC
COMPENDEX see p. 146 (engineering) ——— ESA
IRRD (road research) ——— ESA
ISMEC (mechanical) ——— DIALOG — ESA
TRANSDOC (economics) ——— ESA
TRIS (all modes) ——— DIALOG
URBALINE (GLC news) on **SCICON**

Veterinary sciences see Agriculture and Bioscience

210

Water science and technology see also Biosciences, Environmental sciences, Marine science and Geology

Database	BRS	DATASTAR	DIALOG	DIMDI	ESA	INFOLINE	INKA	SDC
AFFE (freshwater studies)					ESA			
AQUACULTURE			DIALOG					
AQUALINE			DIALOG			INFOLINE		
AQUATIC SCIENCE & FISHERIES ABS.			DIALOG	DIMDI				
COMPENDEX see p. 146 (engineering)	BRS	DATASTAR	DIALOG		ESA	INFOLINE	INKA	SDC
FLUIDEX			DIALOG		ESA			
METEOROLOGICAL & GEOASTROPHYSICAL ABS			DIALOG					
OCEANIC ABSTRACTS			DIALOG		ESA			
WATER RESOURCES ABSTRACTS			DIALOG					
WATERLIT						INFOLINE		SDC
WATERNET (water supply)			DIALOG					

Zoology see Biosciences

Searching: the literature and computer databases

This section assumes that you have a good idea of the information you want, the information you do not want, and your reason for wanting the information, and also that you have selected and have gained access to the appropriate guides or databases. It must not be considered in isolation from other sections, as searching is only part of the process for obtaining useful information. For example, it is particularly important to keep your statement of intent (see p. 4) in mind. Otherwise you may turn a search for useful information into an exercise for finding information for its own sake, useful or not.

There are two main types of use for databases or equivalent hard-copy publications.

(1) *Current awareness* (keeping up to date with the latest information, see also p. 284). Here the emphasis is on rapid information processing, which can be achieved if the information or references can be fed straight into the computer without human indexing or abstracting to slow down the process. The computer can then read or scan this simple input, such as the titles of journal articles, and merely pick out those titles containing certain preselected words or combinations of words, or print out KWIC or KWOC indexes (see p. 222). In the case of computerized searching, this process can be applied with increasing cost to abstracts or even complete articles, and is known as free-text or natural-language searching (or full-text searching in the case of complete documents or articles). One problem with free-text searching is the variability of the personal vocabularies of authors and searchers, and the frequent use of misleading titles, especially in the more news-oriented sources or in the less precise or soft sciences, such as some of the social sciences.

(2) *Retrospective searching* (searching back through time to see what information already exists). Here speed is less relevant as far as the input stage is concerned, but accuracy and permanence are important. With factual or numeric databanks, reliability of the data itself is also especially important. The time-consuming processes of abstracting and indexing can be applied, and the articles indexed by their concepts or meanings rather than by (say) words in titles, which may be ambiguous. This method generally uses what is known as a controlled vocabulary. Both indexers and searchers have to choose their indexing or search terms from a preselected collection of subject headings (usually called a thesaurus, see p. 113). A problem still remains with controlled vocabulary in that if the word the searcher or indexer would

normally use is not 'allowed', a choice of synonyms or compromize words may be offered, and once personal choice comes into the process it becomes unreliable.

General principles

Searching usually presupposes that there are useful items of information in existence, and is mainly concerned with guessing or predicting either some of the contents of useful items, such as titles or authors' names, or which indexing terms or classifications have been given to those items. The final stage is the insertion of those predictions into an effective search strategy and the application of search techniques appropriate to whichever guide or information system is being used.

There are, therefore, three main aspects of searching to be considered. Although capable of being described separately, these interact with one another and cannot be separated in practice.

(1) *The language aspect.* Nearly all information handling depends in some way on the use of languages, from indexing terms to abstracts, and from titles to the full text of original documents. The guessing or predicting procedure is only one side of the language aspect; the other is the ability to recognize relevant or useful information in the items retrieved (including titles, abstracts, full text) or even in indexing terms or classifications thrown up during the search.

(2) *The literature/database aspect.* This concerns the necessity of knowing and understanding the structure of the guides, documents or databases in order to search them effectively.

(3) *The libraries and information-services aspect.* These services provide the facilities for searching, and must become familiar before appropriate guides or databases may be identified, located and used. For online searching, a knowledge of the search commands and special facilities is obviously necessary.

A proper consideration of all three aspects results in a compatible search strategy, which means that the correct language is used in the correct guides/ databases in the correct way, and that the results will be meaningful and useful.

The hunting cycle. Searching is nearly always a cyclic, iterative or repeating procedure. A particular search strategy is tried, and the results fed back into a new, improved strategy. This might mean, for example, a different selection of search terms, or a different way of searching for them, such as different combinations, permutations or truncations (p. 241).

The objective. This cannot be overemphasized, because of the dangers mentioned earlier. Whatever else happens in a search, the searcher must never forget why the information is being sought; unless the information found can be used productively, perhaps to solve a problem, it should be rejected, however relevant it may at first appear in the subject sense. It is very rare for searching to be performed for its own sake, where the search results themselves (hit rates or references) supply the required data. This can happen in citation or literature/data analysis, in the examination of trends in terminology, in such cases as failure-report analysis where frequency of reported failure is a guide to reliability, in search training and in bibliography compilation. Occasionally, of course, people want a search to show that no information exists on a particular topic, such as a patent search to check the novelty of their new invention, or a final search before writing up a doctoral thesis to confirm the originality of the research. People may believe that no information exists and request searches as a form of insurance.

Search efficiency. This is usually expressed in terms of precision (a measure of the amount of rubbish you retrieve), and recall (a measure of the amount of relevant material available that you do not retrieve). In searching, particularly computerized searching, precision is normally inversely related to recall, so it is only rarely possible to perform a perfect search. In practice, this often means that if you wish to be reasonably sure that you have not missed anything, you will have to endure a large amount of irrelevant output as well. Conversely, if you wish to be spared the rubbish, you will also inevitably miss a great deal of useful material.

Relevance is often a subjective evaluation that can easily change or be changed. For example, if a researcher has found relatively few references on a particular topic, he may believe that all references at the same subject level will be useful (everything on lasers). If a search is then performed which produces 500 similar references, the researcher, realizing that he simply cannot cope with that number, will quite possibly narrow his definition of what is relevant to him. On the other hand, if a search produces a reasonable number of apparently useful references, the researcher may be perfectly satisfied, even impressed, because he does not know that a significant number of key papers have been missed owing to an inadequate search strategy. This subjective evaluation of relevance and search efficiency is sometimes referred to as user naivety, and probably applies equally well to searcher intermediaries, some or all of the time.

Choosing search terms

Before searching in any index, contents list or database, you must know which indexing terms, subject headings, words, phrases, codes, word fragments, etc., to look for. We shall refer to these terms, etc., as *search terms*. This process of choosing search terms is considered here, but it may take place at any time. You may make a list days before the search, and you will almost certainly need to introduce new search terms as your search progresses.

What you are trying to do is predict which search terms have been used by the indexers, abstractors or authors to 'label' the information that you want. For example, if you wanted to find the boiling point of ethyl alcohol, it might have been indexed or described in a contents list, abstract or title under Alcohol, Ethyl Alcohol, Ethanol, C_2H_5OH, Boiling Points, Physical Properties, Organic Compounds or just Liquids. This is a problem of language.

The usual practice is to make intelligent guesses at the various possible search terms, and then to search by trial and error. Although this method may frequently work, there are times when your experience will not be wide enough to cover all the possibilities. For example, there are particular problems with

American terminology, such as the use of 'ground effect vehicles' to describe 'hovercraft'.

Aids for identifying search terms

When you have difficulty in predicting suitable search terms, particularly in areas outside your normal subject expertise, the following may help.

Colleagues. It is surprising how frequently people forget other people as a source of information.

Thesauri (these are described on p. 113, and are particularly useful for identifying controlled terms or descriptors). To illustrate the use of these, suppose you are looking for the search term 'hovercraft' in an American index, and find nothing. You could then look up 'hovercraft' in a technical thesaurus, and possibly find the following entry:

Hovercraft   use GROUND EFFECT VEHICLES

This means that the second term is 'preferred'. You then look up 'ground effect vehicles', and find something like this:

GROUND EFFECT VEHICLES

UF Air cushion vehicles
    Hovercraft

NT HOVERCAT MK II
    SR.N6

BT AIRCRAFT

RT AMPHIBIOUS AIRCRAFT
    FLYING PLATFORMS

where UF represents used for, NT narrower terms, BT broader terms and RT related terms.

Some examples of thesauri are included in the section on p. 114. Details of other thesauri may be obtained from the appropriate guides or from Aslib, who maintain a record of thesauri.

Reference books, textbooks, etc. When you are entering a new subject field, you may have to familiarize yourself with the associated terminology. Encyclopaedias, taxonomies, textbooks, and technical dictionaries are particularly useful.

Subject family trees/library classification schemes (the latter are covered on p. 249). It may help to consider how the narrow subject field surrounding your problem fits into the main subject disciplines by constructing a family tree. This is done by writing a description of the narrow subject field at the bottom of a page, followed at the top of the page by all those subject disciplines which have a bearing on the narrow field. You can then complete the family tree by repeatedly breaking down or combining broader subjects to form narrower ones. For example:

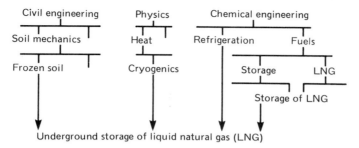

Underground storage of liquid natural gas (LNG)

The family-tree approach helps you to appreciate that most problems are multi-disciplinary in nature, and should prevent you from relying solely on the literature of your own or any single subject. It will also indicate likely broader search terms if you find nothing under a narrow search term.

Library classification schemes have some points in common with subject family trees and thesauri.

Known references. These can be fed into a search strategy by using either the author's name or combinations of title words. If the index or database includes the reference, you will be able to identify which subject section, heading or indexing terms have been used for your known reference, and which can therefore be used to find other items on the same topic (see pp. 222 and 241).

These aids should enable you to compile a list of search terms to cover your subject interests. Remember that this selection may have to be changed during the course of your search. Other points worth considering are:

Synonyms. There are frequently several different terms which mean the same thing or are very similar in meaning. Subject thesauri are usually the best guides to help find all the possible terms via the 'see also', 'used for', and 'use' headings. English-language thesauri and word finders are useful for the more general terms. Examples, apart from those already mentioned, include car and automobile, malnutrition and undernourishment, and sound and acoustics.

Generic terms. When there is a part-whole relationship between terms, such as chlorides-halides, it is (a) unlikely that when the term chloride is used the term halide will automatically be used as well (autoposted), and (b) likely that some information about chlorides will only be indexed under halides. This means that, to perform an exhaustive search for either halides or chlorides, both terms must be used, even though halides may not always include chlorides. Another typical example is information about isocohedrons appearing under polygons. Thesauri are the main aids here with their coverage of 'broader' and 'narrow' terms.

Antonyms. Terms with opposite meanings may be used to describe the same thing. For example, the stability of a material may also be described as its instability. Noise from a machine may be described in terms of silence or quietness. High vacuum may be used to describe conditions of extremely low pressure. Other pairs of terms include: solubility and insolubility, elasticity and plasticity, hardness and softness, flammable and inflammable, and strength and weakness.

Word order. The natural order of words in phrases is sometimes reversed to

produce indexing terms or subject headings. This causes problems in using alphabetical indexes and in using free text searching to cover descriptors (see p. 230). Examples include:
    Heating: solar
    Pulp, wood
    Vehicles—ground effect (hovercraft!)
Only good indexes or thesauri can usually help with such inversions.

Novel (new or unusual), rare, unambiguous or highly specific terms. These are all highly desirable properties to look for when selecting search terms. Normally these give good precision if not always such good recall. The problem with such terms is that they may only be useful for free text searching, not having enough occurrences to warrant inclusion in a thesaurus or classification schedule. Examples are balsa (wood) and Surtsey (a small volcanic island).

Homonyms and homographs. There are sometimes individual search terms which can have several different meanings or different applications. For example: bearing, seal, filter, die (German version of the), van (part of a name), orange (part of Orange Free State), naval (part of establishment name), and weir (also a surname).

Acronyms. These can cause trouble when they form real words, such as SAIL (space project) and CARS (chemical analysis).

Changes in terminology. Classification systems are notorious for changes, but subject headings, search terms, thesaurus terms and descriptors may also change over the years. Sometimes it is mere fashion, such as fused salts and molten salts; sometimes it is a growing subject area splitting up into narrower areas, such as spectroscopy and all its derivatives; and sometimes it is a more fundamental redefining of a subject area or use of more explicit terminology.

'Manual' versus 'computer' searching

A great deal has been written on this subject, but for most people who have had significant experience in using both techniques, the advantages and disadvantages of computer searching are only too clear.

Advantages

• Very fast retrieval of information or references, possibly involving a substantial saving of staff time.
• Sophisticated search techniques, such as overlapping concepts and word proximity searches.
• Interactive searching, giving 'hits' and the ability to make changes as you go along.
• Thorough searching with far less chance of mistakes.
• Very wide-ranging, with access to a greater number of databases than usually available locally in hard-copy form.
• Ability to search titles, and sometimes abstracts or even the full text, instead of just indexing terms.
• Sometimes more up to date than hard-copy collections.
• Access to extra facilities, such as downloading, online document ordering, private files, and electronic mail and publishing.

- Printed lists of references or records, possibly merged or sorted into required order.

Disadvantages

- Relatively high exposed costs.
- Special equipment and telecommunications interface required.
- A wider range and possibly a higher level of skills required by searchers.
- Browsing can be expensive and is unlikely to provide the joys of pure serendipity.
- Only one search can be performed at a time if there is only one set of equipment or one password available.
- Service dependent on reliability of equipment, power supply, telecommunications and computers.
- Congested telecommunications or computers can mean a slow service at a much greater overall cost.
- Limited coverage back over the years (often to only about 1970), which can exclude vital basic information.
- Time-consuming for many isolated quick-reference enquiries.
- Larger amounts of staff time spent on administration, training and maintaining skills.

Of course these will change in the course of time, hopefully for the better, with such improvements as user-friendly equipment and services resulting in reduced staff time, and the input of basic information, initially in the form of encyclopaedias, textbooks and handbooks.

Literature searching

This section must not be considered in isolation from the others, as literature searching is only part of the process of obtaining useful information. This section only covers the stage where you have the literature before you. It includes searching all types of literature, from books and journal articles to directories and abstracting journals. Although it concentrates on searching the more complicated types (e.g. indexing and abstracting journals), most of what is said applies equally well to books, articles, patents, reports, etc.

The plan of action

(1)
> Familiarize yourself with the layout of the document or volume(s) you are about to search, e.g. types of index, subject arrangement, etc. (There may be how-to-use instructions.)

(2)
> Choose which section(s) to search, e.g. contents page, author or subject indexes, etc., and, if appropriate, which period of time to cover (e.g. back to 1960).

(3)

> Search selected sections using search terms (chosen prior to, or during, the search).

(4)

> Record what you find and where you find it.

In practice, this is a cyclic process where it is often necessary to go back to earlier stages as the search proceeds, e.g. to try different indexes, periods in time, or search terms.

Familiarization

A few minutes spent here, having avoided the temptation to rush into the search, may save you a considerable amount of time and effort. Library/information-office staff or experienced colleagues can be particularly helpful at this stage.

Instructions

There is frequently a 'how-to-use' section in the more complicated types of literature, such as reference material and guides. Where these instructions are very detailed (e.g. for *Chemical Abstracts*), it may only be necessary to read one or two appropriate sections, depending on the nature of your search.

The things to look for

(1) Types of index available: in addition to author and subject indexes, there may be corporate, chemical formulae, geographic, taxonomic or other special indexes to aid searching. There may also be an index of publications covered and their abbreviations.

(2) Periods of time covered/cumulated indexes: where a publication is issued in parts over a period of time, the overall layout and indexes may differ, depending on whether the parts are annual volumes, monthly issues, supplements, etc. Monthly indexes may be cumulated into semi-annual or annual indexes, and there are sometimes multi-annual indexes to save searching time. The layout, types of index, indexing terms, and even the title of the publication may change from time to time.

(3) Lists of contents/subject classifications: sometimes the contents of documents or volumes have a structure (e.g. patents have claims, reports normally have conclusions). In the case of books, abstracting journals, etc., there may be a specific subject arrangement for the contents which is described by a separate list of contents or subject layout.

(4) Location of indexes: they may be bound in, as an integral part of the volumes containing abstracts, references or data, or they may take the form of separate volumes. Multi-annual indexes are sometimes kept together, separate from the main sequence.

Choosing which sections to search

Once you have familiarized yourself with the literature you are about to search, this choice is usually a simple one. There are, however, certain points worth mentioning:

(1) Types of index available: where you already have a reference on your search topic, which should have been covered by the literature you are about to use, it is a sound principle to check whether it has been covered via the author index. If it has, you can then try to find out how it has been treated: for example, in the subject index or subject arrangement. Once you find the appropriate indexing term or location, you can usually expect to find other material, on your search topic, in the same place.

(2) Periods of time covered/cumulated indexes: where there are separate indexes covering successive periods in time (e.g. abstracting and indexing journals), you must pick those indexes most likely to cover the period in which your search topic was reported (or, in the case of citation indexes, those most likely to cover the period in which your references were quoted). This is particularly important with very recent inventions and discoveries. The normal practice, in these circumstances, is to start with the latest indexes (e.g. current issues of an abstracting journal) and work back through semi-annual, annual or multi-annual indexes.

Occasionally when looking for information on a rare subject, it is preferable to identify appropriate indexing terms by use of multi-annual indexes, as there is a greater chance of finding something here than in an annual index or current-issue index.

(3) Lists of contents/subject classifications: these frequently supplement subject indexes, and it is often advisable to search both sections. This is particularly true for books, where authors may deliberately leave out subject headings in the subject index, where they have already been covered by the list of contents.

When a subject index is not provided, or for some reason does not appear to help, it may be necessary to scan a whole subject section: for example, in current issues of abstracting journals, where the abstracts are grouped by subject. This grouping can be an advantage, as it brings related items together which makes searching simple. The disadvantage is that each item can only appear in one group, so that there is only one subject approach (whereas there will be multiple entries in ordinary subject indexes).

Choosing search terms

With traditional, manual, or hard-copy literature searching, there are relatively few fields (author, title, classification, subject headings, etc.) which are covered by indexes, especially in the case of subject searching. With subject searching in abstracting and indexing journals, for example, indexes are often restricted to thesaurus terms, that is a fixed or controlled vocabulary of subject headings (or keywords, descriptors, search terms etc.) or, in some indexes, to words in titles. It is unusual to be able to perform retrospective searches by looking under classifications, except in the case of certain guides to books or to patents. Choosing subject search terms is, therefore, usually the prediction of subject headings or possibly title words, by guesswork, backed up mainly by the use or thesauri, trial and error in the actual indexes, and the use of any known references as levers or keys into the indexing system (see p. 215).

Searching procedures

This is the stage where you search the chosen indexes, contents lists, etc., for your selected search terms. As suggested earlier, a flexible approach is always necessary as you may have to introduce new search terms which arise during the search from the indexes, etc., or from the information you find. Bear in mind that mistakes can occur in indexes, and remember that indexing policy can change over the years, and different search terms may apply to the same subject at different periods of time. Always take note of cross-references and of the search terms at either side of the one you are looking for in the alphabetical sequence. Also check all the sub-headings of any potentially useful main headings.

If you are regularly scanning the current numbers of a particular indexing journal, it is helpful to keep an alphabetical list of search terms by your side as an aid to memory. You can underline the most important terms, so that when time is very short, you can restrict your search in a sensible way.

Indexes

These are usually the main link between search terms and the information you seek (or references to that information). You may find all the details you need for using the index in the 'how-to-use' section, if there is one. We shall describe here some of the aspects of indexes which may cause you difficulties. If there is a contents page as well as an index, it sometimes happens that the index specifically excludes headings that appear in the list of contents, presumably because they are taken for granted.

INDEXING RULES

Indexes usually contain lists of search terms arranged in an alphabetical order. Where simple, single words are concerned, there is usually no problem. Where search terms consist of several words, numerals, abbreviations or acronyms, there is more than just one way of arranging them into an alphabetical sequence. For this reason, most indexes are constructed according to sets of rules. These rules can differ from one index to another.

Alphabetization. There are two main ways of placing search terms which consist of two or more words into an alphabetical order. Each group of words may be considered as one continuous string of letters (ignoring spaces), and this is known as letter-by-letter alphabetization. Alternatively, the first words in each group may be considered separately, and the subsequent words only considered when the first words are identical. For example:

	Letter-by-letter	Word-by-word
Groups of words:	Newhaven	New York
	Newton	Newhaven
	New York	Newton

Hyphenated words:	Waterproofing	Water-wheels
	Water-wheels	Waterproofing
Abbreviations, etc.:	Uranium	USA (letters treated as words)
	USA	Uranium

Expansions. Abbreviations, acronyms, numerals, etc., may also appear in an order as if spelt out in letters. For example:

5%	as	Five per cent
Mc	as	Mac
Müller	as	Mueller
NATO	as	North Atlantic Treaty Organization
St	as	Saint
USA	as	United States of America

Ignored words. Certain words, prefixes, prepositions, articles, numerals, etc., may be ignored when placing a group of words in an alphabetical order. For example:

De la Mare $\left.\begin{array}{l}\\\\\end{array}\right\}$ may be inverted $\left\{\begin{array}{l}\text{Braun, Von}\\\text{Mare, De la}\end{array}\right.$
Von Braun

Journal of Inorganic Chemistry $\left.\begin{array}{l}\\\\\end{array}\right\}$ may be inverted $\left\{\begin{array}{l}\textit{Journal American Chemical}\\\quad\textit{Society}\\\textit{Journal Inorganic Chemistry}\end{array}\right.$
Journal of the American Chemical Society

2,4-Dinitrobenzene $\left.\begin{array}{l}\\\\\end{array}\right\}$ may be inverted $\left\{\begin{array}{l}\text{Butane (iso)}\\\text{Dinitrobenzene (2,4-)}\end{array}\right.$
iso-Butane

The parts that are ignored are frequently placed in brackets, either in their natural position or at the end of a word or group of words, in order to preserve meaning.

Although it is unnecessary to learn all the different types of indexing rules, you must again be prepared to adopt a flexible approach when searching an index. If you cannot find a particular search term at the first attempt, consider where else it might have been placed, and try there. Good indexes will not only offer assistance in their 'how-to-use' section, but will also provide appropriate cross-referencing to help you find awkward search terms (e.g. US, *see* United States).

DETAILS SUPPLIED IN AN INDEX

Some indexes look different from the simple layout found in, say, telephone directories. The examples, described below, cover the main types you will encounter. Very specialized indexes such as chemical formulae or structural indexes are beyond the scope of this book.

There are two kinds of information in each index entry, which can cause you difficulty. There is the information which tells you if the entry is worth following up, and this may vary from a single broad search term to a specific search term, backed up by a descriptive title. There is also the information which directs you away from the index to the body of the work or to other volumes or documents. These directions may vary from a complicated code to a complete bibliographical reference. For example, an easy-to-use index might contain the following (imaginary) entry:

ANALYSIS: the chemical analysis of iron in copper. J. Smith. *Chemical Analysis*. Vol. 3, No. 2, March 1963, pp. 5–9

with adequate cross-referencing, such as:

Iron: analysis in copper, *see* ANALYSIS

Compare this with the following entry, which could be all that you would find in a subject index to direct you to the same reference:

ANALYSIS: 2:13:4486

In the first example it would be easy to find the entry on analysis of iron in copper, and the reference which follows is obviously details of the corresponding journal article. In the second example you would have to look under the single broad search term, 'analysis', then decipher the code which follows (this could be 2—1972, 13—issue number, 4486—abstract number in the issue of an abstracting journal covered by the index you are consulting).

The following kinds of index illustrate these problems.

KWIC (Keyword in Context) indexes. These are usually computer-produced indexes which manipulate titles of articles, etc., to construct an index using all the significant words. For example:

in copper	ANALYSIS of iron [details of reference, etc.]
iron in	COPPER. Analysis of [details of reference, etc.]
Analysis of	IRON in copper [details of reference, etc.]

The title is printed out as many times as is necessary to bring each one of its significant words to an appropriate place alphabetically, in the centre of the page (cf. KWOC, Keyword out of Context). A reference, abstract number, etc., is usually given alongside. These indexes rely on short descriptive titles. If the title is too long, it is simply cut to size, often making it difficult to understand.

Chain indexes. These usually consist of entries containing strings of search terms arranged alphabetically, or in some other order, with one of the search terms carrying the reference, etc. For example:

Accidents, Aircraft, *see* AIRCRAFT, Accidents
AIRCRAFT, Accidents, Statistics
   British aircraft accidents. J. Smith. *Statistics Bulletin*. Vol. 3, No. 5, May 1963, pp. 5–9 (an imaginary reference)

Encyclopaedic indexes. With these there are no separate subject indexes, and the contents (e.g. abstracts) and index are combined into one. This usually means that specific information about each entry (e.g. abstract) occurs in only one place—with the entry. The indexes tend to be less specific, with one search term covering a number of different entries, and this can cause difficulties, particularly as you have to rely more heavily on cross-references. Known references may be used to identify appropriate search terms (see p. 215).

Where adjacent search terms can be pages apart, it is especially necessary to check the search terms on either side of the one you choose to look for, as you may miss related search terms or sub-headings. The latter are common in this type of index (e.g. discoloration could be a sub-heading under the search term 'plastics').

Co-ordinate indexes (see also p. 270). With this type of index you will find an alphabetical list of search terms with a group of numbers listed alongside each one. These numbers usually refer to abstracts, references, etc., in the main body of the work. To locate items which require several search terms to describe them, you have to co-ordinate the terms by looking for coincident numbers. For example, if a reference (No. 54) to the analysis of gold was covered by this type of index you might see something like this under analysis and gold in the index:

ANALYSIS 10   73,54   87   59
GOLD   30,51,42   54,65

where the coincident number is 54, meaning that reference No. 54 has been indexed under both analysis and gold. As in this example, the numbers are sometimes arranged so that the last figure fits into the sequence 0–9 across the page. Although this assists searching for coincident numbers, these indexes can still be tedious to use.

Lists of contents/subject classifications

The two main problems with searching these lists are that usually only broad search terms are given, and there is normally little or no cross-referencing. There is usually only one broad subject approach to each item covered, compared with normal subject indexes, where there may be a number of quite specific search terms.

Recording what you do and what you find

At some time in the future, you may wish to extend your search to cover earlier periods or to bring it up to date. You may even wish to broaden the search using different search terms or guides. In either case a record of what was done during previous searches will be required. This record should include the statement of intent, the titles, dates and indexes of the guides which have been searched, and the search terms which you used.

The record of what you find must not only contain sufficient detail to enable you to obtain the original document at a later date, but must also allow you to refer back to the source of reference (i.e. the guide), so that you can either check for accuracy or recall the identity of a useful guide for further searches. Whenever you note down a reference, begin by recording it *exactly* as you find it. Do not expand abbreviated journal titles unless this can be done with absolute certainty, for instance by consulting a 'list of periodicals covered' published as part of an abstracting journal, if that is where you found the reference. Should the latter be thought worthy of long-term storage in your personal record system, the details must be entered as fully as possible, even though this may mean checking various guides in order to complete or enrich the bibliographic data. The source of each reference should be included, too.

We give examples below demonstrating the lengths to which you might go in recording references for personal files. Our intention is not to indicate how such references should be prepared for publication, since this would depend very much on the medium in question (journal and book publishers have individual house styles, as do organizations originating reports or university departments concerned with theses). For each type of document we provide two examples, one relatively simple, the other more complex, but no attempt has been made to illustrate all the complexities which can occur. We have chosen an author's name as the first element wherever we could, because many personal-reference filing systems are author-oriented. Our pairs of examples are preceded by notes of the bibliographic data required for the literature specified.

Books: author(s) or editor(s), title, edition, volume(s), place of publication, publisher, date, ISBN when available, and where the reference was found. If you are only interested in part of a book, give appropriate details. For example:

- Elliott, D. J.
  Integrated Circuit Fabrication Technology
  New York, McGraw-Hill, 1982 (ISBN 0 07 019238 3)
  [Subject Guide to Books in Print 1982–3, listed under Integrated circuits]

- Todd, John
  Numerical analysis. Chapter 7, pp. 1–90 to 1–125, of
  Condon, E. U. and Odishaw, Hugh eds.
  Handbook of Physics. 2nd ed.
  New York, McGraw-Hill, 1967
  [Advanced Engineering Mathematics, by Erwin Kreyszig, 4th ed., 1979, ref. G 16 in Appendix 1]

Journal articles: author(s), title of article, name of periodical, volume and issue numbers, date, range of pagination, and where the reference was found. For example:

- Barger, James E.
  Signaling along elastic plates with wide band acoustic pulses
  Journal of the Acoustical Society of America, v. 76, no. 6, December 1984, pp. 1721–30
  [Copy in Bill Smith's filing cabinet]

- Abt, Steven R. and Ruff, James F.
  Estimating culvert scour in cohesive material
  Proceedings of the American Society of Civil Engineers; Journal of the Hydraulics Division, v. 108, no. HY 1, January 1982, pp. 25–34
  [Engineering Index, v. 81, pt. 2, 1982, abs. no. 024644, recorded as ASCE J Hydraul Div, listed under Culverts—testing]

Reports: author(s), organization(s) associated with the work, title of report, publication details including date and all alpha-numeric serial codes allocated by organizations originating or processing the document, and where the reference was found. For example:

- Kimber, R. M. (Transport and Road Research Laboratory)
  The effects of wheel clamping in central London
  Crowthorne, Berkshire, TRRL, 1984

TRRL Laboratory Report no. LR 1136
[Online search in IRRD database, using the phrase Wheel clamping]

● Tam, L. T. and Singhal, A. K. (CHAM of North America, Inc., Huntsville, Ala.)
Numerical analysis of flow and heat transfer in the VAFB LOX storage Dewar tank. Final report
Oct. 1984
NASA-CR-174028; NAS 1.26:174028; CHAM/4040-6
[STAR, v. 23, no. 2, January 23, 1985, abs. no. N85-11315, listed under Cat. 34 Fluid mechanics and heat transfer]

Standards: standardization organization(s), title of standard, publication details including date and alpha-numeric serial code(s), and where the reference was found. For example:

● British Standards Institution
Specification for powered home lifts
London, BSI, 1980
BS 5900: 1980
[BSI Catalogue 1985, p. 289, indexed under Lifts, home, powered]

● ANSI (American National Standards Institute)/IEEE (Institute of Electrical and Electronics Engineers)
American National Standard Pascal computer programming language
New York, IEEE, 1983
ANSI/IEEE 770X3.97–1983
[1983 Index to IEEE Publications, p. 589, indexed under IEEE standards; document has American National Standard Pascal. . . on the cover, but IEEE Standard Pascal. . . on the title-page]

Patents: inventor(s)/applicant(s) including organization(s) associated with the application, title of patent, alpha-numeric serial code(s) relating to either application or patent, date (of filing, or at which the application/specification was published, as appropriate), note of any equivalents, and where the reference was found. For example:

● Leakey, David Martin (General Electric Company Ltd.)
Telephone systems
UK Patent Application GB 2,047,048 A, 19 November 1980
[Abridgements/Abstracts of Specifications, div. H3-H5, pt. E, 1980, indexed under UK classification H4 K1U1]

● Shah, Rajiv R. (Texas Instruments Inc.)
Laser processing of PSG, oxide and nitride via absorption optimized selective laser annealing
European Patent Application EP 109,499, 30 May 1984, designated states DE (Germany, Federal Republic) and FR (France). Equivalents: JP 59–103346 (Japan, Kokai Tokkyo Koho) and US 4,472,456 (United States Patent)
[Online search in WPIL database, using international classification H 01 L 21/26, retrieved descriptive title only: Smoothing insulator layers in semiconductor device mfr.—by exposing to laser at wavelength absorbed

by layer only and opt. substrate; Chemical Abstracts, v. 101, no. 24, 10 December 1984, abs. no. 101: 220954h, listed under Section 76 Electric phenomena, patent family indexed under EP 109,499]

Conference papers: author(s), title of paper, alpha-numeric serial code (if any), name of conference with its location and date/publication details of proceedings including title, editor, date, range of pagination for the paper in question, and where the reference was found. For example:

- Carson, B. H.
  Fuel efficiency of small aircraft
  AIAA Paper no. 80–1847 for AIAA Aircraft Systems Meeting, Anaheim, California, August 4–6, 1980
  New York, American Institute of Aeronautics and Astronautics, 1980
  [Engineering Index, v. 80, pt. 1, 1981, abs. no. 002868, listed under Aircraft, efficiency]

- Fredriksson, Sverker
  Hello diquark, goodbye gluon!
  Proceedings of the Hadronic Session of the 19th Rencontre de Moriond, La Plagne—Savoie—France, March 4–10, 1984, v. 2: New Particle Production, edited by J. Tran Thanh Van. Gif sur Yvette, Editions Frontières, 1984, pp. 479–85
  [Copy in Southampton University Library at QC 721 REN, indexed in local conference catalogue under New particle production]

Theses: author, title, type of thesis or dissertation (including degree), awarding institution, date, and where the reference was found. For example:

- Norton, C. R.
  Applications of computer graphics to plant science
  M.Sc. Thesis, University of St. Andrews, 1980
  [Aslib Index to Theses, v. 29, pt. 2, 1982, ref. no. 6816, listed under Biological sciences: botany: general, indexed under Computer graphics]

- Klocke, Norman Lee
  Evaluation of evaporation and transpiration for irrigated corn
  Ph.D. Thesis, Colorado State University, 1983
  [Dissertation Abstracts International, v. 44, no. 6, December 1983, p. 1900-B, University Microfilms order no. DA8317806, listed under Engineering, agricultural]

Government publications: author(s)/government department(s) or agent(s), title, chairman of investigating committee (if appropriate), publication details including date, and where the reference was found. For example:

- GB Department of Employment
  Flixborough disaster—Report of the Court of Inquiry. Formal investigation into the accident on June 1 1974 at the Nypro Factory at Flixborough Chairman: R. J. Parker, Q.C.
  London, HMSO, 1975
  [Government Publications 1975, p. 1672, indexed under Flixborough Inquiry]

● US Congress. House of Representatives. Committee on Government Operations. Environment, Energy and Natural Resources Subcommittee Nuclear safety—three years after Three Mile Island. Joint hearing before certain subcommittees of the Committees on Government Operations and Interior and Insular Affairs, House of Representatives, 97th Congress, 2nd session, March 12, 1982
Washington, DC, US Government Printing Office, 1982
[Monthly Catalog of US Government Publications, 1982, ref. no. 82–24901, indexed under Three Mile Island nuclear power plant]

When quoting an abstracting or indexing journal as a source of reference, it is helpful to include volume, cumulation, issue, page or abstract numbers, where appropriate, and the date.

You can also add any extra information which may assist you in the future, either to evaluate the reference or locate the original document. Note, too, your reason for keeping a reference when this is not immediately obvious, especially if you are in any danger of forgetting why it was selected in the first place. See also **Storing references**, p. 269.

Computer database searching

This includes searching all types of database, including databanks and bibliographic, full-text, numeric and directory databases. Most of the details for retrospective searching mentioned apply equally well to current awareness, but any differences are covered in that section (p. 290). This section only describes the stage where you have selected a specific database, and assumes that you know how to log on and enter that database.

The plan of action

(1)

> Familiarize yourself with the structure of the database records. See which fields are searchable and/or printable, and limitable, and how they may be searched.

(2)

> Choose which fields or combinations of fields (such as the basic index) to search, and, if appropriate, what limitations to make (such as language or year of publication).

(3)

> Search selected fields using search terms and strategy (chosen prior to, or during, the search), either singly or in combination. Observe results, and if necessary modify search strategy by using different search terms or different combinations.

(4)

> Print out or download appropriate records or parts of
> records (such as titles and abstract numbers). Note any
> details required for administrative purposes.

Familiarization

With abstracting and indexing journals, the types of index and the methods of
linking index entries with references have to be considered. With databases, it
is nearly always the 'record' which has to be examined, together with the codes
required to search the various parts of the record (fields), the codes necessary
to limit the search results, and the formats available for printing out records.
Consider the fictitious record:

84–063129
The influence of TV programmes on the use of tobacco and the smoking behaviour of
adolescents.
E. T. Teragic and E. Koms.
Slavonic Tobacco Research Institute.
Journal of Brain Measurement, vol. 38, no. 5, May 1983, pp. 184–192.
JOBMMT
A7, D593, T14
Galvanic skin response measurements on twenty-five tobacco-smoking teenagers . . . .
. . . . further
work in this area is recommended.
TELEVISION/Stimulation; ADOLESCENTS; TOBACCO SMOKING.
TV programmes; galvanic skin response measurements; conditioned responses;
behaviour; teenagers; cigarettes.

This particular record is composed of ten fields: accession number; title of
article; authors; corporate source; journal title and issue details; journal
CODEN; classifications; abstract; descriptors (controlled terms); and
identifiers (extra or supplementary uncontrolled or free text terms). Records
may not always contain some of these fields, such as abstracts, or may contain
other fields, such as chemical formulae, report numbers or contract numbers.
Normally most of the fields can be searched either singly or in combination,
and most of the fields can be printed out.

Instructions

These normally take the form of manuals, guides, quick-reference booklets or
sheets, and thesauri supplied by the hosts and database producers. As far as
individual databases are concerned, the host usually supplies short summary
pages for each database showing the contents of a typical record, the search
codes available, and the formats which can be printed. These are generally
backed up by larger database chapters which go into greater detail, especially
dealing with the peculiarities of the individual database and the techniques
and problems associated with particular fields. The database producers often
supply thesauri, and some even provide comprehensive manuals for searching
their own databases on a range of hosts which mount them.

The things to look for

(1) The types of field available. There may be special fields in some databases which are essential for efficient searching, such as CAS Registry Numbers in chemically related files, taxonomic codes in biological files, and document numbers in reports and patents files.

(2) The nature of the 'basic index'. Normally if you simply enter a search term, the computer will effectively search a selection of fields, which might include titles, indexing terms, abstracts or corporate sources. Alternatively it may either search most or all of the fields, or display a range of options and ask you to specify which fields you wish to select. This can cause problems, and these are dealt with in the next section.

(3) Any splitting of the database into separate files covering different periods in time. In this case, if more than one file has to be searched it is essential to save the search strategy rather than have to type it in again for each file.

(4) Check if there is a thesaurus which you can use to identify permitted subject headings.

(5) The contents of the printable formats. There may be standard formats ranging from titles only to full records, or you may be able to choose your own format by specifying selected fields for printing out.

Choosing which fields to search

Field selection can be obvious; for example, in the case of an author search, you select the author field. There can be advantages and disadvantages in the use of the following fields.

Basic index. Most subject searching is probably done in the basic index (with no particular field specified), probably because it confers a certain confidence that nothing will be missed since the basic index normally includes the title, indexing terms, and, if available, an abstract. In certain cases this can result in a serious failure to retrieve relevant references. For example, if the free-text phrase 'carbon adj paper' is entered (retrieves 'carbon' adjacent to 'paper'), it may be assumed that it will pick up carbon paper in the title and in the indexing terms. However, descriptors do not always conform to natural language, and the appropriate controlled term might be 'paper, carbon', which would not be picked up.

   If the basic index includes the abstract, the number of 'hits' from a given strategy can be increased by over 100 per cent, but with many of the extra references being of less relevance.

   If the basic index includes the corporate source names, searching may result in the retrieval of useless hits because of a mechanism known as 'false drops'. This is where a search strategy produces hits by combining search terms which are not directly linked conceptually in a given record, perhaps because they are in different sentences or fields. For example, a search in the basic index for the irrigation of orange trees (using ORANGE and IRRIGATION) might result in false drops from articles on quite different topics by people working at the Institute of Irrigation in the Orange Free State.

Title. This is useful for the free text searching of terms which may not have been suitable candidates for descriptor or thesaurus terms, because either the individual terms were too vague (such as those in the phrases 'information systems for design engineers' or 'current trends'), or they were too new or rare in occurrence.

Journal CODEN. These are useful terms for running a search on a particular journal or journals that your library either does take (for quick accessible information) or does not take (for current awareness of material which is not available locally, but might be very relevant). Conversely, you can use CODEN in the negative sense to remove references from a search that should already have been picked up by reading in your own library (to save print costs on expensive files).

Classification codes. These are mainly used when text words alone are ambiguous, such as 'bearing', which could be classified as a dynamic device under mechanical engineering, or a static device under civil engineering. They can also be used in combination for broad subject searches for current awareness or browsing. Seal, noise and filter are other examples which may need qualification, especially in multidisciplinary databases. Classifications can sometimes be useful in isolating acronyms from their common language equivalents, as in the case of SAIL (a space project) and sail.

Abstracts. These are a useful field to extend a free text search if the hit rate is too low in other fields or if an exhaustive search is required. The comments above, about free text searching under titles, apply here. Sometimes individual sentences may be searched to prevent false drops between terms in different sentences.

Descriptors. If descriptors exist which unambiguously describe your search topic, you may well be spared the problem of finding alternative terms, such as synonyms. Otherwise it is normal practice to include descriptors and free text terms in a search, either mixed together, or in separate sub-searches.

Identifiers. These are extra, informal terms assigned by the indexer to assist in online searching. They are nearly always worth including in a free text search.

Choosing search terms

One of the great advantages of online searching over hard-copy searching is the number and variety of record fields and databases that may be searched. The computer can be commanded to pick out virtually any data element from a database (words, fragments, phrases and codes). This increases the number of types of search term that may be used, and can result in a greater efficiency and depth of searching.

Most of the general advice on choosing search terms (p. 213) applies here, but in addition there are two extra facilities that may help.

(a) If you have already performed an initial online search and have retrieved a set of reasonably relevant records, some hosts offer you the ability to obtain a ranked list of search terms occurring in those records, in descending order of occurrence. This enables you to check if you have missed an obvious search term which could then be fed back into a new improved strategy. It also tells you what the records are about; for example, are they about design (see below)? The command is ZOOM for ESA-IRS and GET for PERGAMON INFOLINE. The following example is the first part of the display of ranked index terms obtained by 'zooming' a set of records retrieved by the term 'hovercraft'.

Text analysis results

Frq	Words/phrases
26	GROUND EFFECT MACHINES
16	HOVERCRAFT
13	AIR CUSHION VEHICLES
12	HOVERCRAFT GROUND EFFECT MACHINES
7	BIBLIOGRAPHIES
7	NTISNERACD
6	NTISDODXA
6	PERFORMANCE TESTS
6	SURFACE EFFECT SHIPS
5	AERODYNAMIC CHARACTERISTICS
5	DESIGN
	. . . .
	. . . .

(b) The advent of cross-file searching (p. 149) has made it practicable to search a wide range of databases using very precise strategies in order to produce a collection of highly relevant records. This is a possible alternative to an exhaustive search in one or two of the most appropriate databases, in which the same total number of references might have been found, but which would probably have included many references of less relevance.

A further sensible precaution is to include any search terms which would be required to pick out references that have been provided as relevant examples: for example, by using combinations of title words.

One other technique is to make up some imaginary but relevant-sounding article titles, and then select terms from those titles and combine them into a strategy which would retrieve all the titles plus, hopefully, some real ones.

Searching procedures

For a computer search to take place, the search terms must be entered (by an end user or an intermediary such as a librarian or information officer) in a form that the computer can both 'read' and recognize. The search queries must be computer- or machine-readable, that is entered manually online by keying-in commands on terminal equipment, or entered in batch mode via punched tape or cards, or magnetic tape, disks or other storage devices (prepared offline and fed in in one batch). Some modern equipment allows a combination of entries from the keyboard, memory devices (such as ROMS) and disks. For the computer to recognize what is keyed in, a command language is used which is usually specific to a given host.

● Search commands are single search statements entered in one step, which may be anything from simple to complex combinations.
● Search strategies are the complete sets of commands or statements that make up a search.

● Search profiles are usually strategies that have been refined until sufficiently reliable to be fed into the computer at regular intervals (possibly automatically) to provide a current-awareness service. A profile usually represents a subject interest of a particular person or group.

Search commands, strategies or profiles may be compiled offline if the equipment and software permit, and then sent online to the computer line by line or in blocks. Search strategies may also be stored temporarily or permanently on suitable local equipment or on the host computer, depending on the facilities available.

There are several ways to understand the search process; two of them are set out below. The first is to take a specific reference and see how various strategies to retrieve the reference (and other similar ones) could be built up and refined. This approach is theoretical and uses simple Boolean logic equations, but does not typify any particular host command language. The second way is to examine a selection of real search commands, as used for two well-known hosts, for retrieving the record shown on p. 228.

Remember that the more planning that can be done before you go online (time permitting), the more money and time may be saved, and the more systematic will be the search. Although interaction is a vital facility with online systems, if it is overused it can result in a muddle of search terms and combinations which become increasingly difficult to re-interpret, either as the search proceeds or else at a later date.

The computerized search

Consider the fictitious reference: 'The Incidence of Lung Cancer in Cigarette Smokers, by O. C. Cabot. *British Journal of Cancer*, Vol. 5, No. 6, pp. 55–71, June 1970.

This reference may be stored on magnetic tape in the same sequence as written above but in machine- or computer-readable form. The computer can be programmed to search a tape containing such references by simply scanning the tape from beginning to end, looking for words or combinations of words (perhaps in a certain order). This is known as a sequential search and is a relatively lengthy process, because the computer must search all the tape or tapes, and all the references or the whole database. Hosts now use more sophisticated methods of storing the searchable parts of the references by constructing inverted files, similar to those used for co-ordinate indexes, on disks instead of tape. The method of searching these is known as random-access, because the computer only has to search through a small 'relevant' part of the database.

Consider a simple sequential search where the computer scans the reference in the same sequence as it is written, as well as all the other references on the tape or tapes. We want the computer to pick out the Cabot reference and any other references on that subject.

Suppose we program the computer to search for all references containing the word 'cancer' in any part of that reference. The computer would usually be programmed to extract the references containing the required word or words (e.g. 'cancer'), store them, and then print them out after the search. These references are sometimes called 'hits'.

In this case we would probably end up with hundreds of references about all kinds of cancers, cancer-inducing substances, cancer in rats, etc., etc. Our reference by Cabot would be picked up on two counts, because 'cancer' occurs in the title of the article and also in the name of the periodical. For the printout to be really useful, it must have greater 'precision', i.e. we need to eliminate the unwanted references or 'noise'.

We can do this by programming the computer to pick out references that contain both the word 'cancer' and the word 'lung'. This would still include our Cabot reference, but it should exclude articles with titles such as 'Cancer of the liver' etc.

In this case we would probably end up with a hundred or so references, which might still include some titles we did not want, e.g. 'Cancer of the lung in horses'. We might also miss useful references that did not have this combination of words in their titles, e.g. 'Cancer in cigarette smokers'. Thus for a printout to be really useful it must also have a reasonably high 'recall', i.e. it must not miss large numbers of useful references. (As mentioned previously, in practice it is virtually impossible to have 100 per cent precision and 100 per cent recall. A compromise is necessary, and most printouts include some unwanted references and miss some useful ones; see p. 213.)

So now we tell the computer to pick out any reference containing either the words 'lung' and 'cancer' or the words 'cancer' and 'smokers', and so on and so on.

What we have been doing in a roundabout way is to construct and refine a search strategy or profile. There is a logical way to approach this, and the method is given below, but remember the result is always a compromize between high recall and high precision (and ultimately between both of these and cost).

STRATEGY/PROFILE CONSTRUCTION

What follows is meant to illustrate the general principles involved in profile construction, *not* the working of any particular service. Suppose you are interested in the relationship between lung cancer and cigarette smoking, and you wish an information service to supply you with relevant references. You also want references to everything published by your rival Dr O. C. Cabot.

(1) Analyze the subject of your search into its separate concepts:

(a)	(b)	(c)	(d)
Lungs	Smoking	Cancer	Dr O. C. Cabot

where (a) is a part of the human body; (b) is a possible cause of cancer; (c) is a disease; and (d) is a person's name.

(2) Expand these concepts into groups of similar or related terms, using either natural language or a thesaurus of permitted terms:

(a)	(b)	(c)	(d)
Lungs	Smoking	Cancer	O. C. Cabot
Respiratory system	Cigarettes	Tumours	
Bronchi	Cigars	Growths	
Alveoli	Pipes	Disease	
Trachea	Tobacco	Damage	
	Nicotine		

(3) Truncate the terms if appropriate:

(a)	(b)	(c)	(d)
Lung*	Smok*	Cancer*	O. C. Cabot
Respiratory system*	Cigar*	Tumour*	
Bronchi	Pipe*	Growth*	
Alveoli	Tobacco	Disease*	
Trachea	Nicotine	Damage	

This means in the case of Smok*, for example, that the computer will pick out any word beginning with Smok, i.e. smoke, smoking, smokers. It is sometimes possible to truncate at both ends of the word, e.g.*chem* could be used for alchemy and chemistry. This simply reduces the number of search terms.

(4) Regroup these terms in 'parameters' by either splitting or combining concepts as necessary. A parameter is simply a number of terms which have been grouped together for convenience as far as the search strategy is concerned. This should become clear in due course.

P1	P2	P3	P4	P5	P6
Lung*	Respiratory system*	Smok*	Cancer*	Disease*	Cabot O C
Alveoli	Bronchi	Cigar*	Tumour*	Damage	
	Trachea	Pipe*	Growth*		
		Tobacco			
		Nicotine			

The reason for splitting the first concept is that while alveoli are parts of the lung, bronchi and trachea only lead to it, and respiratory system is merely a general or broad term. Thus the terms in P1 are more specific to our interest than those in P2. (Similar reasoning applies to the split of concept (c) into P4 and P5.)

(5) Relate the parameters to the required output by writing a 'logic equation' (the mathematical process is sometimes referred to as Boolean algebra or logic). As a first step, we could write:

Output = P1 *and* P3 *and* (P4 *or* P5)

This means that any article which contains any one of the terms in P1 *and* any one of the terms in P3 *and* any one of the terms in *either* P4 *or* P5 will be printed out by the computer. Note that all the terms within a parameter are effectively connected by *or* logic.
  As a second step we could write:

Output = (P2 *and* P3 *and* P4) *or* P6

This means that any article which has been given at least one index term from each of P2, P3 and P4 will be retrieved, and *any* article which has been indexed under P6 will also be retrieved. (Cabot O C is sometimes referred to as an independent term because it will retrieve references without having to be linked with any other terms.)
  Combining these two steps we can write:

Total output = [P1 *and* P3 *and* (P4 *or* P5)]
   or [P2 *and* P3 *and* P4]
   or [P6]

Thus an article indexed under Lungs (P1) and Cigars (P3) and Cancer (P4) will be retrieved, but an article indexed under Bronchi (P2) and Smoking (P3) and Damage (P5) will not be retrieved (it probably relates to a case of chronic bronchitis anyway!)

More sophisticated logic, including *with* and *not* terms, is easy to understand once the basic principles given here have been accepted.

It is also possible to employ a search strategy which uses a system of allocating numerical weights to each search term according to its relevance. This weighting system can be superimposed on a logic equation, so that if the sum of the weights falls below a prearranged limit (the threshold weight), the article will not be retrieved. When two terms in the same parameter occur, only the one with the highest weight contributes to the total sum. Suppose our parameters are the same as our expanded concepts and the weighting is as follows:

Pa	Pb	Pc	Pd
Lung* 10	Smok* 10	Cancer* 10	Cabot O C 30
Respiratory system* 5	Cigar* 10	Tumour* 10	
Bronchi 5	Pipe* 10	Growth* 10	
Alveoli 10	Tobacco 10	Disease* 5	
Trachea 5	Nicotine 10	Damage 5	

Our search strategy is now represented by the logic equation

   Output = Pa *and/or* Pb *and/or* Pc *and/or* Pd

If a threshold weight of 25 is chosen, articles containing

Lungs	*and*	Smoking	*and*	Cancer	
10	+	10	+	10	=30

or Cabot O C = 30 will be retrieved, but articles containing

Bronchi	*and*	Smoking	*and*	Damage	
5	+	10	+	5	=20

will not be retrieved.

The order of references in a printout may be arranged by author, journal, date, language or weight, depending on the system.

One of the advantages of constructing a search profile is that it makes you think more carefully about the precise nature of the information you require.

With online systems it is possible to key your profile directly into the computer terminal, and see how many references would be retrieved, at each stage of the profile construction. It is also possible to display or printout some of the references as you proceed, to see how the profile is performing, making adjustments where necessary (interactive searching).

Command languages

The two command languages chosen to illustrate searching are those used by DIALOG and DATA-STAR. DIALOG is a derivation of the original Lockheed system language and is similar to that used by ESA-IRS. DATA-STAR's command language is a derivation of IBM's STAIRS and is similar to that used by BRS. Note that with DATA-STAR there are different 'modes', and it is necessary to be in the search mode, rather than the print mode, for searching. DIALOG is, effectively, always in the search mode. When a computer is ready to receive a command, it will normally print or display a 'prompt' using a symbol such as * or ?, a statement number, or some other message. When you wish to send a command to the computer, you must type it in and then press the enter, return, carriage return or CR key to transmit the command. For error correction, see p. 239.

In most cases the examples get progressively more selective, and they all retrieve the record shown on p. 228. DATA-STAR commands are shown in italics.

Search commands	All retrieved references contain:
SELECT TOBACCO (or S TOBACCO) *TOBACCO*	tobacco in the basic index
S TOBACCO/TI *TOBACCO.TI.*	tobacco in title (this is called field qualification, and different suffixes or prefixes are used for each field required)
TOBACCO..AB.	tobacco anywhere except occurrences only in the abstract
S TOBACCO OR SMOKING *TOBACCO OR SMOKING*	either tobacco or smoking or both in the basic index
S TOBACCO AND TELEVISION S TOBACCO(C)TELEVISION *TOBACCO AND TELEVISION*	both tobacco and also television in the basic index
S TOBACCO(F)(TELEVISION OR TV) *TOBACCO SAME (TELEVISION OR TV)*	tobacco and also either television or TV (or both) in the same field
S TOBACCO(F)TELEVISION/DE S TOBACCO(C)TELEVISION/DE S TOBACCO/DE AND TELEVISION/DE *TOBACCO.DE. SAME TELEVISION.DE.* *TOBACCO.DE. AND TELEVISION.DE.*	both tobacco and television in the descriptor field (/CT for ESA-IRS); (C) is equivalent to AND
S TOBACCO/DE AND TV/TI,AB,ID *TOBACCO.DE. AND TV.TI,AB,ID.*	tobacco in the descriptor field and also TV in one or more of the other fields indicated; ID means 'identifier'
S TOBACCO(S)TEENAGERS *TOBACCO WITH TEENAGERS*	tobacco and teenagers in the same sentence; (S) is not always available and may mean sub-field
S TOBACCO(2N)SMOKING	tobacco within 2 words of smoking, in any order
S TOBACCO (NOT 2N) CHEWING	tobacco unless chewing is within 2 words of it

S TOBACCO(2W)TEENAGERS	tobacco within 2 words of teenagers in the order indicated
S TOBACCO(W)SMOKING *TOBACCO ADJ SMOKING*	tobacco adjacent to smoking in the order given
S TOBACCO(W)SMOKING/AB *TOBACCO ADJ SMOKING.AB.*	tobacco adjacent to smoking in the abstract in the order given
S CS=SLAVONIC(W)CS=TOBACCO *SLAVONIC.IN. ADJ TOBACCO.IN.*	Slavonic adjacent to tobacco in the corporate source field in the order given

AND, OR (and NOT, see below) are called Boolean operators, and (2W) and ADJ are called word proximity, free text or full text operators. Impossible commands, such as those using word proximity operators between terms in different fields, are obviously not permitted.

More than just two terms may be used to produce an even narrower search statement:

S SMOKING(W)BEHAVIOUR(1W)ADOLESCENTS/TI	smoking adjacent to behaviour within one word of adolescents in the title
(SMOKING ADJ BEHAVIOUR) WITH ADOLESCENTS.TI.	smoking adjacent to behaviour in the same sentence as adolescents in the title

This last example illustrates two points: first, it may not be possible to translate exactly a strategy from one command language to another; second, brackets or parentheses show which operator should be applied first. The use of brackets is not always necessary if you know the computer's order of priority for the operators (see p. 239), but it is good practice to use brackets all the time, so that you do not forget them when they really are needed, and also so that the strategy is more obvious or explicit.

The NOT operator. This may be used to negate unwanted references, but it can be a blunt instrument and must be used with care. It is best to use it only when the term being negated is very unlikely to occur in relevant references. For example, if you want information on chlorides, and decide to use the term halides as well, so as to be sure of an exhaustive search, it would be unwise to negate (say) bromides from the halides group. This is because there might be references to bromides which included information about chlorides under the umbrella term of halides, although the word chlorides was not used in the title or indexing terms (such as an article on the 'solubility of halides using bromides as a model'). Of course if you negated bromides from the chlorides group you would miss, for example, references comparing or covering both chlorides and bromides.

The NOT operator is very good for negating search terms such as journal coden, authors, or corporate sources (independent terms).

S TELEVISION(F)SMOKING NOT ADULTS *(TELEVISION SAME SMOKING) NOT ADULTS*	television in the same field as smoking unless adults appears in the basic index

SET NUMBERS

When the computer responds to a search command, it normally gives a running 'set' (or command/statement) number, the number of hits produced by the command, a copy of the command itself for verification purposes, and a prompt (a code such as '?' which tells you that the computer is waiting for you to enter a command). In the following examples, the entered commands are in **bold** type, and the computer prompts and responses are in ordinary type.

For DIALOG:

?S TOBACCO(F)TELEVISION

	1567	TOBACCO
	4783	TELEVISION
S1	434	TOBACCO(F)TELEVISION

?S TOBACCO(1W)TEENAGERS

	1567	TOBACCO
	4213	TEENAGERS
S2	211	TOBACCO(1W)TEENAGERS

These set or statement numbers may be incorporated into search commands.

?S S1 AND S2

	434	S1	references satisfying both statement S1 and
	211	S2	also statement S2
S3	53	S1 AND S2	

?S S3 NOT ADULTS

	53	S3	references satisfying statement S3 but not if
	5301	ADULTS	they contain adults
S4	39	S3 NOT ADULTS	

A 'select steps' command (SS) will provide separate search statement numbers for each search element.

?SS TOBACCO(F)TELEVISION

S5	1567	TOBACCO
S6	4783	TELEVISION
S7	434	TOBACCO(F)TELEVISION

Brackets or parentheses should be used for more complicated search statements, such as:

S (TOBACCO OR SMOKING) AND (TV OR TELEVISION)

to tell the computer the correct order for performing the combinations (but see also next page), and to keep the strategy easy to interpret.

Both DIALOG and ESA-IRS have a 'combine' command which may be used with just set numbers on their own. For example:

COMBINE (1 AND 2) NOT 3   or abbreviated to C(1*2)-3

DIALOG no longer promote the use of the combine command or logic symbols (*, + and −), and do not permit the use of the minus sign as an abbreviation for NOT logic in S or SS select commands.

More complicated statements involving different levels of parentheses (clustering or nesting) are usually possible, such as:

S (S1 OR S2) AND S3 AND ((S4 OR S5) AND (S6 OR S7)) OR S8

or

C(1+2)*3*((4+5)*(6+7))+8

The main problems with complex statements are that they may take a long time to execute, and on some systems you cannot tell whether or not the computer is still processing the commands; after several minutes with nothing happening, you do not know whether to wait a bit longer or to try to log on again in case you have been cut off. Furthermore, on some systems you do not get intermediate results, so that if zero hits is the result, you do not know why.

The computer prompt may include the set or statement number, and the response may just consist of the number of records retrieved.

For DATA-STAR:

11–: **TOBACCO SAME TELEVISION**
RESULT   434
12–: **11 AND TEENAGERS.AB.**
RESULT   64
13–: **12 NOT ADULTS**
RESULT   41
14–: **(TOBACCO OR SMOKING) AND (TV OR TELEVISION)**
RESULT   591
15–: **((1 OR 2) AND 3 AND ((4 OR 5) AND (6 OR 7))) OR 8**
RESULT   1

COMPUTER PRIORITIES IN SEARCH STATEMENTS

The statement 1 OR 2 AND 3 could mean (1 OR 2) AND 3, but is usually interpreted by the computer as 1 OR (2 AND 3). This is because the computer is programmed to perform different operations in a particular order. The normal order of priority is:

( ) before NOT before AND before OR

SEARCH STATEMENT ORDER IN STRATEGIES

It is often helpful to put as many of the subject search statements or terms as you can at the beginning of a strategy, in groups (broad terms first) if appropriate, and all combinations at the end. This is particularly important for saved searches where it may be necessary to make subsequent changes or to prevent muddled strategies that are difficult to interpret at a later date. There has to be a balance between interactive searching (including combining as you go along) and systematic searching strategies. Sometimes interactive searching does have to be used right at the beginning, especially if you are unsure of the terminology or likely hit rates. In this case, it may be preferable to re-enter a tidied-up version if it is required to repeat the search in another database or to save it.

OTHER POINTS ON SEARCHING

Errors. Where these are noticed as typed, they may usually be corrected by back-spacing using 'Control H', that is by pressing the H key while the Control or CTRL key is also depressed. A whole statement may be cancelled, before transmitting, by pressing the Escape key and then the Return key.

Directory displays. These permit a searcher to see a selected part of an alphanumeric listing of search terms, names, numbers or codes, either from the basic index or from a particular field in some cases. This is effectively an online dictionary, index or thesaurus. Typical commands are EXPAND, ROOT and NEIGHBOR. These are particularly good for checking the existence of a term or terms; the spelling of terms, especially authors' and chemical names; the abbreviations for terms, such as corporate sources; and the saving of typing time for alphabetical or numerical ranges of terms, for instance the latest updates to a file, or a run of terms beginning with the same word stem. These ranges may be transferred into search statements by a simple command on some hosts. A typical example of a display of part of an author directory is shown below:

```
                EXPAND AU=SMITH, C. W.
REF  INDEX-TERM           TYPE ITEMS
E1      AU=SMITH, C. D._____1
E2      AU=SMITH, C. F. C._____11
E3      AU=SMITH, C. L._____3
E4      AU=SMITH, C. S._____2
E5      AU=SMITH, C. V._____1
E6    −AU=SMITH, C. W._____15
E7      AU=SMITH, C. WAYNE_____12
E8      AU=SMITH, CARL MAYN_____2
E9      AU=SMITH, CHARLES B._____7
E10     AU=SMITH, D._____10
. . . . . . . . . . . . . . . . . . . . . . . . . . . . . . . . . . . . . . . .
```

In this case the command SE6-E7 would result in a sensible selection.

File openings. These are the commands to start or restart in a new or existing file, erasing all previous search strategies. Typical examples are: BEGIN29, B3, . .C/INSP, FILE PSYC. The command should normally be given before starting a search that is to be saved, or before executing a saved search so as to reset the statement number to one. Because these commands erase existing statements, care must be taken before using them to save any strategy that is required for repeat searches.

Record displays. It is often helpful to display parts of retrieved records to see how the search strategy is working. Titles can be displayed to check for relevance. Indexing terms can be displayed to obtain new terms to feed back into the strategy (see also ZOOM and GET, p. 231). The commands used are the same as those described under local display and printing on p. 242. Care must be taken, sometimes, to obtain a representative selection of hits for evaluation over the total period of time covered, rather than just the latest half-dozen.

Strategy display. These commands allow all or part of the ongoing search strategy to be displayed. This can be essential for people without printers running in parallel, or with obscured paper feeds. Examples include DISPLAY SETS, DS, D8-25, . .D ALL, HISTORY, and RECAP.

Too few (or missed) hits. This is usually due to either an over-restrictive search or a lack of relevant records in the file being searched. In the former case, the search may be widened by using broader terms, using both free text terms and descriptors, reducing any restrictions introduced by proximity operators, truncating (see below), or starting again with a different selection of search terms, possibly obtained via a GET or ZOOM command. Different databases may be used, selected perhaps via a cross-file search on whatever part of your strategy actually produced hits. You may also use the known reference method mentioned under literature searching. The known reference may be obtained as a hit by entering a very specific word proximity search on its title, or sometimes by entering the accession number, perhaps available via the hard-copy equivalent to the database. The full record can then be displayed, and any classifications, descriptors or identifiers noted and fed back into the strategy. Any search strategy can then be checked for its ability to retrieve the known reference by simply applying NOT logic (negating the known reference from the strategy). If the number of hits goes down by one, the strategy is retrieving the known reference. As mentioned previously, it is sensible to compile your search strategy so that it does retrieve any known references that you may have, perhaps by proximity search statements to cover the titles.

Too many hits. This may be due to unavoidably vague terminology, too broad a search strategy, or a very popular topic. In the first two cases it may be possible to use classifications, narrower terms, or more restrictive search strategies, such as proximity operators. In the case of a prolific subject area, individual search terms or whole search strategies may be limited by year, language or field. If the strategy and host permit it, retrospective limiting to title words is a crude, but often effective, way of extracting many of the most relevant records (for example, to limit set no. 2 to titles only, the command is S2/TI for DIALOG or 2.TI. for DATA-STAR). As an alternative, the results may be limited by intersection (ANDing) with terms like review, survey, bibliography, state, current, trend, etc., and their plural forms.

Truncation. This can be both a time-saving and an insurance device to capture a group of terms (via a search command) which all contain the same word stem or fragment. Truncation is usually right-handed, rarely left-handed, and is depicted by symbols such as ?, #, $, and :, depending on the host. For example, ENGINEER? might pick up ENGINEER, ENGINEERS and ENGINEERING, while *CAR* might pick up AUTOCARS, CARRIAGE, CAR, etc. This unlimited truncation, when applied to popular stems, or when applied in a extreme way, such as SPEC? or VIBR$, can cause severe problems: lengthy response or computer-processing times; irrelevant retrieval, such as special instead of spectra; and computer cut-offs if more than, say, 100 terms or phrases begin with the stem in question. If the terms you require are few in number and have a popular stem, it is advisable to type them all out separately (for example, spectra or spectrum or spectroscopy or spectroscopic).

Truncation can normally be restricted to a given number of characters, such as two in the cases of VIBRATION?? or VIBRATION$2. Some hosts permit truncation symbols inside words, to represent one universal character (or more with some systems) for coping with alternative spellings, such as organi?ation, wom?n, colo$r, or sul$ur.

Recording and preserving searches

There are three parts to a search: the strategy, the results in terms of the numbers of hits, and the results in terms of actual records or references. Depending on your equipment and software, and also on particular hosts, some or all of these parts may be output, saved or stored, either locally and/or on the host computer, temporarily or permanently.

The search strategy may be saved on the host computer by using commands like END/SAVE, END/SAVE TEMP, SAVE, SAVE OLD, or . .SAVE. As mentioned earlier, it is best to clear away any old strategies before starting a new one which is to be saved, and it is essential to give the save command before opening a new file or database, or the entire strategy will be lost.

The executed search (with hits) is automatically purged on logoff by some hosts; kept until space is needed, by other hosts; or kept for a limited time, on the receipt of the appropriate command, by at least one other host. If you are accidentally cut off or 'thrown off' because of a computer or network failure, the host usually allows a short interval of time for you to re-establish a connection, before deleting the search.

Local display or printing

The records may be displayed on a VDU and/or printed out on your own equipment, either partially or completely, by commands which normally state the set number, the format (selection of fields), and the number or range of items to be displayed or printed. Examples are:

T14/4/1–20
. .P 14 ALL/ALL
PRT SS 14 FULL 20

which all mean: print from set 14, the full records of all (20) items retrieved (hits). It is possible to use other predetermined formats, or to make up your own choice of fields with some systems, or to print out a selection of items.

Be very careful not to make a mistake with these commands, because some hosts charge for the number of records requested in the command transmitted or sent, not for what is actually printed out (you might use the break key to stop the printing, or you might be accidentally cut off). If you are at all unsure of your search strategy, or the reliability of the telecommunications connection, it is best to print out in batches of no more than 20 to 40 references at once, especially with databases that have high print royalties.

Records are normally printed out in chronological order (latest first) unless otherwise requested, but there are some databases, such as *ERIC* and *BIOSIS*, which have more than one chronological sequence, and therefore do not provide all the latest references together.

Local printing is immediate and is possibly less expensive than remote offline printing, provided that a high-speed printer is used, and only short formats (just references) are printed.

Offline printing may be done by the host using very high-speed printers, either at the host's computer installation or at regional centres, and then sent to the searcher or end user by mail. Commands are usually similar to the local display/print commands, and under certain conditions, may be cancelled after having been sent. (Some hosts have files which are unavailable online, but which may be searched offline using an appropriate command, and printed out offline in the normal way.) Typical commands are:

PR23/5/1–200
. .PRINTOFF 23 AU,TI,SO/1–200

Offline printing can be less expensive than online printing for full-format records (especially for long abstracts), particularly if your own printer is slow. The quality of the offline prints may be significantly better than that from local printers. It may also be inconvenient to tie up local equipment for long periods of time in printing out large numbers of references.

Downloading

This is the electronic capture of records for editing, sorting or merging, prior to printing; or for input into a different information storage and retrieval system. This can only be done with the specific permission (and possibly involving special charges) of some database producers, and not at all with a few others at the present time. The main worry for database producers is a loss of revenue caused by piracy and copyright infringement. The loss could result from customers being able to re-search records (previously retrieved in the normal way) on their own computer system without further online connection to the host, or (more seriously) selling repackaged information to third parties. The problem has been partially solved by some producers permitting downloading under certain contractual conditions, and by one host (only ESA-IRS, so far) introducing a downloading command, which incurs increased royalties but provides flagged fields to make the records easier for processing on a local system. Capture of records for editing is technically quite easy to do, whereas capturing significant numbers of records for large-scale information storage and retrieval systems can be both expensive and difficult. If you wish to download, it is essential to check with both the host and database producer that it is possible both technically and legally.

Guides to searching

Many of the subject guides (p. 42) contain sections on searching the literature, and some of the guides to information services (p. 186) cover online searching. Online searching techniques are also to be found in hosts' and database producers' manuals, guides and newsletters, and also in the various online periodicals such as *Online*, *Online Review* and *Database*.

Interpretation of references

Assuming that you have obtained a collection of references as a result of a hard-copy literature search or an online computer search, the next step is to work out what types of original document or source material have been cited. This is necessary because you will either have to search for them in an appropriate catalogue or shelf sequence in a library, or give appropriate details on an inter-library loan form (unless you are using an online ordering service). It should be understood that the boundary lines between the different types of literature are sometimes blurred, and that there may be considerable overlap between different kinds, such as a conference paper published as a journal article (and different libraries may treat certain types in different ways).

Interpreting references is almost the reverse of the process for recording references covered in an earlier section, except that with interpretation it is necessary to spot those features which typify different kinds of original document. As can be seen by checking the references given in the earlier section (p. 224), there are certain attributes (apart from any obvious indications) which can be used to classify the original document type.

- *Books:* a publisher, year, possibly an edition, and usually no range of pages, unless only a chapter is being cited.
- *Periodical articles:* a journal name or abbreviation, issue details giving volume, issue number and date, and a range of page numbers.
- *Conference papers:* a conference or publication name, a location, and a date that is often a range of days.
- *Reports:* a report number, usually a mixture of letters and figures often containing abbreviations for laboratory, research or technical report, note or memorandum, or of the originating organization's name.
- *Patents:* a patent number (often in millions for the UK and the US), normally preceded by initial letters or abbreviations for country names, and one or more specific dates.
- *Standards:* a standard number, sometimes preceded by the name or initials of the issuing body, and year of publication.
- *Theses:* the type of degree, name of awarding institution (usually a university), and year of award.

Many libraries find that conference papers, conference proceedings and reports give most problems.

Obtaining literature in a usable form

You may eventually reach a stage where you have references to relevant literature; then comes the problem of locating and obtaining it in a usable form. Libraries offer the principal means of tackling this problem. They are, of course, important at all stages: from providing the guides at the beginning of your search to producing the required references to journal articles, reports, books and so on at the end. You may also be able to use them for obtaining photocopies, microform reading facilities, or even translations of foreign-language material. Alternatively, commercial services are available to cover some of these needs.

Various possibilities are dealt with in this section, beginning with libraries and their services, and concluding with a word or two about the book trade. However, some modern methods of obtaining literature, which will become increasingly important in the future, are mentioned elsewhere. For the online document-ordering facility provided by certain information services, see p. 151. Note, too, that the section on **Information transmission systems** (p. 183) covers topics such as facsimile transmission, telex and teletex, electronic publishing, and electronic document delivery.

Library systems and services

The main function of a library, as far as you are concerned, is to acquire and maintain a collection of useful literature in such a way that any document forming part of the stock can be located quickly and easily, normally by means of the library's catalogues.

However, it is important to remember that items not held as part of the stock can generally be obtained either as a loan from another library (through the inter-library lending system), or by purchase if this is thought desirable.

Bear in mind, too, that library staff are trained to help you in your quest for information; they may not be experts in your subject, but they should have a good working knowledge both of their own collection and of information sources in general, backed up by experience in the art of finding answers to subject enquiries. Do not be afraid to ask, because you think librarians are incapable of dealing with technical questions. At worst they may be able to suggest further sources of information in cases where an immediate answer is not available; at best they may be able to save you time and effort in locating and searching the literature and its guides. Be reasonable in your expectations and demands, though. A librarian responsible for a large collection is unlikely to remember every single book and journal held in stock; so acquire the habit of assisting yourself whenever possible, in this case by using the library's catalogues.

Using a library

See the flow diagram below for a summary of this section.

> Familiarize yourself with the main areas of the library, its regulations and facilities. Make yourself known to a member of staff on your first visit and ask to see any library guides or leaflets.

> Use the catalogues (author, subject, periodicals, and so on, not forgetting any special ones which may be unique to that library) to see what is held and where it is kept. Note details of items you wish to see.

> Consult library guides, plans or staff to determine the physical location of specific material. Proceed to the shelves, or complete a requisition slip for items not on open access.

> Search the shelves for the material you require, remembering it may be mis-shelved, in use elsewhere, on loan, missing, etc. Consult the staff if you have difficulty in finding what you want. Browse, where necessary, to discover alternative items.

> Use the appropriate library facilities and services, such as borrowing, inter-library loans, microforms, photocopying, reservations, and translations. Ask the library/information staff whenever you need help and advice.

Familiarization

On your first visit to a library you should familiarize yourself with the layout of the building or room; take note of the regulations governing its use and the facilities and services offered. Make yourself known to a member of staff (it may be necessary to comply with certain formalities, such as signing a visitors' book, before you can use the collections) and ask to see any library guides or leaflets which might enable you to make more efficent use of the material and services available.

Catalogues

When you go into a library you are generally in search of something specific, for instance a book by a given author, an article in a particular journal, or information about a certain subject. The initial step, therefore, is to find out whether the library holds the relevant material and, if so, where it is kept. The answers to these questions may be obtained from the catalogues, which function both as a record of stock held and as an aid to locating items on the shelves. Catalogues can take any of the following (physical) forms: cards in cabinets, slips in looseleaf binders, files of computer printout, cloth-bound volumes, cassetted microfilms, and microfiche in folders; or there may be online access via computer terminals. There is considerable variation from one library to another, especially in format and the nature/amount of the information contained, but you will nearly always find:

(a) An author, or name, catalogue (which may include titles)—where you look for books by a given author, or the name of an organization where this is responsible for the item in question, or a specific title if these are included.
(b) A subject catalogue—where you look for material on a particular topic.
(c) A periodicals catalogue—which lists journals held by title, subject, associated organization, and so on.

Sometimes these will be combined (a name catalogue which includes journal titles and/or subject headings, for example), or they may appear in different forms (a card catalogue for books, but a microfiche catalogue for periodicals), or there might even be variations based on acquisition/publication dates (for instance, a card catalogue for material acquired before 1980 and a microfiche catalogue for stock added thereafter). A recent trend towards computerizing current library records often raises a financially difficult problem: retrospective conversion of existing catalogues to machine-readable form. Thus the need to choose between alternative kinds of catalogue, depending mainly if not entirely on the date of what is sought, will probably inconvenience those using certain larger UK libraries (at least) for quite a while.

Many libraries have special catalogues relating to particular sections of their stock. For example, you may find separate catalogues of reports, reference material, theses (in academic libraries) and translations. You may further encounter individual catalogues for departmental and branch libraries. There are also union catalogues which cover all libraries in a particular organization (a university or company, for instance), or the combined resources of several independent establishments (as in the case of a group containing both public and academic libraries).

Remember that library catalogues are produced primarily for the benefit of *you*, the user, not just for the convenience of library staff, so take advantage of these records as often as you can.

Various sets of rules have been drawn up to govern the compilation of catalogues; however, different systems are used in different libraries, ranging from the *Anglo-American Cataloguing Rules* (2nd ed., London, The Library Association; Chicago, American Library Association, 1978—known as AACR2) to simpler 'homemade' schemes.

There are certain basic details which you can expect to find in most catalogue entries:

● *Author:* the name of the author or authors, including initials, sometimes with forenames given in full. This can take the form of an organization (e.g. Institution of Mechanical Engineers) and may be subdivided (e.g. Royal Society of London. Mathematical Tables Committee).
● *Title:* the title of the work, which may be abbreviated.
● *Edition:* a statement of the edition, if not the first.
● *Imprint:* some, or all, of the following details: place of publication, name of the publisher and date of publication.
● *Physical description:* some, or all, of the following details (appropriately modified in the case of non-book items): number of pages, number of volumes if more than one, size, and a note of any accompanying material (such as charts or microfiche supplements).
● *Location:* a code (usually consisting of letters and figures) indicating where the item is shelved in the library. If a subject classification scheme is in use (see below) this code will probably be based on it.

Of course, many catalogue entries contain more information, but these are the main points to observe. Periodicals generally have simpler entries giving title, length of the 'run', imprint and location.

Nothing has been said about the layout of the entry, as this again varies from one library to another. If in doubt, ask the staff.

How are entries arranged in the various catalogues?

Author (or name)/title catalogue. Entries appear under alphabetically filed headings derived either from the names of authors, editors, compilers, organizations and conferences, or from the titles of books, publishers' series and journals. Multi-author works can often be traced using the name of any author involved. Nowadays one of the entries for a book will generally be found under its title; this was formerly the exception rather than the rule, although certain libraries regularly made such title entries. Conference proceedings are notoriously difficult to find in this kind of catalogue: they are often entered under the official name of the conference (where one exists); otherwise under the title given on the title-page of the published proceedings or, alternatively, under the name of the organization responsible (extra entries for editors may sometimes be provided). Government departments can also be difficult to find in an author catalogue, because they form a subdivision after the name of a country (e.g. Great Britain. Department of the Environment); cross-references, if present, should assist here. Otherwise the golden rule is: if you cannot find what you want, ask the staff.

Subject catalogue. There are two main ways of presenting a subject catalogue. The first involves allocating a subject heading (or headings) to each entry, and catalogues produced in this way are as easy to use as a conventional index. The second method commonly adopted is to allocate a subject code or classification instead of a subject heading. In order to use this kind of catalogue it is first necessary to discover the code/classification which has been given to the subject you have in mind, which may be done via published guides to the various classification schemes (see below) or through the library's own subject index, if it has one. The principal advantage of the second method, accounting for its wide use, is that the subject code can be lettered on the spine of the book and used as a location symbol for placing items on the shelves. Naturally, more than one classification can be given to a multi-subject book, enabling additional entries to be made in the subject catalogue.

Periodicals catalogue. Entries are normally arranged alphabetically by title. This is less simple than it sounds, because catalogue 'filing rules' may direct that insignificant words (such as *and, of* and *the*) be ignored; also that some types of journal be entered under the name of their associated organization (irrespective of journal title). A good periodicals catalogue provides more information than the basic title, length of run, imprint and location: it cross-references changes of title (which are fairly frequent among scientific and technical journals) as well as giving additional entries for associated organizations.

There are some general points to bear in mind when using any library catalogue.

● Be prepared to look under more than one heading; remember that there are generally several access points for each book in a library's catalogue provided by its author, title and subject entries.
● When you find a relevant entry, read it carefully and then look at the adjacent entries (especially in the case of card catalogues)—you should do this because, for example, different editions of the same book may appear as separate entries on adjacent cards.
● Make a brief note of the relevant entry on a scrap of paper (author, short title, location) before you proceed to the shelves; it is all too easy to forget the precise details of the items you require if the catalogue is in one part of the library building and the books are in another.

Classification schemes

The purpose of such schemes is to provide a structured arrangement of subjects covering all (or some) fields of human knowledge, generally in such a way that a code is allocated to each individual topic.

Classification is thus the assignment of subject codes to individual documents in order to facilitate their manipulation, and that of their records. Books, for example, can then be shelved in subject code order—whereas alphabetical subject order ('elephants' next to 'electrons') would be difficult to arrange and would inconvenience most scientists and engineers.

Users of libraries in Britain and North America will normally encounter one of three standard schemes:

Universal Decimal Classification (UDC)
Dewey Decimal Classification (of US origin)
Library of Congress Classification (of US origin)

possibly modified to suit local needs, although occasionally they may find an *entirely* 'home-made' variety.

Each classification, or classmark, assigned to a particular topic is usually obtained by subdividing broader subjects. To illustrate this, the way in which a classmark is arrived at is shown for the three standard schemes mentioned, the subject in question being 'cigarettes'.

Universal Decimal Classification

6	Applied sciences. Medicine. Technology
66	Chemical industry and technology
663	Beverages. Stimulants. Narcotics
663.9	Cocoa. Chocolate. Coffee. Tea. Tobacco. Stimulants. Narcotics
663.97	Tobacco industry. Cigars. Cigarettes
663.974	Cigarettes
663.974.6	Cork-tipped. Gold-tipped, etc.

Dewey Decimal Classification

600	Technology (applied sciences)
670	Manufactures
679	Other products of specific materials
679.7	Of tobacco
679.73	Cigarettes

Library of Congress Classification

T	Technology
TS	Manufactures
TS 1950 (-2301)	Miscellaneous industries
TS 2220 (-2283)	Tobacco industry
TS 2260	Cigars and cigarettes

From this simple example it will be seen that UDC makes provision for very fine subdivision, for which reason it is often employed in libraries having a large collection of material in a relatively small subject area (libraries of commercial organizations, for example).

The Dewey Decimal scheme, despite a superficial resemblance to UDC (technology classifications begin with the digit 6 in both schemes, for example) departs from it as the subject divisions become narrower (663.97 is equivalent to 'Tobacco industry' in UDC, but denotes 'Coffee substitutes'—a subsection of nonalcoholic brewed beverages—according to Dewey). Both Dewey and the Library of Congress classification schemes adopt a somewhat broader viewpoint, which makes them well suited for libraries where the material covers most areas of human knowledge without undue specialization: the former will be found in many public libraries, the latter in some university libraries. The chart on p. 252 illustrates in outline how these classification schemes treat the two main areas of interest to scientists and engineers.

All classification schemes suffer from disadvantages. One of these is the tendency to separate different aspects of a given topic. For example, the UDC scheme allocates class 535.33 (in the physics section) to 'Spectra in general. Emission spectra', but the spectroscopist may also find relevant material at class 543.42 (in the chemistry section), which is intended for 'Spectrum analysis. Spectroscopy. Spectrography, etc.' Different classification numbers in the scheme lead to separation of material on the shelves, so library users *must* be prepared to look in more than one place. Another major disadvantage of classification schemes is that they soon become out of date, especially in most areas of science and technology, where knowledge is rapidly advancing. An attempt to combat this is made by publishing new editions of the schemes, or lists of amendments, from time to time, but these often arrive too late to be of immediate use. Librarians try to classify and shelve their stock as soon as it is acquired: it is not desirable that they should hold back books on a new subject until the official classification amendment has arrived). The generation of 'home-made' amendments leads to the local irregularities in classification found in many libraries (there is rarely time available to undertake re-classification of existing stock) and emphasizes the need to consult whatever subject indexes/guides the library which you are using has compiled.

Finally, for those who may wish to know something of the published guides to the major classification schemes (for example, to find out which code has been allocated to a particular subject) the following are available:

Universal Decimal Classification International Medium Edition—English Text. Part 1: systematic tables; part 2: index. 2 vols. London, British Standards Institution, 1985. British Standard BS 1000M: 1985.

○ BSI is responsible for publishing the UDC schedules. They appear as parts of British Standard 1000, a method of production which allows small sections of the scheme to be easily updated. BS 1000C: 1963, *Guide to the Universal Decimal Classification (UDC)*, forms a useful synopsis of the scheme, but has no subject index. Each individual part, however, such as BS 1000 [54]: 1972 (*UDC 54 Chemistry. Crystallography. Mineralogy*), is provided with its own detailed subject index. Amendments are issued from time to time, as with any other British Standard.

Dewey Decimal Classification and Relative Index, devised by M. Dewey. 19th ed. 3 vols. Albany, NY, Forest Press, 1979.

○ Volume 1 consists of an introduction followed by auxiliary tables, which are used in conjunction with the schedules forming the second volume. Volume 3 contains the detailed index linking subjects (arranged alphabetically) with the appropriate classification code.

Library of Congress Subject Headings, prepared by the Subject Cataloging Division. 9th ed. 2 vols. Washington, DC, Library of Congress, 1979.

○ This is an alphabetical list of subject headings, cross-referenced rather like a thesaurus (see p. 113) giving some, but by no means all, associated class numbers. It is kept up to date by quarterly supplements cumulating in annual volumes. More detail can be obtained from the individual classification schedules (such as that covering class Q: science), each of which has its own alphabetical subject index.

Finding the shelves

Having noted, from the library catalogues, brief details of the items you wish to see, you can proceed to the shelves (in an open-access library). There will generally be plans, leaflets or some other form of library signposting to help you find the sections you need to search. If in doubt, ask a member of staff to direct you. In closed-access libraries (and for certain categories of material in other libraries) you will have to complete a requisition slip for each item

Chart 11. Synopses of library classification schemes covering the two main areas of interest to scientists and engineers

Universal Decimal Classification

5   Mathematics and natural sciences
51 Mathematics
52 Astronomy. Geodesy and surveying
53 Physics and mechanics
54 Chemistry
55 Geology and associated sciences. Meteorology
56 Palaeontology. Fossils
57 Biological sciences
58 Botany. Plant biology and taxonomy
59 Zoology. Animal biology and taxonomy

6   Applied sciences. Medicine. Technology
61 Medical sciences. Health and safety
62 Engineering and technology generally
63 Agriculture. Forestry. Livestock. Fisheries
64 Domestic science and economy. Household management
65 Management. Organization of industry, business, communication and transport
66 Chemical industry and technology
67 Industries and crafts based on processable materials
68 Specialized trades and industries for finished articles
69 Building: materials, construction, trades

Dewey Decimal Classification

500 Pure sciences
510 Mathematics
520 Astronomy and allied sciences
530 Physics
540 Chemistry and allied sciences
550 Sciences of earth and other worlds
560 Paleontology
570 Life sciences
580 Botanical sciences
590 Zoological sciences

600 Technology (applied sciences)
610 Medical sciences
620 Engineering and allied operations
630 Agriculture and related technologies
640 Domestic arts and sciences
650 Managerial services
660 Chemical and related technologies
670 Manufactures
680 Miscellaneous manufactures
690 Buildings

Library of Congress Classification

Q   Science
QA   Mathematics
QB   Astronomy
QC   Physics
QD   Chemistry
QE   Geology
QH   Natural history
QK   Botany
QL   Zoology
QM   Human anatomy
QP   Physiology
QR   Bacteriology

T   Technology—general
TA   Engineering—general. Civil engineering—general
TC   Hydraulic engineering
TD   Environmental technology. Sanitary engineering
TE   Highway engineering. Roads and pavements
TF   Railroad engineering and operation
TG   Bridge engineering
TH   Building construction
TJ   Mechanical engineering and machinery
TK   Electrical engineering. Electronics. Nuclear engineering
TL   Motor vehicles. Aeronautics. Astronautics
TN   Mining engineering. Metallurgy
TP   Chemical technology
TR   Photography
TS   Manufactures
TT   Handicrafts. Arts and crafts
TX   Home economics

required. It is in your own interest to write clearly, giving all the information required; otherwise the wrong item may be delivered to you or the waiting time may be unduly prolonged.

Searching the shelves

When you get to the appropriate shelves, be prepared to make a really thorough search. In an open-access library it is impossible to keep everything exactly in its correct place (library users themselves are sometimes responsible for misplacing items they have consulted), so it often pays to look around. Of course, if borrowing is permitted, some of the material you want may be on loan; the library staff will have a record of this and sometimes, in the case of a popular item, it may be worthwhile enquiring whether it is on loan before you go to the shelves. If the specific material you wanted to see is not there, or if you are not quite sure which book might contain the information you require, it generally pays to browse among what is available in that subject area. Remember that not all books of the same subject-classification may be shelved in the same place: most libraries have separate sequences for large books, reference books, less up-to-date books, and so on. This will cause you no trouble if you have made proper use of the catalogues.

If a particular volume is not found in its correct place on the shelves, it may be: in use, on loan, misshelved, being photocopied, at binding, or lost. Library staff will usually be able to determine which of these categories apply and advise on appropriate action (for instance an inter-library loan request, or a visit to a neighbouring library).

Searchers for periodicals in libraries seem to encounter two main difficulties:

(1) the inability to locate a specific title in an alphabetical sequence, generally because they have not realized that special filing rules may apply (ignoring, or not ignoring, words such as *and*, *of* and *the*) or, perhaps, because they have remembered the title incorrectly;
(2) the failure to appreciate that the run of a particular title may have to be divided among different 'bays' in a given stack of shelves (even to the extent of being distributed between opposite sides of a double-sided stack, or opposite bays of adjacent stacks): the only advice that can be given is—look around.

Library facilities and services

Apart from acquiring, cataloguing and storing material, libraries offer various supporting facilities and services. Naturally, the scope and availability of these varies from one library to another; the principal ones are listed below.

BORROWING

Most libraries offer some kind of borrowing facility if you can satisfy appropriate conditions or become a member. The amount of material you can borrow and the length of the loan depend on local library regulations.

INFORMATION SERVICES

Some libraries have members of staff responsible for comprehensive information services (as opposed to reference enquiries—see below). These information officers may undertake literature searches, advise on the use of mechanized information services and on sources of information in general. They can also be involved in running local current-awareness or abstracting services.

INTER-LIBRARY LOANS

You cannot expect a given library to hold *every* document (book, report, journal article, etc.) you need to consult. However, items not available from stock can nearly always be borrowed by your library using the national (or international) inter-library lending network. In the UK most scientific/ technical (and much other) material is borrowed from the BLLD (British Library Lending Division) at Boston Spa in Yorkshire, and generally arrives within a week to 10 days of the request being made. Certain types of material may present problems, however:

● Very recently published volumes take time to get into the library system, and there can be waiting lists owing to popular demand.
● Reports (especially those issued by small organizations and some international bodies).
● Theses (particularly those of foreign origin, unless available through schemes like that operated by University Microfilms International).
● Conference proceedings (which may not be published until long after the conference was held).

If you know that you will have to make use of material in these categories, identify and request specific items at the earliest possible stage. It is most important, too, that inter-library loan application forms should be completed as fully and as accurately as possible. Many references cited in the literature are bibliographically inadequate for this purpose; often they require verification or enhancement, which is the responsibility of either the person making the request or staff at the initiating library.

Material obtained on inter-library loan usually comes in one of the following forms:

(1) The original document (book, report, volume or issue of a journal, etc.) which you may be permitted to borrow, but which has to be returned to the issuing library (photocopies can sometimes be made, however—see below).
(2) A photocopy of the original document, for which a charge may be made (though you will probably be allowed to retain the photocopy if you have paid for it).
(3) A microform version of the original document (see below), which has to be read using special equipment (generally available in larger libraries).

In the future, though, greater use may be made of facsimile transmission, teletex, and electronic document delivery. Whilst inter-loan requests sent by telex have been commonplace among libraries for years, online ordering now seems likely to become of at least comparable importance during the next decade.

No charge is made (other, perhaps, than a nominal reservation or request fee) for original documents obtained on inter-library loan via a public library in the UK. Photocopies, however, may be charged for, although some libraries bear a portion of the cost so that the requester does not have to pay the full amount. As such photocopies can be expensive, ask for an estimate first.

Some inter-library loans librarians may find the following publication of value: *Document Retrieval: sources and services*, edited by B. W. Champany and S. M. Hotz (2nd ed., San Francisco, The Information Store, 1982). It is international in scope, including libraries which have developed services beyond the needs of their own communities, information centres of nonprofit organizations which will similarly help non-members, and commercial enterprises undertaking document retrieval as a business.

MICROFORMS

There are three main types of microform.

(1) Microfilm: roll film (negative or positive) either 35mm or 16mm in width.
(2) Microfiche: sheet film (negative or positive) about the size of a postcard, with rows of frames representing pages of the original document. Most currently produced microfiche conform to one of the following American standards: (a) COSATI (Committee on Scientific and Technical Information) 105 × 148.75mm (approximately 4 × 6 in) carrying up to 72 frames arranged in 6 rows of 12; (b) NMA (National Microfilm Association) 105 × 148.75mm (approximately 4 × 6 in) carrying up to 98 frames arranged in 7 rows of 14. The former is now being gradually dropped in favour of the latter.
(3) Microcard: photographic print (usually positive) on card, consisting of rows of frames as in a microfiche.

Types 1 and 2 are transparent, so enlarged 'hard copy' can be made from them without too much trouble, but type 3 is opaque and very difficult to copy.

Microfilms are used particularly for the reproduction of newspapers, journals and theses. Microfiche are extensively used for reports (the average report occupies one or two fiche). Microcards are mainly used for reproducing journals and books.

Details of commercial microfilm services in the UK may be obtained from a directory produced by Cimtech—the National Centre for Information Media and Technology (formerly the National Reprographic Centre for Documentation—NRCd), which is available only via the telephone/television information retrieval system PRESTEL, on page 2886 at the time of writing.

Cimtech—the National Centre for Information Media and Technology, Hatfield Polytechnic, PO Box 109, College Lane, Hatfield, Herts AL10 9AB (tel. 070-72 68100) offers, among other things, an enquiry and consulting service on microform and reprographic systems and techniques, including word and information processing, videotex and optical disks. This organization also issues specialized reports, and publishes the magazine *Information Media and Technology* (formerly *Reprographics Quarterly: the journal of NRCd)*, which features useful equipment surveys and reviews.

PHOTOCOPYING

Most libraries offer some form of photocopying service. Modern photocopiers are capable of supplying excellent reproductions of text, line drawings and photographs. Many machines offer additional features, which can include an enlarging/reducing facility, double-sided copying, the production of transparencies for overhead projection, and the ability to print (in relatively small quantities) documents such as reports and theses, where the use of larger-scale printing processes is not justified. The availability and cost of these services vary considerably from one library to another, as does the time taken for processing an order. You will sometimes find coin- or token-operated machines provided for instant photocopies.

Copyright law, which varies from country to country, has serious implications not only for students, researchers and librarians, but for authors (of reports, theses, conference papers, journal articles . . .) and publishers as well. The law sometimes applies in less than obvious circumstances. For example, in the UK, technical drawings may be protected by copyright, and it is possible for that copyright to be infringed by the unauthorized manufacture of whatever they represent (even, in an extreme case, where one component has been 'copied' from another, with no direct use made of the original drawings by the infringer at all). This comes about because copyright legislation is concerned with the reproduction of certain things in material form, including three-dimensional objects from two-dimensional drawings; so if a sufficiently strong link can be established between an original protected by copyright and something derived from it without authority, infringement may be proved to have occurred.

The photocopying practices of some library users have helped to create considerable concern among copyright owners in recent years, concern which may well precipitate changes in UK copyright law and/or the spread of licensing agreements between educational (and other) organizations and publishers' copyright licensing agencies. Until law and practice are clarified further, most institutions and libraries will probably adopt a more cautious attitude towards photocopying, and thus avoid exposing themselves unnecessarily to the risk of copyright-infringement litigation.

In the UK, when you ask a library to photocopy a document on your behalf, you may be asked to sign a form of declaration and undertaking, to comply with the provisions of the Copyright Act (1956). The effects of copyright legislation on photocopying cannot be dealt with fully here. A booklet entitled *Photocopying and the Law*, published by the British Copyright Council in 1970, which provided guidelines as to what might be considered acceptable in practice by a number of publishers, was withdrawn in 1984; it has been replaced by *Reprographic Copying of Books and Journals* (London, British Copyright Council, 1985). In general terms, a library may be permitted to supply you with a *single* copy of a journal article (or, providing certain conditions are met, part of a book) for the purposes of *research and private study*, but only if you have not previously been supplied with a copy of the item concerned by any library.

Some libraries also have access to photographic services for the production of prints, slides, microfilms, etc. Again, copyright laws must be observed in connection with such work, and it is often necessary to obtain the permission of the copyright owner before photographing a book illustration (including graphs, charts, and technical drawings).

Details of commercially available reprographic services will generally be found in the commercial sections of local telephone directories, or in guides such as *KOMPASS: United Kingdom* (see p. 21).

UK readers wishing to learn more about legal aspects of copyright may find the following publications of interest.

A User's Guide to Copyright, by M. F. Flint. 2nd ed. London, Butterworths, 1985.
Copinger and Skone James on Copyright, by E. P. Skone James. 12th ed. London, Sweet & Maxwell, 1980.
The Modern Law of Copyright, by H. Laddie, P. Prescott and M. Vitoria. London, Butterworths, 1980.
Copyright for Librarians, by L. J. Taylor. Hastings, Tamarisk Books, 1980.
  For information regarding copyright law in the USA, see the following.
Nimmer on Copyright, by M. B. Nimmer. 4 vols. (looseleaf, with updating service covering the most recent legislation). New York, Matthew Bender, 1963– .
Copyright Handbook, by D. F. Johnston. New York, Bowker, 1978.

REFERENCE ENQUIRIES

These are not to be confused with *information services* (see above). Most libraries make provision for staff to answer straightforward enquiries, including those made by telephone, but you must not expect long or complicated questions to be tackled in this way (ask yourself whether someone could *reasonably* be expected to find the information you want in about 5 minutes).

RESERVATIONS

If a library offers borrowing facilities, it will almost certainly run a reservations service whereby you can be notified when a book which was out on loan is returned to the library; the book being held for you to collect within a few days.

TRANSLATIONS

Some foreign-language material is automatically translated into English within several months of publication; for example, the so-called cover-to-cover translations of Russian journals. Other material may be translated on an ad hoc basis, and the fact recorded in one of several indexes or publications.

  Cover-to-cover translations of journals are listed in *Journals in Translation* (3rd ed., Boston Spa, British Library Lending Division, 1982) and *A Guide to Scientific and Technical Journals in Translation*, compiled by C. J. Himmelsbach and G. E. Brociner (2nd ed., New York, Special Libraries Association, 1972), as well as in some of the **Guides to periodicals** (see p. 97). It may also be worth consulting *Journals with Translations held by the Science Reference Library*, by B. A. Alexander (London, Science Reference Library, 1985).

  Translations of individual articles are covered by *World Transindex*, 1978– (whose predecessors include *World Index of Scientific Translations*, 1967–77, and *Transatom Bulletin*, 1960–77) published in Delft, Netherlands, by the International Translations Centre, and *Translations Register-Index*, 1968– (formerly *Technical Translations*, 1959–67), a publication of the National Translations Center in Chicago. Both of these arrange entries by subject and index them by journal (etc.) citation. The *WTI* database, equivalent to *World Transindex*, can be searched online via the European Space Agency information retrieval service (see p. 154). Another guide in this field is *British Reports, Translations and Theses*, 1981– (see p. 124), which was preceded by *BLL(D) Announcement Bulletin*, 1973–80 (formerly *NLL Announcement Bulletin*, 1971–72) and *NLL Translations Bulletin*, 1961–70 (formerly *LLU Translations Bulletin*, 1959–60).

However, the main source of information in the UK about existing translations is a card index maintained by the BLLD (for details of this organization see p. 259). The BLLD will answer enquiries about translations from borrowers and non-borrowers alike. Virtually all the material listed is held in stock, and retention photocopies can be made of most items. Enquiries about translations are generally made because someone wishes to obtain a copy: for this reason requests on BLLD loan/photocopy request forms are preferred.

If you find no translation of a document in which you are interested is available (and you cannot translate it yourself with the aid of a dictionary), consider whether an English abstract is sufficient. If so, consult an appropriate abstracting journal; if not, consider having a full, or partial, translation made. This can be done in three ways:

(1) Your organization may offer a translations service.
(2) You may have colleagues prepared to undertake this work, for a suitable fee.
(3) In general, commercial services are fast, but can be quite expensive. As a very rough guide, a survey carried out in November 1983 by the London-based Translators' Guild indicated that the rate for translating 1000 words (about 2½ pages of double-spaced typescript on A4 paper) from one of the main European languages into English varied from £25 to £40 depending on the technical content of the original material. Translations into English from Slavonic and Oriental languages would have cost considerably more than this. For further details of translation services consult an appropriate professional association: for example the Translators' Guild of The Institute of Linguists (24a Highbury Grove, London N5 2EA, tel. 01-359 7445), or the American Translators Association (109 Croton Avenue, Ossining, NY 10562, tel. (914) 941-1500). It may also be worth looking in *Translation & Translators: an international directory and guide* by S. Congrat-Butlar (New York, Bowker, 1979).

It is often best to go through the article with a translator at your side, as it may be that you only need to have one or two sections fully translated. Your librarian or information officer will usually be able to advise on the availability of local translators.

Further information about translations can be obtained from *Aslib Proceedings*, Vol 31, No. 11, November 1979, which contains several articles on the subject, including 'Tracking down translations' by B. J. Birch on pp. 500–11, and from two papers by J. P. Chillag (of the BLLD): 'Translations and their guides', published in *NLL Review*, Vol. 1, No. 2, April 1971, pp. 46–53, and 'Translations and translating services at the Lending Division' in *Interlending Review*, Vol. 8, No. 4, October 1980, pp. 136–7.

Principal libraries in the UK

Most towns have a public library (small or large) to which local residents may belong. Cities and towns where there is substantial industrial activity may support extensive library (including reference and information) facilities.

The libraries of universities, polytechnics, technical colleges, etc., may offer some of their facilities to the public, so it is worth investigating this possibility in your area. Many professional organizations (e.g. the Royal Society of

Chemistry or the Furniture Industry Research Association) have library and information services available for the use of their members.

Commercial organizations may have library services: these will normally be for the use of employees, but you may be able to make indirect contact with them through one of the **Regional services** (see p. 181). Official organizations may also have library facilities for staff (and sometimes public) use.

There are commercially operated lending libraries, such as that run by Messrs H. K. Lewis, PO Box 66, 136 Gower Street, London WC1E 6BS (postal service only). Membership of this is open to both individuals and libraries on payment of an annual subscription (related to the number of books which can be on loan to the subscriber at any one time). Advantages of this system include long-term loans and relatively rapid access to recently published works on (in the case of Messrs H. K. Lewis) science, technology and medicine.

Apart from the types of library described above, there are two examples of sufficient importance to be mentioned individually.

British Library Science Reference and Information Service (SRIS)

| Holborn Reading Room: | 25 Southampton Buildings, Chancery Lane, London WC2A 1AW Tel. 01-405 8721 |
| Aldwych Reading Room: | 9 Kean Street, Drury Lane, London WC2B 4AT Tel. 01-636 1544, ext. 229 |

Please see 'Publisher's note' after Preface.

This began as the Patent Office Library; it subsequently became known as the National Reference Library of Science and Invention, and then as the Science Reference Library (SRL). The collections are freely available to all for reference without formality (other than signing the visitors book).

The Holborn Reading Room has literature on physics, chemistry, engineering, technology and commerce, together with all patents and trademark publications (there is a separate reading room for foreign patents in Chancery House). Although recent material is on open access, time must be allowed for older items to be fetched from a store. The British Library's Business Information Service is also based here.

The Aldwych Reading Room has literature on the life sciences, biotechnology, medicine, earth sciences, astronomy and pure mathematics. Books and recent issues of periodicals are readily available, but time should be allowed for earlier periodicals to be brought from a store.

Reference, enquiry and information services are provided at both branches, together with while-you-wait (and postal) photocopying. Help with foreign languages is also obtainable by prior arrangement.

British Library Document Supply Centre (BLDSC)

Boston Spa, Wetherby, West Yorkshire LS23 7BQ
Tel. Boston Spa (0937) 843434
Please see 'Publisher's note' after Preface.

This began as the National Lending Library for Science and Technology, better known as the NLL. It later became the British Library Lending Division, or BLLD. Most of the scientific/technical material supplied by inter-library loan in the UK comes from this source. (Librarians concerned with inter-library loans will find the *Users' Handbook* of particular value.) In addition, it is worth noting that the BLDSC does make provision for visitors, who may use the Reading Room (which contains an extensive collection of abstracting/indexing journals, bibliographies and reference books), and items from the main collections can also be made available. If you have to undertake a fullscale manual literature search it may be worth spending a few days at the BLDSC, especially if you live within reach of Boston Spa.

Those in search of business information should not overlook the commercial sections of reference libraries in major cities such as Aberdeen, Birmingham, Glasgow, Leeds, Liverpool, Manchester, Newcastle, Nottingham and Sheffield. Another example, this time in central London, may be cited as typical.

The City Business Library
Gillett House, 55 Basinghall Street, London EC2V 5BX
Tel. 01-638 8215/6

This has developed from the Guildhall Library. Basically a public reference library with material on open access, the collections are particularly strong in directories (including trade directories) both UK and overseas, company data, market data, newspapers, official publications and other material associated with business and commerce. Photocopying and reference enquiry services are also available. Enquiries are handled in the strictest confidence.

It may also be worth noting the existence of European Documentation Centres, perhaps the most easily accessible sources of EEC information. There are 45 of them, mainly housed in university and polytechnic libraries, and they receive the major official publications and documentation of the various Community institutions. The London office of the Commission of the European Communities (8 Storey's Gate, London SW1P 3AT, tel. 01-222 8122) will supply addresses of EDCs, and also has its own information unit/ library.

Guides to libraries

While many libraries produce guides (leaflets, plans, and so on) describing their own layout and services, this section is concerned with guides which cover a number of libraries, giving basic information such as addresses and telephone numbers. This may be supplemented by notes on size and scope of stock, an indication of special facilities, and lists of publications produced by the libraries in question.

Uses

● To find out which libraries specialize in a given subject field.
● To see what library resources are available in a particular town, region or country.

● To obtain information about specific libraries.

Access

Via libraries.

Caution

Most of the guides are not particularly up to date, so bear in mind that certain details (e.g. telephone numbers, opening hours, name of librarian) may be inaccurate.

Examples

Aslib Directory of Information Sources in the United Kingdom, edited by E. M. Codlin. 5th ed. London, Aslib, 1982–84.
○ Volume 1 covers science, technology and commerce, while volume 2 deals with social sciences, medicine and the humanities. Entries are arranged alphabetically by organization, all kinds of library being included. Address and telephone number are supplemented by: person to whom enquiries should be addressed, subject coverage, special collections and/or services available, and publications produced. A detailed subject index is provided in each volume.

Libraries in the United Kingdom and the Republic of Ireland. 11th ed. London, The Library Association, 1984.
○ Basically a list of libraries with addresses, telephone numbers, and names of librarians (no indication of subject specialities), this has sections which include public libraries, university libraries, polytechnic libraries, and selected national, government and special libraries. An index of places and institutions is provided. The Library Association has also produced a series of guides to various regions of the UK containing rather more information in each entry; however, these have not been updated recently.
Regional information services (see p. 181) may also produce directories of resources in their area (as does HATRICS, for example).

Guide to Government Department and other Libraries. London, The British Library, Science Reference Library, biennially.
○ Entries are arranged alphabetically by name of organization, under fairly broad subject divisions. Address and telephone number are augmented by name of librarian and notes on stock and subject coverage, availability to the public, whether loans are permitted, opening hours, services, and publications. An index of organizations is provided.

The Libraries, Museums and Art Galleries Year Book, 1978–79, edited by A. Brink. Cambridge, James Clarke & Co, 1981.
○ This publication is divided into five sections, covering the British Library; public libraries (UK, followed by the Channel Islands and the Isle of Man); special libraries (for example, those associated with academic or commerical organizations); museums, art galleries and stately homes; and the Republic of Ireland (public libraries, special libraries, museums, etc.). For libraries, addresses, opening hours, and details of subject specialities are provided. The index includes subject entries. A new edition has been announced as *The Libraries Yearbook*, omitting coverage of museums and art galleries.

American Library Directory, compiled and edited by Jaques Cattell Press. New York, Bowker, annually.
○ Lists over 30000 public, academic, special, government and armed-forces libraries in the US, its possessions, and Canada. The arrangement is geographical (by town within state for the USA) with entries which include, in addition to addresses and telephone numbers, the following: names of key personnel, size and nature of stock, subject coverage, special collections and/or services available, and publications produced. An index of organizations is provided.

Subject Collections, compiled by L. Ash. 5th ed. New York, Bowker, 1978.
○ Covers academic, public, and special libraries and museums in the US and Canada. Entries, which are arranged under fairly narrow subject headings, give address and telephone number for each library, enhanced by notes describing the strength of its stock on the subject concerned. The same publisher is also responsible for *Subject Collections in European Libraries*, compiled by R. C. Lewanski (2nd ed., 1978). Here entries are arranged by country under subject headings based on the Dewey Decimal Classification scheme, an index of subjects being provided.

Directory of Special Libraries and Information Centers, edited by B. T. Darnay. 8th ed. 3 vols. Detroit, Gale, 1983.
○ The first volume (which is in two parts) contains over 16000 entries describing special libraries and information centres in the US and Canada, including computerized services; a subject index is provided. In volume 2 a geographic index lists by state or province the institutions described in volume 1, whilst a personnel index does likewise for the library staff there mentioned. Volume 3, *New Special Libraries*, forms a periodical supplement to the first volume, with cumulating indexes. The same publisher also issues *Subject Directory of Special Libraries*, edited by B. T. Darnay (8th ed., 1983). This is in five volumes (the fifth dealing with science and technology) and contains the same information found in volume 1 of the main *Directory*, but here arranged in subject sections.

World Guide to Libraries, edited by H. Lengenfelder. 6th ed. Munich, K. G. Saur, 1983.
○ Covers national libraries, university and college libraries, general research libraries and public libraries with holdings of 30000 volumes or more, and special libraries with 3000 volumes or more. Entries are arranged according to category of library (such as national, university, special or public) by country within continent. Addresses and telephone numbers are supplemented by some or all of the following: year of foundation, name of librarian, size of stock, a note on participation in the inter-library loan system, and membership of local, national and international library associations. A subject index is provided. The same publisher has also produced *World Guide to Special Libraries*, edited by H. Lengenfelder (1983), with entries arranged by subject.

It may also be worth noting that *The World of Learning* (see p. 27) serves to some extent as another international library guide.

Guides to the book trade

Occasionally you may want to buy books and other kinds of literature for your reference shelves: it is more convenient to own some of the items which you use frequently over long periods of time than to keep borrowing them from, or consulting them in, a library.

Apart from being a place where you buy books, a good bookshop will have (perhaps on special display) a selection of the very latest publications in your field. It may also have on its shelves items which, for one reason or another, are not available in your library. Books not in stock, but in print, can be ordered via a bookshop, though it is wise to ask for estimates of price and delivery time in the case of material coming from overseas.

There are comparatively few guides to bookshops. The 'yellow pages' of a telephone directory will indicate those in a particular region of the UK. The Booksellers Association of Great Britain and Ireland (154 Buckingham Palace Road, London SW1W 9TZ, tel. 01-730 8214) produces a list of its members, and Peter Marcan (31 Rowliff Road, High Wycombe, Bucks) has compiled and published a *Directory of Specialist Bookdealers in the United Kingdom handling mainly New Books* (2nd ed., 1982), with entries arranged under broad subject headings. The *American Book Trade Directory*, compiled and edited by Jaques Cattell Press (New York, Bowker, annually), lists over 20000 US and Canadian bookstores and book wholesalers, with entries arranged geographically. A companion volume, *International Book Trade Directory* (New York, Bowker, 1979), provides information on some 30000 booksellers and wholesalers in 170 countries outside North America, especially those handling foreign publications; again the entries are geographically arranged. Finally an *International Directory of Booksellers*, published in London by the Library Association, and in Munich by K. G. Saur, in 1978, lists more than 63000 booksellers and wholesalers from 134 countries around the world, with entries arranged geographically, including an indication of subject specialities.

If you want details of booksellers specializing in particular material, consider asking the staff of an appropriate library, society, institution or museum for advice.

It may sometimes be necessary to contact publishers (e.g. to obtain permission to quote from, or copy, a copyright work; or when you want information about the availability of a particular book, especially new editions and forthcoming publications; or if you are looking for someone to publish *your* book). Some of the **Guides to books** (see p. 62), **Guides to newspapers** (see p. 76) and **Guides to periodicals** (see p. 97) contain useful lists of publishers' names, addresses and telephone numbers. Similar information can also be obtained from the following:

Whitaker's Publishers in the United Kingdom and their Addresses. London, J. Whitaker & Sons, annually.
○ Entries, which are arranged alphabetically by publisher, generally include telephone/telex numbers and ISBN prefixes. (International Standard Book Number—a unique ten-digit number with elements identifying language/ country of origin, publisher and publication, used by the book trade for ordering purposes and by librarians for retrieving bibliographic records.) A numerical index linking ISBN prefixes with publishers is also provided.

Cassell & the Publishers Association Directory of Publishing in Great Britain, the Commonwealth, Ireland, Pakistan & South Africa. 10th ed. London, Cassell, 1982.
○ In the first part, book publishers and publishing agents, the arrangement is alphabetical by publisher within country. As well as addresses, ISBN prefixes, telephone and telex numbers, the entries give names of senior executives, an indication of subject coverage, and details of companies acting as representatives and agents. The remaining parts cover: overseas publishers represented in Great Britain, British publishers classified by fields of activity, authors' agents, trade and allied associations and societies, and trade and allied services for Great Britain. There is a combined index to the last two parts, followed by separate indexes of personal names, ISBN prefixes, and publishers, imprints and agents.

5001 Hard-to-Find Publishers and their Addresses. 2nd ed. Reading, Berks, Alan Armstrong & Associates, 1984.
○ An alphabetical listing of over 7000 publishers with their addresses and telephone/telex numbers where possible.

Publishers, Distributors & Wholesalers of the United States. New York, Bowker, annually.
○ The 1984 edition listed some 35000 firms alphabetically, giving ISBN prefixes, addresses and telephone numbers, among other data. A key to the publishers' abbreviations used in *Books in Print* (see p. 64) is provided, as are indexes of ISBN prefixes and wholesalers. A related Bowker title is *Who Distributes What and Where* (3rd ed., 1983), an international guide to distributors of US publications, with entries arranged alphabetically by publisher; a geographical index is also provided.

Publishers' International Directory with ISBN Index. 10th ed. 2 vols. Munich, K. G. Saur, 1983.
○ Volume 1 lists nearly 134000 publishers (providing ISBN prefixes for almost 60000 of these) with entries arranged alphabetically by publisher within country. Other details which may be given include addresses, telephone/telex numbers and, sometimes, subject specialities. The second volume consists of a numerical index linking ISBN prefixes with their publishers.

International ISBN Publishers' Directory, compiled and edited by the International ISBN Agency, Berlin. 4th ed. New York, Bowker, 1983.
○ Coverage is wider than the title suggests, publishers' addresses from some countries outside the ISBN system being included. This guide has three sections: an alphabetical sequence of about 100000 publishers (with their addresses, and ISBN prefixes where relevant) from 50 countries, a numerical directory linking ISBN prefixes with publishers (again addresses are provided), and a geographical index with publishers listed alphabetically under their respective countries.

Apart from their own stock lists, some publishers/booksellers produce reference catalogues which can be used as guides to a wide range of material. One example is *The Top 3000 Directories & Annuals* (see p. 116) published by specialist bookseller Alan Armstrong & Associates. Another is:

Guide to International Journals and Periodicals. Folkestone, Wm. Dawson & Sons, annually.

○ Better known as Dawson's 'little red book', this contains subscription details (country, frequency, price) of those titles most frequently demanded by the firm's customers. There is a classified section, with titles arranged under fairly broad subject headings, and a separate listing of yearbooks and directories.

You may find it helpful to be included on the mailing lists of booksellers and publishers specializing in your field, as you would then receive catalogues, details of new publications, announcements of forthcoming books, and so on, which act as an aid to current awareness.

Organizing and presenting information

We spend a large part of our lives receiving, processing and transmitting information, and it is a sad fact that relatively little is generally known about this subject and that it is not widely taught in schools or colleges.

The main reason for handling information at work is usually an objective which involves influencing or affecting other people in a desired way. One of the main problems is to integrate all the aspects of information handling with the objective. It is all too easy to isolate the various tasks and aims and forget that they are interdependent. For example, some people might be tempted to acquire a comprehensive set of manufacturers' literature on a given product, when a small selection would have proved perfectly satisfactory. Another example is the piles of 'relevant' xerox copies and pamphlets which grow on some peoples' desks, accumulating dust and waiting for the magic moment when there will be time to read them or to transfer their contents to a report. And there are those talks where the speakers 'read' their papers and show unreadable slides because they have forgotten what it is like to be on the receiving end of this type of presentation.

Objectives

In our real-life situations, objectives are often mixed and complicated. The best approach is probably to look at a given situation from a number of different points of view.

Information. This should always have a subject and an object—information is never required for its own sake. Put another way, it is what you do with the information that counts, or what the information (if you communicate it) can do for you. You should consider how information can help you in the short term in a number of ways:

- Stimulation, e.g. to help you to be creative and motivated or interested.
- Current awareness, e.g. to keep you up-to-date with the 'competition' and development.
- Methodology, e.g. to tell you how to do something.
- Theory, e.g. to predict different or future phenomena.
- Data, e.g. to compare your own results or use in calculations.

- Solutions, e.g. to show how other people dealt with similar problems.
- Education, e.g. to give background details in a new field or to pass an exam.

Communication. You should consider the method you will use to communicate the results of your work:

- Proposals, e.g. to sell a product or service.
- Reports, e.g. to satisfy an employer of progress.
- Talks, e.g. to persuade clients to adopt a design solution.
- Plans, e.g. to satisfy authorities that regulations have been obeyed.
- Articles, e.g. to demonstrate professional expertise.
- Drawings, e.g. to express a design to clients and contractors.

You also need to consider your abilities, timing and production services in relation to whatever form of communication you intend to use.

The recipient. The last, and for some the most important, consideration should be of the recipient of your communication. What are their objectives and how are they able and willing to attain them? Can their objectives be compatible with yours? Again a process of integration is necessary. The points to consider are:

- How do you wish to influence them?
- What level do they understand?
- What will they use the information for?
- How much time will they spare to read it?
- How would they prefer to receive information—a long essay, drawings, or brief, clear notes?

Of course, an appreciation of all these considerations should influence literature searching and personal records. Literature searching is dealt with in an earlier section, but a few examples of these influences are:

- If you are in a hurry for information do not bother to look in indexes to remote documents which cannot be obtained quickly.
- Make sure that 'relevance' applies to your objectives and not just the subject of your search.

Personal records of references are also dealt with in the section on literature searching, but their relationship to objectives is covered here, where it will be shown how they should be far more than mere records of what you have found, but working documents which will play a direct role in helping you to achieve your objectives, develop working drawings, write your report, etc.

To summarize, the key concept is the integration of all aspects of your objectives and the method of achieving those objectives. In particular:

- Decide exactly who you wish to influence and in what way.
- Imagine yourself in their position and decide what would influence you in the desired way.

Do not forget your priorities and your timing, as you may find that the advice that follows is just a counsel of perfection, that time has evaporated and that you must make the inevitable compromises.

Gathering and recording information

Apart from the concept of integrating the gathering and organizing of information with objectives and presentation of information, the most important thing to remember is that this stage reflects the way you work, and tends to develop into a habit very quickly.

The problem for some people is that it is easy to drift into sloppy habits and very difficult to break out of them. Develop a sound procedure and then stick to it if at all possible. Some examples of where things go wrong are:

● It is time to write your report, you are in a desperate rush and you are not sure which items in the pile of literature on your desk are really relevant, without re-reading the lot.
● You remember lending a vital article to someone, but to whom?
● A conference proceedings arrives two months after you requested it on loan, and you cannot remember why you ever wanted it.
● A really useful photocopy of an article contains no details of its source, so now you do not know how to quote it.
● The substance of the good idea you had, and which seemed so vivid that you'd never forget, has evaporated by the next day.

First of all, it is usually best to deal with references and documents or copies as they appear. This can save time in reorientation after an interlude of forgetfulness and can prevent a demoralizing back-log of the kind mentioned earlier.

Secondly, at the most basic level it is essential to keep a proper record of what you have done and what you have to do. A simple way to deal with these is to use the following.

Day book or journal. In this you record those aspects of what you have done which you may need to recall: for example, who you rang and what information they gave you; where you searched for a reference to a new material and what you found; a possible solution to design problems. In other words, the day book should be the equivalent of a scientist's laboratory notebook. If duplicate copies of entries are required for cross-filing or sending elsewhere, carbon or xerox copies can be made; if copies are going to be needed frequently, special notebooks with duplicate pages are available.

Diary. Preferably of the large desk variety, this can be a very helpful workhorse if used properly as an extensive *aide-mémoire*. Apart from appointments you can note, for example the names of those people you have to contact tomorrow or next week and their telephone numbers; regular chores like checking accounts or pending trays, personal deadlines and even running or continuous tasks which have to be done 'some time'—these can be kept in the weekend sections.

Pocket notebook. For some people this is a vital 'mobile memory'. It can be divided into sections such as books or references to be looked up, things to do and 'ideas'. The last suggestion is essential for the many people who remember or think of things to do while at home and forget them before they get back to work.

Of course, the habit of using these recording devices must include cross-references, transferring appropriate details from one to another and also to other records, e.g. ideas from the pocket notebook to the diary for action, or information from the day book to reference cards or other working documents. The contents of all of them should be regularly reviewed, and pages crossed through when 'dead' as far as future use or action is concerned.

Evaluating references

All stages in the process of finding information involve some degree of evaluation and storage, especially when handling search terms, references, abstracts, original documents, etc. You will normally have to rely on your own personal judgement and previous experience, but always take advantage of other people's advice if it is available. Search terms have already been discussed on p. 213, and the recording of references on p. 223.

You may obtain references from journal articles, books, reports, indexing journals, abstracting journals, etc., or from colleagues. In the latter case, evaluation can be simpler because you will have the advice of your colleague to assist you. With abstracting journals, the information in the abstract may offer valuable help. In the case of references at the end of journal articles or in books, reports or similar documents, you can be assisted to some extent by that part of the text which refers you to the reference. With indexing journals, the only additional help you get is from the subject headings, where these are not taken directly from the titles of articles covered.

The reference itself contains the title, which is the main indicator of usefulness; the author, who may be known to you, or who may be investigated (see **People**, p. 14); and, in the case of a journal reference, the journal title, which may help to indicate the subject coverage of its contents.

Frequently the only way to tell if a reference is of use is to consult the original document. If the document is not easily accessible, this process can be time-consuming and costly, especially if inter-library loans are involved. Supplementary information can sometimes be obtained by checking whether the reference has been covered by an abstracting journal. If it has, the abstract may provide sufficient detail either to give you the information you want or to help you decide whether the original document should be obtained.

The accuracy or completeness of a reference, especially to a journal article, can frequently be checked by use of abstracting/indexing journals (*Science Citation Index* can be useful here). Book references and journal titles can be validated in either the **Guides to books** (p. 62) or **Guides to periodicals** (p. 97). Bear in mind that inaccuracies can also occur in these aids!

Storing references

Many people use some form of card index for their collection of references, with one reference recorded on each card. When the collection is small, it is usually convenient to keep them in alphabetical author order, and when a search has to be made, the whole batch of cards can be scanned. With larger collections, it may be necessary to group the cards into subject sections. Sometimes an easy way to do this is to use the same subject division occurring in a comprehensive textbook in your field (if one exists). The index to the book can then be used, after a fashion, as an index to your own collection of references.

A problem associated with having one card to represent each reference is what to do when you have, say, a reference on the analysis of iron in copper. Do you write out identical cards for each subject section, or do you put one card under 'analysis' and make cross-reference cards for 'iron' and 'copper'?

One solution is an inverted file, with cards representing subjects rather than references in a subject file. Imagine you obtain your 50th reference and it is on the analysis of iron in copper. You make out one card with full details of the reference, writing the number '50' at the top. This card is filed in numerical order with your other 49 reference cards. You now take from your set of subject cards the 'analysis' card, the 'iron' card and the 'copper' card (these are kept in alphabetical subject order). On each of these you write the number '50' and then return them to their alphabetical order. Conversely, if you wished to know whether you had any references on the analysis of iron in copper, you could take out the three appropriate subject cards, look for any common number (e.g. 50), and then look up these numbers in the numerical sequence of reference cards. (This is a form of co-ordinate index—see p. 223).

This imaginary reference could be used to create the following records:

Analysis of iron in copper. J. Smith.
J. Chemistry. Vol. 5, No. 2. May 1967, pp. 1–2.

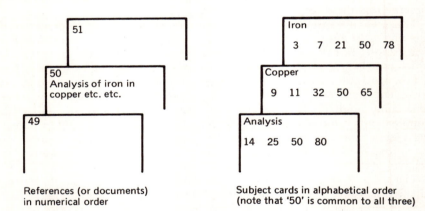

References (or documents)   Subject cards in alphabetical order
in numerical order   (note that '50' is common to all three)

A refinement of this system is the use of optical-coincidence cards, as used in the Geodex System (p. 176) or as available for personal or company information retrieval from Anson Systems of Liverpool. These and other types of card indexes are described in detail in the books below.

Computerized record keeping is growing in popularity, ranging from software packages for home computers to sophisticated software suites mounted on mainframe computers or minicomputers and available via your own organization or an online host (p. 151). For personal use, a typical facility would allow you to design your own record layout (such as that shown on p. 271), and would then provide you with that layout, on the screen, every time you wished to enter a new reference, prompting you to fill in each section (or field) in turn. You would be able to organize those records into any order required, or search them for particular items of your choice, and finally print out a selected listing.

For further details of personal records consult:

Information Storage and Retrieval Systems for Individual Researchers, by G. Jahoda. London, Wiley, 1970.
A Guide to Personal Indexes using Edge Notched and Peek-a-boo Cards, by A. C. Foskett. 2nd ed. London, Bingley, 1970.
Personal Documentation for Professionals: means and methods, by V. Stibic. Amsterdam, North-Holland, 1980.

Some of the subject guides also offer advice.

Reference extension

Some aspects of recording references and constructing personal reference collections having been dealt with above and in the section on searching, this section shows how these records may be extended to increase their usefulness.

Card for reference extension

Author	Subject	Document
Ref 1		
Ref 2		
Full ref		
Location	Ref source	
Requested	Use	
Reason		

The figure above shows a reference card designed to do just this. The three boxes along the top edge are to enable easy filing under author and/or subject and/or end document. For example, if you were interested in a particular subject but with no end document (such as a report) in mind, all the reference cards pertaining to that subject could be labelled accordingly and stored in alphabetical author order in the section of your card index system for that subject.

● 'Ref 1' is for recording a reference in a particular house-style, e.g. where the reference is to be cited in a particular journal or company report and where there are strict rules for writing references (see p. 277).
● 'Ref 2' allows for a second, different house-style for the same reference—you

may want to cite a reference in two different documents which have different editorial rules.

- 'Full ref' contains any missing details from 'Ref 1' or 'Ref 2' so that there is always sufficient detail to accommodate any other house-style. If no end document is in view the complete reference is entered in this box. The point of 'Ref 1' and 'Ref 2' is to save time when compiling a bibliography or list of references for a report, article, etc. The appropriate set of cards can be handed to a typist in the correct order with the instruction to type only 'Ref 1' items, for instance.
- 'Location' is for locating the document or copy. This can work in various ways: it may indicate a place in your main xerox file, a shelfmark and book number in your local library, or a colleague's desk drawer. Alternatively it may show a temporary location in a file specially compiled to write a report, or even a loan to a colleague in another office and the date of the loan.
- 'Ref source' is to record your source of reference so that you or anyone else can check on the accuracy of your reference or so that you can extend your search if appropriate. Accuracy here is very important for identification and for inter-library loan purposes.
- 'Requested' is for recording attempts to obtain the original document or a copy. This may involve a book order or reservation, a request via an inter-library lending department, a direct request for a reprint to an author or an online document order following a computer search—all noting the date requested. All cards referring to awaited items can be stored in a separate part of the file, reviewed regularly for action, the date noted on arrival and the card ultimately refiled in an appropriate section.
- 'Use' is for a list of the actual or potential applications for the information in the original document. This might include specific end documents or more general 'education'.
- 'Reason' is for noting what it was about the information in the original document that made it worth a record. Perhaps it was an excellent example of a previous practice which would be suitable for introduction into a new design. Perhaps it was a new technique which should be borne in mind by the contractor when making tenders for a particular kind of construction work.
- Lastly, if an end document is to be written, the reverse of the card may be used to write an essential extract of the original, so that when reporting or writing commences there is something to build on, or even use if appropriate, without starting again from scratch.

Evaluating documents

You should always try to adopt a critical approach to recorded information, as with any information, since it may be wrong. Even some of the well-established properties of chemical compounds have been found inaccurate (e.g. the properties of the so-called inert gases). If your work depends, for example, on certain basic data, the origin of the data should be checked, i.e. original journal article, measurement method, etc. Data from handbooks may be vaguely defined and misleading; for example, properties of materials may vary with environmental conditions. Ideally the data should also be cross-checked with other sources of information (such as independent calculations, measurements, theories, etc.) through your own findings or via the advice of a colleague or an expert in the subject.

When evaluating documents or their component parts (e.g. articles in journals, chapters in books), publishers, authors and their place of work, dates and journals may be critically considered where appropriate. People frequently rely on the date more than anything else, assuming that the most recent information is the best. Although this is frequently found to be a correct assumption in the case of scientific research, the critical approach should never be neglected. In the case of books you should always check in the appropriate guides, or with the publisher, that you have the latest edition. You can sometimes find helpful book reviews in appropriate journals. It is always a sound practice to obtain any review articles in your field, and these will frequently contain a comparative or critical evaluation of journal articles.

Some types of publication contain sections which assist evaluation. These include contents pages, abstracts, indexes, introductions, conclusions, lists of references, illustrations, tables, graphs, etc. Patents are a special case (see p. 81).

If you find that certain journals or report series contain a number of useful items, it is often worth consulting the annual indexes to the journals or the complete report series, for similar items. If you find particular authors or organizations produce useful work, it is worth using their names as search terms in appropriate guides.

Storing documents

Accessibility is of great importance to users of libraries. Perhaps for this reason a good personal library plays a very important role in the work of many researchers. The main problems that exist with these personal collections are size and cost. It is easy to fall into the trap of thinking that an article which has been photocopied and filed away is an article which has been read. The result is thousands of photocopies, etc., stored for posterity at enormous cost. At the same time, it has to be admitted that some of these papers will be referred to, partly because they are immediately accessible, and may provide really useful information. A sensible compromise is best. If a document is of obvious use, a copy should be kept where feasible. If it is of possible use, it should only be kept or copied if it has been particularly difficult to obtain. Otherwise a reference (indicating where the original can be quickly located) should be a sufficient record.

Larger documents such as books and reports usually present only minor problems with storage and retrieval. Papers, photocopies, reprints, etc., may cause some difficulty. Like references, they may be stored in alphabetical author order, or grouped together in subject files or boxes. Alternatively, they may be stored in the 'inverted' fashion (see section above) where each item is numbered, and a subject index or set of subject cards is used for identification purposes.

When dealing with originals, reprints or copies it pays to treat them as working documents (provided they are yours and not office or library stock). Extend them in just the same way as the reference cards, to enhance their usefulness and to make them work for you rather than against you. For example, do not be afraid to mark up the margins with pencil comments or to highlight lines using fluorescent marker pens. You may agree or disagree, you may think something is a good new idea or you may be stimulated to think of something quite different—in all cases note it down. If you do have the misfortune to have to go back through a number of documents at reporting time, at least you will not have to read every line to see what it was that originally caught your eye.

After making a photocopy you *must* ensure that full details of the original document appear somewhere on the copy, if necessary writing them in by hand. Otherwise you may later be faced with tracing the source of an unidentified photocopy, which is often a time-consuming, if not impossible, task.

Communicating information

You may cause a greater impact with verbal communication—for example, the great political speeches—but with written communications you have much greater control. If you make a mistake when talking to an audience, it is difficult to correct it, even if you spot it. When talking, you may suffer from nerves and perhaps lack of experience, and also from the effects of your surroundings, including your audience. With writing, you can normally correct the finished product before it reaches the reader.

Written communication usually takes the form of a letter, memorandum, report, thesis or periodical article. We shall briefly consider reports as an example.

Report writing

There is no rigid procedure, but the notes below coupled with a flexible approach should offer some help. You should check whether your organization issues any instructions on layout, house-style, etc.

Objective

Prepare a statement of intent. This can prove an invaluable yardstick when deciding whether or not to include items of information.

Timing

Work out a time schedule. This means fixing or acknowledging a deadline for completion of the report. Then check the availability and speed of production facilities, typing, reproduction, photography, binding, etc. This will then enable you to work out how long you have to actually write the report. Sometimes the production of the report causes more problems than the writing.

Information

Assemble the results from your work and any other relevant material, bearing in mind your statement of intent. Some people find that it helps to have a card or folder, for example, for each section of the report or for each subject-division of the work they are reporting. As you acquire information or ideas, you can add them to the appropriate collection. Ideas can be very fleeting; you should note them down as they occur. As a part of your work, you should have examined all the ideas and information with a critical eye, particularly checking for accuracy and comparing your results and ideas with previous work. Time spent in simply thinking around the material you have and how you might present it is never wasted. At this stage it can help to write out a provisional list of section headings covering the work you have done or the information you wish to present. These headings can then be arranged into a logical order, and shown to colleagues for comment. Even at this early stage criticism can be helpful. If you have problems in deciding on the structure of the report as far as the work done is concerned, or how to organize your work into sensible chapters, then consider the alternatives and take the one that fits best with your objectives.

Each heading can be expanded by adding brief notes about your own work, ideas and information from other sources which relate to that heading. Use your objectives to guide you as far as selecting which topics require special emphasis, and try to predict questions that may arise in the minds of the people you expect to read your report.

Writing

Writing the report can best be described by considering its various parts. A typical layout is given next and you should try to arrange the material and the ideas that you have to fit this type of pattern. Reports are usually best written over a short period of time, to preserve continuity. Do not wait for perfect sentences to form in your mind, but write whatever comes to mind. This way you will find it easier to start, and you will find it is always easier to criticize and correct something on paper.

A word processor may be used for writing almost anything that is to be composed of textual or numeric information. It is essentially a device for typing in the data, storing it in a memory, recalling all or parts of it for display, checking, correcting, editing, altering, or printing out copies. The equipment available ranges from electronic typewriters with limited WP facilities, to microcomputers with WP software packages and dedicated word processors. Some are very sophisticated, permitting all kinds of advanced data manipulation, such as automatic spelling checking, movement of whole columns and rows of figures, 'cutting and pasting', merging address lists with standard letters for 'mailing shots', and searching for specified words which are then altered. Most systems allow new data to be inserted anywhere in a text file (such as a report), automatically moving up the following text to make room. Word processors may suffer from poor screen displays (glare, inadequate size, and inability to display italics, bold etc.), and lack of compatibility with other equipment, such as memory disks from other word processors. They can be excellent for composing text, major or repeated corrections, regular updating, or standard letters, but they can be fiddly for making a few small corrections, especially when security demands the keeping of several back-up copies on disk. Record keeping for all the files in their varying stages of correction and corresponding back-up copies can be a headache, and with some systems, it is easy to wipe off several hours work by mistake.

Layout

The accepted page size is now A4 (210 × 297mm), and it is usual to put page numbers centrally at the bottom of each sheet. The various parts of the report are considered below.

FRONT COVER

- Report number: may be used for physically storing and retrieving the report, if it is part of a series.
- Title: must be descriptive, as it may be used to retrieve the report by subject, or it may be used to decide if the report is worth reading.
- Author's name.
- Name of organization producing the report.
- Date when writing was completed.
- Security classification if necessary.

TITLE PAGE

Usually contains the information already given on the front cover, perhaps with a little more detail, such as names of sponsoring bodies, contract numbers and acknowledgements. The signature of the writer and his supervisor, if required, should always appear here.

SUMMARY PAGE

The summary or abstract of the report is best kept to one page. It should be written as though it might be the only part of the report that a busy executive will have time to read. It must contain the 'message' that the report is trying to communicate. At the very least, the summary is the bridge between the title and the body of the report.

TABLE OF CONTENTS

This should list all the principal headings in the report in the order in which they appear, giving appropriate page numbers. The table should allow a reader to see at a glance how the report is laid out and to be able quickly to locate any section which he wishes to read. Contents pages are usually written last because they have to cover the entire contents. It is usual to number (or letter) sections as well as pages. Sections may be systematically broken down by numbers and letters or by decimals. For example:

Principal section	1		1
Main section	a	*or*	1.1
Sub-section	(i)		1.11

INTRODUCTION

This should give the reason behind the report, the objectives, terms of reference, background, state-of-the-art, previous work and appropriate discussion which should justify the work which follows.

THE WORK DONE

The work done is usually split into two main parts: (1) experimental or investigatory, method and results; (2) discussion and theoretical interpretation. This is the vital section when it comes to details and accuracy. Make sure you have covered all the influencing variables. If you have trouble in breaking the work down into logical sections see **Information** (p. 274). Diagrams, pictures, graphs, tables, etc. should only be used when they represent the best way of communicating information. If one of your 'figures' is referred to in detail from just one part of the text, or it is essential to the understanding of the next, it should be positioned adjacent to the corresponding text. If neither of these apply, or the figure is referred to from numerous parts of the text, or the size and number of illustrations would unbalance the text, making it difficult to follow, the figures should be placed in the appendices section.

The text should refer to the figures via figure numbers or codes and/or page numbers. Codes can be a simple running number right through the report or a compound number made up of chapter number/section number, etc. For example, 'Table 4.12.2' could mean a table in Chapter 4 section 12 paragraph 2, whereas 'Fig. 45' could simply refer to the 45th figure. Make sure that you explain in the text why you are referring to a figure, and that the figure itself backs up the point you have made in the text, that its title, labelling, symbols and units are all crystal clear, well defined, and are in a form that your reader can easily use if he has to. Use footnotes or annotations if necessary to avoid overcrowding a drawing. Lastly do not forget that figures cause the most problems in their making and reproducing—so check on what can or cannot be done first.

CONCLUSIONS AND RECOMMENDATIONS

These form the end-product of your work and will hopefully justify your efforts. Like the summary, these may be the only parts of your report that are actually read by busy managers or customers, so again special care is required.

APPENDICES

- Tables, graphs, photographs, diagrams (only when it is not appropriate to include these in the body of the report).
- The list of references is often a neglected area and as a result causes all sorts of problems. There are two points to remember. First, the whole reason for having references is for other people to obtain the original documents, and therefore it is vital to be accurate and to give as much information as possible. For example the title will help them to decide if the original is worth having; the issue date should be included, even if it duplicates volume and issue numbers, as a double-check in case you make a mistake and to give an idea of currency; the complete pagination will give an idea of the length of the item and hence one measure of its usefulness; the complete title of a journal will prevent misinterpretation of abbreviated journal titles. Second, you should check whether there is an imposed house-style for references, as mentioned earlier. Whether or not there is a house-style, you should be consistent in the way you record references; for example you can indicate 'volume' by volume,

Volume, vol., Vol., vol, Vol, v., V., v, V, or by underlining or by using bold print!

There are two main ways of linking the text of your report with individual references in the list; these are the Harvard System, and the numeric system as frequently found in journal articles. Quite a lot of detail may be found in the appropriate British Standards, but briefly the two systems work as follows.

Harvard system. This uses authors' surnames and years as the link, e.g. the text may contain inserts such as 'the method used by Jones (1980) was also . . .' or 'the established method (Brown 1960) was also . . .'. In this system the list of references would be in alphabetical/date order and of the form

> BROWN, J. 1960. Title of the article. *Journal Name*, Vol. 6, No. 3, March. pp. 542–584.
> JONES, D. 1980. Title of the article. *Journal Name*, Vol. 28, No. 6, June. pp. 33–34.

Numeric system. This uses numbers as the link, e.g. the text may contain inserts such as 'the method used by Jones[4]. . .' or 'the method (5) used by us . . .'. In this system the references are listed in number order in the same order as they appear in the text, e.g.

> 4. JONES, D. Title of the article. *Journal Name*, Vol. 28, No. 6, June 1980. pp. 33–34.
> 5. BROWN, J. Title of the article. *Journal Name*, Vol. 6, No. 3, March 1960. pp. 542–584.

Both systems have their advantages and disadvantages and, depending on what you are writing, you may find one better than the other, or alternatively you may find a hybrid or extended system the best for your purposes.

If you think that new references may have to be added as you go along, the pure numeric system may prove awkward if you have to re-number all the text/reference numbers. If you want to be able to refer from your references to your text, the numeric system is the best, although any list of references may be annotated by adding those page or paragraph numbers* where the text refers to the references, e.g.

> BROWN, J. 1960. Title of the article. *Journal Name*, Vol. 6, No. 3, March. pp. 542–584 [42, 43, 80]*

Where such annotations are used, an appropriate explanatory note should be given at the beginning of the list. If you think it is important that the reader should know who wrote a particular reference and when, as it is referred to in the text, the Harvard system is better; but it can be bulky, and upset reading flow if a number of references are cited close together.

● Various other addenda may include any acknowledgements, a glossary, list of abbreviations and symbols with explanations, and distribution list. It is most important to clearly define any units which are used in the report. If the report is long enough to justify an index, it should always be given one.

REVISION

This is an essential part of report writing. It is a good idea to give the proofs to a colleague for comment, and also to put the report aside for a few days before revising it yourself.

Further reading

The details here are merely an introduction to written communication. The serious writer should take the time to study a few of the many books, pamphlets or journal articles on this subject. For example:

Scientists must Write: A Guide to Better Writing for Scientists, Engineers and Students, by R. Barrass. London, Chapman and Hall, 1978.
Writing Technical Reports, by B. M. Cooper. London, Penguin, 1964.
How to Write and Publish a Scientific Paper, by R. A. Day. 2nd ed. Philadelphia, ISI, 1983.
So You Have to Write a Technical Report: Elements of Technical Report Writing, by D. E. Gray. Washington, DC, Information Resources Press, 1970.
Good Style for Scientific and Engineering Writing, by J. Kirkman. London, Pitman, 1980.
How to Write Reports, by J. Mitchell. London, Fontana, 1974.
IEEE Transactions on Professional Communication. New York, IEEE, quarterly.
'Writing your thesis', by J. M. Pratt. *Chemistry in Britain,* Vol. 20, No. 12. December 1984, pp. 1114–1115.

There are numerous British Standards which are helpful; the following is just a selection.

BS 4811: 1972 (1979) The presentation of research and development reports
BS 4821: 1972 (1980) The presentation of theses
BS 1629: 1976   Bibliographical references
BS 5605: 1978   Citing publications by bibliographic references
BS 3700: 1976 (1983) Recommendations for the preparation of indexes for books, periodicals, and other publications
BS 5775: parts 1–13   Specifications for quantities, units and symbols

See also *BSI Catalogue,* especially for standards giving glossaries of subject terms. Other countries may have their own standards, such as those of the American National Standards Institute (ANSI):

ANSI Z39.18 1974 Guidelines for format and production of scientific and technical reports
ANSI Z39.29 1974 Standard for bibliographic references
ANSI Z39.16 1979 Standard for scientific papers for written or oral presentation
ANSI Z39.14 1979 Standard for writing abstracts

One should always have a good English dictionary close at hand when writing a report. Three additional reference books which help with language difficulties are:

Thesaurus of English Words and Phrases, by P. M. Roget. Harlow, Longman, 1982.

Fowler's Modern English Usage, by E. Gowers. 2nd ed. Oxford, OUP, 1965.
Mind the Stop, by G. V. Carey. London, Penguin, 1971.

Giving a lecture

Just as with written communication the main thing is to consider your readers, so with verbal communication the main thing is to consider your audience: what are the listeners going to get out of your talk? Will they be bored? Are they going to understand what you say? Will they be able to interpret any slides you may show? Is there enough 'new' information to make the talk seem worthwhile? Giving a talk or lecture ought to be an enjoyable experience both for the audience and for you. Your enthusiasm for the subject (or lack of it) will be communicated to your listeners. Unlike written communication, a speaker and his audience are in a 'live' situation where it is essential to take into account the interest and comfort of the listeners at every stage. If one or two members of the audience become restless, this will be communicated to the others: watch for the warning signs (excessive yawning, undue movement of chairs, expressions of boredom or non-comprehension, muttering, people walking out!)—but you should be able to control this situation to some extent by varying your presentation, perhaps even departing from your original intention in extreme cases.

Objective

Decide exactly what your talk is to achieve, remembering that you will be able to impart only a comparatively limited amount of 'new' information during the length of an average talk (do not cram too much in—you can always distribute a handout covering the details which have to be omitted). Give careful consideration to the nature of your audience and the reason for the lecture, which will determine your approach: are you trying to sell something, describe your own work, review the work of others, instruct your audience on a particular subject, or what? Estimate, as accurately as you can, the amount of knowledge possessed by your listeners. You must not waste time unnecessarily by telling them what they know already, or pitch the talk at so high a level that you will not be understood. If you have to address a mixed group of people, in the sense that they have different subject backgrounds or levels of ability to understand, try to make sure each section of the audience is catered for, so that everybody will get *something* out of your talk. Bear in mind that if you had to prepare talks based on given material to (say) your company's sales staff, and its research staff, the approaches adopted should be rather different. In the former case, for example, lengthy attention to technical minutiae would be inappropriate, while in the latter it might even be welcomed. It may help to write a short statement of your objective to keep before you throughout the time spent on preparation.

Timing

Work out a time schedule. You will generally know the date of the talk, and think there is plenty of time in hand. However, it takes very much longer to prepare a lecture than to deliver it (a rough estimate is at least 10 hours to prepare a 1-hour talk, but for a prestigious event it could take a *great* deal longer). Also, visual aids take time to produce, especially if made professionally, so allow for this in your schedule.

Information

Gather your information as described in the section on **Report writing** above. You will generally acquire rather more than you intend to use, making selection of material possible.

Preparation

Your lecture should consist of a logical progression of ideas leading, if the material permits, to a climax. In other words, it must have a well-defined structure designed to ensure that each point is grasped by the audience before you proceed to the next.

Some people prefer to work out what they are going to say and then write brief notes for use in the lecture; others prepare a complete text which they intend to read (this method should be avoided if at all possible, as it generally creates a very bad effect). In either case think in terms of composing a speech from a play rather than a written report. Do not be afraid of adopting a slightly dramatic approach to your audience, but avoid overdoing it.

Your listeners will assume you are an expert, since you presume to address them. Let them retain this impression. You must have confidence in your own knowledge of the subject, backed by thorough preparation of your material. Give the audience something to look at from time to time. The importance of *good* visual aids cannot be overestimated, but there are pitfalls for the unwary. For example: does the slide contain too much, or not enough information? Will the people at the back of the room be able to read any writing without difficulty? Are there too many slides, or too few? The answers to these questions are best obtained by consulting visual aids experts: your organization may have staff trained to advise on or prepare slides, transparencies for overhead projection, working models, large-scale diagrams, and so on, or the items mentioned in the 'Further reading' section below may help.

Organization

If you are responsible for the organization of your talk, make a list of everything that needs to be done and see to it in good time (e.g. booking a room; making sure a projector is available, if needed; arranging publicity material; signposting the room on the day, if it is hard to find). Be prepared for last-minute hitches, especially where other people are involved. Even if you are not organizing the talk yourself, make sure that the organizer is aware of your need for special equipment (projectors, for instance) and has arranged for it to be provided. You should also familiarize yourself with the place where the lecture will be given, covering such points as layout of the room, location of switches for lights and projector, position of blackboard and chalk. At least plan to arrive well before the lecture is due to start; you may waste time looking for the room if you have not used it previously.

Delivery

You will, of course, have rehearsed the lecture, perhaps with the aid of a tape-recorder, before delivering it. This helps to give you confidence, find the best way of making certain points, and ensure the talk is of the right length.

What is the secret of good lecturing technique? It is the ability to exercise self-control coupled with a relaxed approach. Try to speak clearly, naturally and not too quickly, guarding against verbal mannerisms. '*Well*, Ladies and Gentlemen—*um, er—Well*, Mr. Chairman, I must first thank you for your—*um*—opening remarks and—*er*—the kind—*er*—things you have said about my—*um*—work. *Well*, Ladies and Gentlemen, this evening I want to . . .' This sort of thing is maddeningly hard for inexperienced speakers to avoid, especially at the start of a talk, when they seem particularly vulnerable, but a determined effort should be made to begin your lecture clearly and decisively: 'Mr. Chairman, thank you for you kind remarks! Ladies and Gentlemen, this evening I am going to . . .'

Stand still, or relatively still: do not juggle with the chalk, your notes, or anything else—you are not a circus act.

Look the audience in its face, if you can: after all, there is no need to be ashamed, unless you have skimped the preparation of your material. Your listeners will hear you better if you face them (those in the back row also have a right to hear what is said) and will feel more involved if you look at them, individually, from time to time.

Try to judge audience reaction as you go along, and respond when the need arises. If you make a mistake, correct it without a lot of fuss—and carry on: do not be thrown.

Humour in a 'serious' lecture can be overdone. Your audience will no doubt appreciate a brief, relevant, anecdote (or humorous turn of phrase) occasionally, but forced humour and irrelevant jokes are out of place. Leave this sort of thing alone unless it comes naturally to you. Allow time to answer questions; perhaps at the end of the lecture or, better still, during its course, which increases the sense of audience involvement. Remember that even an expert cannot be expected to know everything about a subject, so it is better to give no answer than one which is incorrect ('I need notice of that question!'). Sometimes you can stall for time ('I am afraid I do not understand your question. Would you mind rephrasing it?') or even stimulate another member of the audience to reply—the possibilities are endless.

Finally, your audience deserves to be thanked for listening to you.

Further reading

First, a selection of titles on public speaking and lecturing.

A Handbook of Public Speaking for Scientists and Engineers, by P. Kenny. Bristol, Adam Hilger, 1982.
Lecturing, by P. Borrell. Keele, University of Keele, 1977. Keele University Library Occasional Publications No. 14.
Effective Technical Speeches and Sessions: a guide for speakers and program chairmen, by H. H. Manko. New York, McGraw-Hill, 1969.
Presenting Technical Ideas: a guide to audience communication, by W. A. Mambert. New York, Wiley, 1968.

Effective Presentation: the communication of ideas by words and visual aids, by A. Jay. London, British Institute of Management, 1971.

Next, a few references on audiovisual aids.

Presentation of Data in Science: publications, slides, posters, overhead projections, tape-slides, television, by L. Reynolds and D. Simmonds. The Hague, Nijhoff, 1981.
A Manager's Guide to Audiovisuals, by S. Allen. New York, McGraw-Hill, 1979.
OHP: a guide to the use of the overhead projector, by R. W. Rowatt. Glasgow, Scottish Council for Educational Technology, 1980.

The British Institute of Management has issued a series of brief checklists which includes *Report Writing* (no. 2), *Giving a Talk* (no. 50), and *Using Visual Aids Effectively* (no. 61). Enquiries about these should be addressed to the British Institute of Management Foundation, Management House, Parker Street, London WC2B 5PT.

Finally, a couple of audiovisual presentations can be listed here. The first is concerned with public speaking, and the second with the design and use of slides.

The Floor is Yours, produced by Management Training Ltd, in association with the British Institute of Management; film or video, distributed in the UK by The Guild Organisation, Woodston House, Oundle Road, Peterborough PE2 9PY (tel. Peterborough 0733 63122). A *Business Speaker's Guide* and a gramophone record are available as part of this training package.

Can We Please Have That The Right Way Round?, produced by Video Arts and Kodak; film or video, distributed in the UK by Video Arts Ltd, Dumbarton House, 68 Oxford Street, London W1N 9LA (tel. 01-637 7288). A booklet entitled *Slide Rules* by A. Jay is also available with this programme.

Current awareness

By current awareness we mean keeping up to date with information about recent developments, particularly those in your own subject field. This process, which may result in the acquisition of substantial amounts of useful information, or no information at all (when it acts as a kind of insurance), is necessary for three main reasons:

- to see whether you are duplicating someone else's work,
- to keep an eye on the activities of your competitors,
- to ensure that your own work does not suffer through ignorance of the latest advances.

Apart from these, you may find that information obtained from current awareness stimulates new lines of enquiry, as well as contributing to the extension of your previous subject knowledge (the 'educational' aspect).

Even if you have completed a project and submitted a report with recommendations, something may be published or reported which could affect those recommendations. You cannot prevent this sort of thing from happening, but you can usually arrange to be informed about it at the earliest opportunity. The following methods are available to you.

Scanning current literature

Only brief notes of appropriate literature and means of access are given here, but further details can be obtained from the sections on **The literature**.

Periodicals. These include the core journals for your specific subject interests and also news-oriented publications such as *New Scientist*, which serve to keep you aware of developments in other fields. Regular access to these is achieved by frequent visits to libraries (some have separate displays for current issues), personal subscriptions, or circulation systems within your organizations. (See also **Newspapers** and **Current-awareness publications** below.)

Books. Access is usually by regular visits to bookshops and libraries (some have separate displays for new material and/or issue lists of their recent acquisitions). Some publishers and bookshops will supply you at regular intervals with brochures or cards describing their latest books.

Patents, trade literature and other types of literature. May be scanned if appropriate for your work. Arrangements can be made for both patents and trade literature to be sent to you on publication (see pp. 80 and 138).

Newspapers. Many of the daily newspapers have regular science or technical features, and some have whole pages devoted to this type of news. (See also **Services** below.)

Current-awareness publications

Current issues of abstracting and indexing journals may be scanned, but these can be months or even years behind with their reporting. There is no sharp dividing line between these and current-awareness publications, but the latter are usually produced expressly for current-awareness purposes, and are not normally intended for retrospective searching. Details of these may be found in **Guides to periodicals** (p. 97) and **Guides to abstracting and indexing journals** (p. 54).

Current Contents

These are little booklets, published each week by the Institute for Scientific Information, which contain facsimile copies of the current contents pages of selected journals in a particular field. The indexes include a list of journal issues covered, a weekly subject index, and an author index and address directory.

Current Contents/Life Sciences (quarterly subject index also available)
Current Contents/Agriculture, Biology, and Environmental Sciences
Current Contents/Engineering, Technology and Applied Sciences
Current Contents/Physical, Chemical and Earth Sciences
Current Contents/Social and Behavioral Sciences
Current Contents/Clinical Practice
Current Contents/CompuMath (contents pages only, monthly)
Current Contents/GeoSciTech (contents pages only, monthly)

Contents pages

Some libraries provide a regular photocopying service for journal contents pages, such as the British Library Lending Division Journal Contents Page Service, which costs about £10 per year per journal title.

Current-awareness journals

Similar in appearance to indexing journals, but usually ephemeral and often with no cumulating printed indexes. Individual or custom-tailored computerized SDIs and standard SDI profiles are covered in **Information services** (p. 140). In practice, there may be very little difference between certain standard profiles and some current-awareness journals. The following is a brief selection.

Current Papers in Physics (CPP). London, INSPEC, fortnightly.
○ One of a series of three titles published by INSPEC (see below). The

contents are arranged in subject order according to a classification scheme which is always shown on the back cover. Each entry normally consists of title, author(s), author's affiliation and complete bibliographical reference. These publications have the same coverage as the corresponding abstracting journals (see *Physics Abstracts*, p. 52), but contain no indexes and do not cumulate.

Current Papers in Electrical and Electronics Engineering (CPE). London, INSPEC, monthly.
Current Papers on Computers and Control (CPC). London, INSPEC, monthly.
○ These are similar to *CPP* above, being companions to *Electrical and Electronics Abstracts* (see p. 50) and *Computer and Control Abstracts*, respectively.

Chemical Titles. Washington DC, Chemical Abstracts Service, fortnightly.
○ Each issue consist of three parts. The first contains a KWIC index (see p. 222) compiled from the titles of current chemical research papers selected from about 700 journals in pure and applied chemistry and chemical engineering. The second part is a bibliographic listing of the titles of papers, arranged in order of journal titles. The third part consists of an author index. The links between the indexes and the bibliographic listing are journal CODEN (code letters and numbers—see p. 101), volume and page numbers. A list of periodicals covered is shown on the inside covers of each issue. See also *Chemical Abstracts* (p. 49).

Health Service Abstracts, London, DHSS, monthly.
○ One of several monthly bulletins, available from the DHSS, which are produced from the *DHSS-DATA* database. Each issue contains entries under broad subject headings, and there are subject and author indexes.

BIOSIS/CAS SELECTS. Philadelphia, BIOSIS, biweekly.
○ A group of bulletins covering 23 specific research topics in the fields of biochemistry, pharmacology and physiology, selected from the BIOSIS and CAS databases and containing abstracts but no indexes. *BioResearch Today* is a similar group of bulletins covering 14 specific research topics selected from *Biological Abstracts*. (See also BioSciences Information Service, p. 176, and *BIOSIS Previews*, p. 145.

CA SELECTS. Columbus, Chemical Abstracts Service, biweekly.
○ A series of about 150 bulletins, each covering a different topic, with selections from *Chemical Abstracts*, and including abstracts but no index.

Chemical Engineering Abstracts Bulletin. Nottingham, Royal Society of Chemistry, monthly.
○ World coverage of chemical and process engineering journals, with abstracts and keyword subject indexes. It is also available as an online database.

Chemical Industry Notes. Columbus, Chemical Abstracts Service, weekly.
○ Aimed at management, and covering chemical and related industrial fields from over 80 leading business and trade jounals, with abstracts, keyword and author indexes. A separate annual volume index is published. This is also available as an online database.

Current Awareness in Biological Sciences—CABS (formerly *International Abstracts of Biological Sciences*). Oxford, Pergamon, monthly.
○ Issues give titles of journal articles, full references, and the authors' addresses, under subject sections and sub-headings, and with sequential accession numbers. There is a separate author index but no subject index. An online database version is available via PERGAMON INFOLINE (p. 194), and Standard Topic Profiles derived from *CABS* are also available from Pergamon.

Current Awareness in Particle Technology (formerly *Particulate Information*). Loughborough, University of Technology, monthly.
○ Worldwide coverage of particle science and technology in most types of literature. References are arranged under a broad subject classification. Author and subject indexes are published quarterly, cumulating annually.

Current Biotechnology Abstracts. Nottingham, Royal Society of Chemistry, monthly.
○ Worldwide coverage of relevant patents and journals; includes abstracts, subject, substance and company indexes. It is also available as an online database.

Current Mathematical Publications (formerly *Contents of Contemporary Mathematical Journals and New Publications*). Rhode Island, American Mathematical Society, biweekly.
○ International coverage of mathematical journals subsequently covered in depth by *Mathematical Reviews*, with individual papers arranged in the same subject classification, but normally with major sections only. Each issue includes a list of journals covered in that issue, a list of new journals, an author index and a 'key' index (for items without authors). The author and key indexes are cumulated every six months.

Key Abstracts. Hitchin, INSPEC, monthly.
○ A group of bulletins covering eight specific topics derived from the INSPEC database, including a contents page, abstracts, but no index. Effectively subsets of the INSPEC abstracts journals (see also p. 146).

Invisible colleges

These are informal groups of people working in related subject fields, who keep each other informed of relevant current developments. They do this by correspondence, exchange of reports, preprints, etc., and through contact at conferences, private meetings, and so on. Many of these groups show no outward indications of their existence, and even their members may not think of themselves as belonging to a specific group.

The advantages of such a system for communicating information are that it is rapid, efficient and uninhibited. There are no publishing delays or editorial restrictions, and friendly advice is frequently only a 'phone call away. Members of these groups also tend to be well established in their field.

The disadvantages are usually experienced by non-members, especially those new to their work, who find difficulty in identifying and joining these groups. Some groups become more formalized, with regular meetings, newsletters, etc., and may even evolve into professional organizations, or give birth to new periodicals.

The best means of entering or forming one of these groups are: correspondence with people who are publishing papers in your field, visits to research groups working in your field, and regular attendance at conferences for the purpose of making contacts (see below).

Conferences (symposia, seminars, meetings), exhibitions, etc.
See also Conference proceedings, p. 65

These perform two main functions, as far as your subject interests are concerned:

(1) To listen to details of the latest work or to see the latest equipment.
(2) To meet people, make contacts and have discussions. Where lists of delegates are issued (sometimes prior to the conferences), you have the opportunity to work out who you should try to meet. Some people find the contacts more important than the papers!

You should bear in mind that residential conferences may prove relatively expensive, especially when you include travelling and subsistence costs.

British Telecom International offer 'International Videoconferencing' currently between the UK and North America. The service can be from public studios or from facilities on a customer's premises. It offers two-way full-motion colour video and audio communication in two modes: face-to-face and high-resolution graphic display, which means that genuine meetings can take place. Enquiries to 01-936 3075 or 2488.

Should you ever wish to organize a conference or meeting, there are many books and journal articles on the subject which can be searched for in the normal way, such as *How to Organize Effective Conferences and Meetings*, by D. Seekings (2nd ed., London, Kogan Page, 1984).

Guides to conferences, exhibitions, etc.

These are by no means comprehensive in their coverage. Access is usually via libraries. Examples include:

Conference Papers Index (formerly *Current Programs*). Bethesda, Cambridge Scientific Abstracts, monthly.
○ Covers scientific, technical, engineering and medical conference papers presented at professional meetings throughout the world. The entries for each meeting are arranged under 17 broad subject sections. Each entry includes the full title of the conference, dates, location, language, sponsors, ordering information, and a list of papers giving titles, authors and their affiliations. Each paper has a running citation number which is used in the indexes. There are monthly author and subject indexes, and separate cumulated annual indexes covering author, subject, conference topic, and conference date. This is also available as an online database.

Forthcoming Conferences, Meetings, and Colloquia in Physics and Electronics. Chelmsford, EEV, quarterly or monthly.

○ Covers the UK and worldwide meetings for the ensuing 12 months, particularly those of the IEE, IEEE, IERE and IOP. Some training courses and exhibitions are included. The bulletin is arranged both chronologically and by subject keywords.

Forthcoming International Scientific and Technical Conferences. London, Aslib, quarterly.
○ The main issue of each year is published in February; this supersedes and updates previous issues. It is followed by a supplement in May and cumulative supplements in August and November. British national meetings are included. The conference entries are arranged in chronological order, giving date, title, location and an address for enquiries. Subject, geographical location and organization indexes are provided.

MInd: the meetings index. New York, InterDok, bimonthly.
○ A new publication covering forthcoming meetings in science, engineering, medicine and technology. The main section contains details of the meetings, listed under keywords. There are separate sponsor, location and date indexes, and a list for contacts.

World Meetings: outside United States and Canada. New York, Macmillan Publishing, quarterly.
○ A 2-year register of all important future medical, scientific and technical meetings to be held outside the USA and Canada which is completely revised and updated quarterly. Each entry gives official title of meeting, location, dates, name(s) of sponsor(s), general information such as name and address for enquiries, brief details of topics in papers, estimated attendance, deadlines for submission of papers, availability of abstracts or papers, and details of any associated exhibitions. There are indexes to subject keywords, locations, dates, deadlines, and sponsors (giving addresses). Macmillan publish companion registries: *World Meetings: United States and Canada; World Meetings: medicine* and *World Meetings: social and behavioral sciences, human services, and management.*

Scientific Meetings. San Diego, Scientific Meetings Publications, quarterly.
○ This publication is a listing of technical, scientific, medical and management organizations and universities that are sponsoring future national, international and regional meetings, symposia, or colloquia. The coverage is international, but with a strong US bias. There are separate chronological and keyword listings. Contact addresses are included in the main listing.

Exhibition Bulletin. London, The London Bureau, monthly.
○ A diary of forthcoming exhibitions and trade fairs at home and overseas, giving details of date, title, venue, and organizers in country, town and date order, except for Great Britain and London, which come first. There is a subject classified listing giving town and date, and an exhibition and conference services directory.

In addition, many journals include details of forthcoming conferences, etc., in their respective fields and, sometimes, brief reports on those that have taken place.

Services

Many of the information services and suppliers described on pp. 139–191 provide current-awareness facilities, and, for convenience, these are described in that section. Remember that many other organizations of this kind offer similar services, such as the British Hydromechanics Research Association (BHRA) at Cranfield, with their SIPs service, a fortnightly current-awareness service based on the generation of their database *FLUIDEX*. Other examples are the monthly SDI bulletins issued by the Transport and Road Research Laboratory (TRRL) at Crowthorne, based mainly on the *International Road Research Documentation (IRRD)* database, and the monthly *Geoprofiles* from Geosystems in London. You should always check your own library/ information service (if you have one) to see if they offer a current-awareness service. Other services include:

Online Services. Most hosts have a facility for automatic SDIs. Profiles are constructed using the methods given in the section on searching (p. 211), except that broader search strategies are normally used because the numbers of references produced per month are so small (relatively), compared with standard retrospective searches, that the resulting higher ratio of irrelevant material is worth accepting in order to ensure that little is missed. It is simply a matter of casting the net more widely. It is particularly important with current-awareness profiles to leave the combinations and permutations until the end of the strategy, so that changes, if facilities permit, are easier to make. Commands like END/SDI or . .SDI 15 ALL will save the strategy, and the host computer will then execute it as an SDI profile automatically each time the database is updated. The results are then printed out and mailed to the user (a print command is included at the end of a profile if it is not already incorporated by the host into the SDI command).

If the searcher wishes to perform the updates at his own convenience, then provided the files are printable in one strict chronological order, and have not been reloaded between searches, an ordinary search save command may be used, and the latest hits since the last run printed out, using simple subtraction to work out how many there should be. Alternatively, with some databases, each update has its own code which can be ANDed with the last set of the saved search.

International Press-Cutting Bureau. This organization will supply you, every week, with current newspaper and magazine cuttings which cover your specific subject interests, or mention specific products or organizations. This service can be particularly useful in commercially competitive areas. The minimum service charges are £80.50 inc. VAT for one topic—100 cuttings or 6 months' service, whichever comes first. Address: Lancaster House, 70 Newington Causeway, London SE1 6DG (tel. 01-403 0608).

TechAlert. This service distributes summaries of the most important exploitable technical information produced by publicly funded research and development to engineers and technologists in industry, via regular published descriptions in certain leading specialist journals (see also OTIU, p. 179). Contact TechAlert Unit, Department of Trade and Industry, Ashdown House (Room 373), 123 Victoria Street, London SW1E 6RB (tel. 01-212 6762 or 6152).

Appendix: Helping the library user

Our brief survey of this topic in a book called *Information Sources in Science and Technology* requires a few words of explanation. We have both been concerned with helping library users for around 20 years; indeed the present book is just one by-product of our activity here. Libraries and library networks can be large, complex organizations. Certain of the materials they contain are difficult to use and/or hard to find, at least for the inexperienced. Library organizers (librarians and information officers) are responsible for providing as much assistance as they can to ensure that library users make the most of whatever stock and facilities are available. Furthermore, if library users themselves are aware of what can be done to help them, they may make sensible suggestions for improvement where the provision of such assistance in their own libraries seems inadequate.

This appendix is written primarily for library organizers who possess little knowledge of 'user education' (as the subject is generally, if not very happily, known) in its broadest sense. However, our comments may also interest those library users anxious to discover techniques for exploiting libraries, the materials held and the services offered, as fully as possible. We outline, with comment, the various forms of help that can be provided, concluding with a few references for anyone wishing to pursue the matter further. In several places, however, we adopt a personal approach and describe in greater depth our experience at Southampton University Library.

Personal assistance

In an ideal world, every library user would find a helpful librarian, instantly accessible, answering questions, taking people to the right shelves, and resolving difficulties of whatever complexity whenever they arose. There would, of course, be no diffident users, shy about approaching library staff or afraid to ask even simple questions. Neither would there be financial obstacles in the way of employing sufficient staff to provide a comprehensive service at all times.

Real life is rather different. Libraries are quite often short-staffed, librarians occasionally fall ill or insist on taking the holidays to which they are entitled, queues form at enquiry desks, telephones are engaged when information officers are wanted urgently or, worse still, there is only a mechanical answering device. Despite all this, personal assistance (when you can get it) is one of the best forms of help encountered by library users.

You may find librarians conveniently posted at reference, enquiry or help desks. More senior library staff sometimes work in open-plan areas adjacent to the stock for which they are responsible, or operate from nearby offices with ever-open doors. However, these people may only be on duty during normal office hours, making specialist assistance hard to come by in the evening and at weekends. Librarians and information officers can give a wide range of information services, depending on local circumstances, from answering basic questions about library layout and stock, through offering individual guidance on the use of less straightforward library materials, to the provision of online information retrieval at a highly sophisticated level.

Their attitudes towards this work vary. Some are willing to carry out almost every task for their users (if they have the time as well as the inclination), others see their role more as helping users to help themselves. In the latter case, for example, the emphasis will be on showing an individual how to use *Science Citation Index* or *Chemical Abstracts* rather than performing a manual literature search on the enquirer's behalf. In libraries where this attitude prevails, attention will probably have been given to devising a substantial library user education programme, involving many of the aids or techniques discussed below. The philosophy is one of persuading people to get more out of libraries than they ever have before by encouraging the active use of library materials and services.

New users present a particular challenge. They may arrive knowing little of the complexities inherent in a library bigger than a single room. On the other hand they may be well versed in the mysteries of a library using a completely different cataloguing and/or classification system, and thus need to unlearn much of what they know. It is important to provide some kind of initiation routine for these users during which at least they see, or better still are formally introduced to, the members of library staff responsible for helping them. Other things for consideration at this stage, including leaflets, talks and tours, are mentioned later; but establishing a measure of personal contact between librarian or information officer and user is the vital factor, as this should 'break the ice' and assist in forming a worthwhile future working relationship.

Library staff situated more or less permanently on open access must bear with fortitude the frustration of having their regular work interrupted for trivial enquiries. They, and their reference desk colleagues, must also be prepared to deal kindly with users who do not know what they want, or cannot express themselves adequately (perhaps because of foreign language difficulties).

It might be argued that there is *no* substitute for personal assistance. This may be true for some library users, though certainly not for all; and even if it is true for all, we must recognize that library staff time has become an increasingly scarce resource in many libraries, which may remain open at times when appropriate staff are not available to assist users, so the possibilities afforded by alternative methods simply cannot be overlooked.

Library signposting

Next time you enter a library, pause and look around. What do you see in the way of signs and notices? Some libraries get carried away with this sort of thing, others fail to supply what might be thought of as the bare minimum. When you are familiar with a particular system, signposting is of little consequence and can generally be ignored. However the inexperienced user is in a completely different situation, and may rely heavily on signs, notices, plans, models, and shelf guiding, especially when there is nobody around to answer queries. Let us consider each of these in turn.

Signs should convey distinctly limited amounts of information very clearly: for instance, directing library users to specific (broad) subject areas or major facilities such as cloakrooms/toilets, photocopying, inter-library loans, book issuing, and reference service points. Notices are employed for larger quantities of information: for example, the opening hours of an academic library having several branches, with differences between term and vacation, or the official regulations governing its use. Some libraries display far too many notices, and these are often neither as well designed nor as carefully executed as they should be for effective presentation of the information concerned. Library floor plans located at strategic positions can be of great assistance; again care should be exercised in their design for maximum effect, resisting the temptation to overburden the user with detail. The value of take-away floor plans is mentioned in the next section. If a library building is really awkward to understand, with numerous levels and staircases that do not interrelate as might be expected, a simplified, cut-away, three-dimensional model may prove of service when prominently sited near the main entrance. Finally, guiding is needed at the shelves, indicating, for example, book classifications/subjects, or the titles of journals housed in a given stack.

Devising and maintaining a good system of library signposting is generally a time-consuming and expensive task, especially if the aid of professionals is enlisted, which is why standards are frequently lower than they might be. If adequate resources are devoted to this where there is a real need (i.e. a relatively complex library, open when staff are not always available for consultation) the potential benefits for users are immense.

Library guides

Probably next in importance to adequate signposting is the provision of printed library guides, which for convenience we will divide into three categories: leaflets, plans, and manuals.

Leaflets should convey relatively limited amounts of information, perhaps concentrating on just a few aspects of the library and its services. For instance, one for new users giving opening hours, an outline of the library's structure and/or layout, names of relevant staff contacts, and highlights from the regulations; or a leaflet showing how to complete an inter-library loan application form originally designed for books and journal articles when other types of literature are required (such as reports, conference papers, patents or theses).

Although floor plans may be displayed at strategic points or appear in leaflets and manuals, the single-sheet take-away variety is valuable too. Librarians who cannot abandon their desks to conduct users around the building may produce an appropriate plan, show the enquirer where to go (marking destination and route where necessary), and present this plan to the user, who will then navigate a course through the library with its aid. More detail may be included in these plans than in fixed-location map displays covered in the previous section.

A collection of leaflets and plans can go towards forming a loose-leaf guide or manual, though manuals are also produced in conventional booklet form. In fact the latter may be quite comprehensive, running to many pages and dealing with the use of library materials as well as the layout/facilities of the library itself. They can even boast floor plans overlaid by a grid, like a commercial street atlas, so that map references are given as well as library shelfmarks to help users find their own way about easily. Sometimes manuals are issued as a handout in connection with formal library instruction courses (see below).

There are several points to remember when preparing leaflets, plans and manuals. The amount and nature of the information for inclusion must be vetted carefully. The presentation of that information should be as attractive as possible; otherwise, however good the guide it could be discarded unread. Consider alternative methods of production before making a final choice (inexpensive short-run copying from your own typescript, high-quality work from a commercial printer, or what?). Bear in mind, too, that publications of this kind go out of date sooner or later; the potential application of word processors should not be overlooked here. It is normally quicker and cheaper to update individual leaflets in a series, or a loose-leaf guide, than to revise and reissue a commercially printed manual. Professional help with design, draughtsmanship, and printing should be sought for all but the least ambitious library publications, since their final appearance is as significant as their information content.

Point-of-use guidance

There are occasions when guidance is most appropriately provided at the place where publications or services are situated and used: for instance, instruction on methods of extracting references from a particular abstracting or indexing journal, and advice on how the library's catalogues or microfilm readers work. We would also include under this heading microform or online catalogues distributed around the different subject areas in a large library, and lists of journals posted at the ends of stacks where they are shelved.

Leaving aside personal tuition, point-of-use guidance media include explanatory posters (and leaflets) as well as visual, audio, or audiovisual displays. Some material of this kind is available commerically, though not always from the producer of the product described, varying from tape-slide guides covering publications such as *Index Medicus* to posters demonstrating the use of *Science Citation Index*. These can be sited and viewed alongside the material to which they refer, but might take up prime reader space. Alternatively, librarians may devise their own point-of-use guidance, especially when dealing with topics like using the local catalogues correctly. In this case, the remarks made earlier about the need for information to be presented both attractively and professionally are equally applicable.

Library talks and tours

If there is a universal way of introducing new users to a library, surely it must be through an orientation talk and/or tour? Variations on this theme are numerous, the best method being determined after considering local objectives and circumstances. Much depends on numbers. If new users are few and far between, individual or small group tours are entirely practicable. Where large numbers of them, all anxious to start using the library immediately, arrive at the same time, tours are probably out of the question unless the library is very generously staffed. In our experience, although small group tours can be highly successful, this is not generally true of large ones. When demonstrating catalogue use, for example, only three or four people nearest the guide can see what is happening, while those on the group's periphery become increasingly bored. In these circumstances there seems no option but to abandon tours in favour of well-illustrated talks, assuming literally hundreds of new users must be inducted within a week or two.

We say talks rather than tape-slides, films or videos, because it is vital to establish *personal* contacts between library staff and users at an early stage (as discussed in the section on personal assistance). We say well-illustrated talks, because learning about libraries is neither a fascinating nor important subject as far as most users are concerned, especially those who perceive no immediate library need and hence enjoy little motivation; so every effort should be made by speakers to devise presentations that are as attractive and interesting as possible. Of course, audiovisual programmes may have a place in these introductory sessions, but we would still consider the physical presence of appropriate library staff essential. Any slides or overhead transparencies illustrating a library talk should be designed and executed carefully, perhaps with professional aid for the best results. Relevant library leaflets can be distributed during induction sessions; indeed much factual information is better disseminated in this form than through the talks and tours themselves.

Before leaving tours, we might say a word about the self-guided, as opposed to the conducted, variety. The former are prospectively less staff-intensive, but we have to admit that if there is a successful formula it has so far eluded us. We have tried two approaches at Southampton:

(1) issuing leaflets which, in association with take-away floor plans, guide the new user through selected parts of the library, giving an opportunity to explore the facilities or stock located in specific areas, and
(2) posting a series of displays at strategic points around the library, each describing the services or materials in its vicinity, followed by directions on reaching the next stage in the tour.

Neither of these proved popular with users. We have not yet experimented with prerecorded audio-cassette guided tours, which might form an acceptable alternative if a sufficient number of portable cassette players could be made available.

Library courses and exercises

It is often quite impossible to impart more than a little of the detailed knowledge about library systems, services and materials required by users during a single orientation talk or tour, especially in educational organizations such as universities, polytechnics and technical colleges (with which this section is almost entirely concerned). Indeed, these users will have different needs at various stages in their academic careers, a situation calling for the development of flexible library instruction programmes. It is essential that formal instruction of this kind be given at the 'right' time, arising naturally in the context of the students' subject courses. Otherwise the library lecture or whatever will probably seem irrelevant, and may do more harm than good. There follows a requirement for sympathetic working relationships between staff from the library and those in academic departments. Where this exists, library instruction modules can be planned in association with course organizers, if not actually integrated with the academic courses themselves (held by many to be the best approach).

It is essential, too, that there are library staff willing and able to become deeply involved with user education programmes, for the latter are most unlikely to flourish where staff are dragooned into work for which they have neither enthusiasm nor particular capability. A certain flair is desirable, coupled with persistence, because these programmes evolve over a period of years rather than emerge in all their glory overnight. Thus, it must be stressed, much will depend on the personalities of the library staff and their academic departmental contacts. When there is mutal understanding of what each can offer, students will derive the greatest benefit. Of course, what has been developed and is working well in one environment will not necessarily do so in another, partly on account of the personal factor.

At Southampton, that part of the instructional programme specifically designed for physical scientists and engineers currently consists of three phases for most students. An introductory talk at the beginning of the first year is followed by an intermediate lecture within a term or so, and a half-day course completes the programme at the end of the second year or the beginning of the third. The introductory talk takes the place of an orientation tour (numbers render the latter impractical), all students being invited on a faculty-wide basis. The intermediate lecture, attended by most engineers in departmental groups, is aimed at helping students to find a few references in the library for tutors' essay assignments. The half-day course, again given on a departmental basis, coincides with the commencement of third-year project work, when students are generally expected to undertake full-scale manual literature searches.

Finding so much factual material worth including in our advanced courses, we were obliged to compile a substantial handout, from which incidentally the present book has grown. The handout has evolved, too, from an unsophisticated set of typewritten, duplicated sheets to a professionally designed and printed 48-page manual (which is solely for internal consumption, being distributed only as part of our library instruction package). In its earliest version the handout possessed a fairly general character, covering resources other than those available at Southampton, but following the initial publication of this book it became more specifically concerned with our own library. We can now concentrate mainly on ideas during the courses, knowing that students will extract detailed factual information from their handouts afterwards when the need arises.

We have also developed various lecture materials in connection with our courses, though these are more conveniently discussed in the next section on audiovisual aids. Nevertheless, the importance of employing attractive and interesting presentations with students who do not immediately appreciate the value of library instruction must be stressed here yet again. We monitor the effects of our courses by means of questionnaires, noting among other things whether or not the students became bored.

Lastly, a little attention may be devoted to library exercises, the practical element in many courses. These vary from simple seek, find and record question/answer sheets devised by library or academic staff, testing basic understanding of the library, its cataloguing, classification and shelving systems, to more elaborate tasks for the advanced student. The latter can include the following, which we have used with reasonable success over the years.

(1) Sample issues of abstracting and indexing journals are set out in a room, accompanied by instruction cards bearing questions to be answered with the aid of each publication. (Laboratory experiments are sometimes conducted in a similar manner.) The students examine several different examples, spending no more than a few minutes on any of them. The main disadvantage is that the materials are not seen in their natural library context, but at least attention can be focused on actually using various types of index.

(2) The students are divided into groups, with designated leaders, each group being assigned to one of perhaps half a dozen rather broad topics. The students receive question/answer sheets that suggest three abstracting or indexing journals (or other guides) per subject, and these are then searched for relevant references, every student covering only one specified year to prevent overcrowding. At the end of the exercise, group leaders collate and report on the findings of their colleagues, raising any difficulties encountered. In this case library materials are used in the proper context.

(3) In connection with their third-year projects, the majority of our undergraduate engineering students perform literature searches that are formally assessed, with marks contributing directly towards degree totals. They are provided with worksheet booklets consisting of two main sections. The first (white paper) is where they enter details of the sources searched, such as abstracting and indexing journals, including the search terms chosen and the period of time covered in each case. The second (coloured paper) is for recording references found, but this is subdivided for different categories of literature, such as journal articles, reports, conference proceedings, theses and books. Students are encouraged to examine a wide range of sources, seeking examples of references in all categories—the educational aspect of the exercise—as well as discovering useful information for the projects themselves. A cross-referencing system in the worksheet booklets shows which references came from a given source, and which source produced a particular reference. The assessment procedure has two stages. First the library staff see all completed worksheets from every group, allocating a grade to each (and comments as well when time permits); then project supervisors receive those for their own students, converting library grades into marks using a previously agreed scale. This two-stage process works well, because library staff are capable of assessing overall search strategies and the accuracy with which references have been recorded, whereas project supervisors can best judge the

relevance of material found, and should detect any serious omissions. Students participating in this scheme are thus highly motivated to learn about the library and use it extensively. The effectiveness of our course, handout and worksheet booklet is also evaluated by a questionnaire, completed after the students have finished their project literature searches.

Although this section has been primarily concerned with the academic environment, there is no reason at all why instructional courses should not take place in other kinds of library where there is sufficient need or demand among the users. For example, a public library with a good stock of patents and their guides might run a course on the exploitation of this literature for its local industrial users.

Audiovisual aids

For convenience we are dealing with audiovisual (a-v) aids under a single heading, but they have also been mentioned elsewhere in view of their point-of-use guidance, library talk and library course applications. There are even online databases covering them. The most commonly encountered forms include sound recordings, tape-slide programmes, films and videos, which will be glanced at in turn.

Sound recordings are readily made and can be used for simple instructional purposes. For instance, we might take a sample issue of an abstracting journal and record a commentary describing how it works, which involves (say) looking up a subject in the index, then turning the pages to find the corresponding abstract. During playback the user physically handles the material described, gaining some familiarity with it in the process. Sound recordings may also form the basis of self-guided library tours.

Before the advent of video, tape-slide was the preferred audiovisual medium for library education, and still has much to commend it. Programmes are generally both cheaper and faster to produce, and capable of being updated easily. They may be intended for either individual/small-group or large-group viewing. One of us completed his first tape-slide programme within a month, from original idea to finished product; it has proved popular with students for more than a decade. However, his second programme was still not quite ready for commercial distribution after seven years.

The first programme, featuring about 30 slides and lasting for approximately 12 minutes, showed someone carrying out a manual literature search. This was unsuitable for use elsewhere because it was designed specifically with our library (including layout, shelf guiding, etc.) and course handout in mind.

The second programme is more general in character, demonstrating the structure of scientific and technical literature by means of a case study. This substantial presentation, with 80 slides and a playing time of nearly three-quarters of an hour, evolved from a lecture opening our advanced library courses. It is based on a genuine research project, the Talking Brooch communication aid for dumb or speech-impaired persons developed in the Electronics Department at Southampton University. Despite the existence of this tape-slide, we normally prefer to deal with its subject 'live' during our courses. A selection of 35mm slides from the Talking Brooch programme is shown on one screen, illustrating the case study's storyline, while a structure of the literature chart is built up on a second screen with the aid of an overhead projector. The structure chart remains visible throughout the lecture, and this somewhat unusual dual-screen approach has been found stimulating by both presenter and audiences alike.

Not only may the Talking Brooch tape-slide be purchased; a book containing a slightly extended version of the same material has been published as well (see *Understanding the Structure of Scientific and Technical Literature*, p. 40). Thus library educators can choose between generating their own tape-slide programmes or making use of those produced by others.

Films are less fashionable now than was once the case, though they remain a satisfactory medium when catering for large lecture-theatre audiences. Various academic library introductory films have been made, but these usually prove difficult and/or costly to update (libraries are often changing) so their popularity has declined. Educational material in this medium is available commerically; check the catalogues of relevant distributors if you wish to explore the possibilities further.

For many instructional purposes, video has now replaced film. This does not mean that it should be the automatic choice in every situation, because tape-slide remains unsurpassed in some respects (cheaper and quicker to produce, easily updated, more appropriate where fine detail must be resolved in the pictures, generally better suited to the large lecture-theatre environment). Video comes into its own when recording active rather than static subjects, so may be ideal for an introductory feature programme about a library, but less suitable for an in-depth account of *Chemical Abstracts* (say) where the viewer is expected to read the small print.

Video also has certain advantages for self-instruction. These days the use of videotape recorders should present no great difficulty, and in theory there is less risk of anything going wrong than with some other forms of student-operated audiovisual equipment. Students can work with video programmes in an 'active' way, as well, leaving a programme in order to undertake exercises (say) and easily resuming later from wherever the interruption occurred.

Before indulging in do-it-yourself audiovisual aids, remember the importance of enlisting professional help if you want a high-quality result. Many organizations have a 'media' department providing assistance with artwork preparation, graphic design, photography, film-making, sound and video recording/editing facilities, staffed by experts willing to advise on technical aspects of programmes as well as their intellectual content. Overlook such a valuable resource at your peril, although it must be admitted that calling upon these professional services may delay substantially the completion of your assignment.

One other factor is worth bearing in mind with audiovisual aids. They may solve problems but they can create them too. For example, perhaps additional equipment is required for the lecture room, which may not be set up (say) for video. Then, the more complicated the equipment or show (we could cite multi-projector tape-slide extravaganzas), the greater the risk of something breaking down or going wrong, leading, in extreme cases, to disaster when the presentation is totally a-v dependent. Our practical advice is, keep things as simple as possible to be on the safe side.

Computer-assisted instruction

Computer-assisted instruction (CAI) or computer-assisted learning (CAL) is a relative newcomer on the education scene. It may be thought of, currently, as a mechanized extension of conventionally printed programmed-learning texts. Basically, the user interacts with a computer which, for example, may display small amounts of information followed by questions, perhaps of the multiple-choice answer variety. When these questions have been answered correctly, or mistakes pointed out and the right answers supplied, the user enters the next stage in the programme.

The popularity of CAI may well increase with the spread of personally owned and publicly accessible micros. Library instructional material of this nature is far from common in the UK at the time of writing, more development having taken place in the USA than anywhere else. However, the technique provides a potentially valuable addition to other educational methods, and can cover topics such as library structure and layout, finding items in catalogues (especially online catalogues), using abstracting/indexing journals, and online searching. CAI, whose applications include point-of-use guidance as well as general orientation, may more readily engage the user's attention, at least until the novelty wears off.

Here we would sound a note of caution. Programmes should be imaginatively conceived and executed skilfully, otherwise interest in them will soon evaporate. Transferring mediocre educational material onto a computer is just not good enough; let us have worthwhile programmes, properly thought out and well designed, making the most of the chosen medium. Combining computers with videos creates an attractive hybrid, interactive video teaching systems, where the programmed-learning approach of CAI can be punctuated by selected optional video extracts reinforcing appropriate key points. Provided that such systems are user-friendly, and their programmes easily updated, they offer one exciting prospect for library instruction in the future.

Further reading

We make no attempt to be comprehensive, or even representative, in this section. We merely suggest a handful of references from the very many available which may interest those wanting more information about methods for helping library users.

First, a couple of titles dealing with library signposting. *Sign Systems for Libraries: solving the way-finding problem*, by D. Pollet and P. C. Haskell, was published in New York by Bowker in 1979. *Signs and Guiding for Libraries*, by L. Reynolds and S. Barrett (London, Clive Bingley, 1981), also touches on publications and stationery.

There is a vast literature on library user education. Several bibliographies have been produced, among which are numbered: *Library Instruction: a bibliography*, by D. Lockwood (Westport, Connecticut, Greenwood Press, 1979); *Bibliographic Instruction in the USA, 1883–1982: an annotated bibliography* and *User Education in the UK, 1912–1982: an annotated bibliography*, both by I. Malley (Loughborough, INFUSE Publications, 1983). Two valuable collections of papers have been edited by J. Lubans: *Educating the Library User* and *Progress in Educating the Library User* (New York, Bowker, 1974 and 1978 respectively).

Mention may also be made of *User Education in Libraries*, by N. Fjällbrant and M. Stevenson (London, Clive Bingley, 1978).

The proceedings of three international conferences on the subject have appeared.

Library User Education: are new approaches needed?, edited by P. Fox, British Library Research and Development Report no. 5503 (London, British Library, 1980).

Second International Conference on Library User Education; Proceedings, edited by P. Fox (Loughborough, INFUSE Publications, 1981).

Third International Conference on Library User Education; Proceedings, edited by P. Fox and I. Malley (Loughborough, INFUSE Publications, 1983).

Finally, for anyone requiring further details of the Talking Brooch presentation developed at Southampton, which was described briefly under audiovisual aids above, the relevant reference is 'Teaching student engineers the structure of scientific and technical literature using an electronics case study', by R. V. Turley, *IEEE Transactions on Education*, Vol. E-26, No. 1, February 1983, pp. 11–16.

Index